## Michel Guisnet

Michel Guisnet was born in Cambrai, France in 1939. He holds a graduate degree in Chemistry from the University of Lille and a Doctorate in Physical Chemistry from the University of Poitiers (1970). He was Assistant Professor in Organic Chemistry (Lille), Assistant Professor and later Professor in Physical Chemistry (1973) in Chemical Engineering (Chemical Reactors-Catalysis) at Poitiers Engineering School (ESIP). He assumed the position of Director of the Laboratoire de Catalyse en Chimie Organique (Poitiers University – CNRS) from 1981 to 1990. He is currently the head of the Chemical, Biological and Geological Engineering Doctoral School. The research interest of his group is in the field of acid and bifunctional zeolite catalysts for Refining, Petrochemical and Fine Chemical Industries, the aim being strongly related to the development of more selective and cleaner processes. Modification of zeolites and characterization, reaction mechanisms (isomerization, cracking, alkylation, hydrocracking, acetylation, etc.) and prevention of deactivation by coking and regeneration are his main research areas. He has been involved with some 400 international publications, 8 patents, Chairman of the first three International Symposia Heterogeneous Catalysis and Fine Chemicals (1988, 1990, 1993)

## Jean-Pierre Gilson

Jean-Pierre Gilson was born in Namur, Belgium, in 1955. He studied Physical Chemistry at the University of Namur and received his Ph.D in 1982 for his work on the reactivity and characterization on the MFI zeolite. From 1982 to 1984, he worked in the Process Research Department of UOP, Des Plaines, USA, on reforming and aromatization catalysts. He then moved to Grace Davison, Columbia, USA and worked on octane enhancing catalysts from 1984 to 1987. Afterwards he joined Shell Research Laboratories in Amsterdam, The Netherlands, where he worked from 1987 till 1996 on various catalysis related subjects such as paraffins and aromatics isomerization, solid acid alkylation, lube oil manufacture and processing and Fischer-Tropsch liquids upgrading. He joined the "Laboratoire de Catalyse et Spectrochimie" at the ENSICAEN-CNRS, France in 1996, and has headed this laboratory from 1997 to 2002. Since 2003 he has been seconded to Total (Refining Marketing) where he directs their Academic Research Program.

# CATALYTIC SCIENCE SERIES

**Series Editor:** Graham J. Hutchings *(Cardiff University)*

---



**Forthcoming:**
Supported Metals in Catalysis
*edited by J. A. Anderson and M. F. Garcia*

CATALYTIC SCIENCE SERIES — VOL. 3

Series Editor: Graham J. Hutchings

# Zeolites for Cleaner Technologies


edited by

**Michel Guisnet**

*Université de Poitiers, France*

**Jean-Pierre Gilson**

*ENSICAEN and Université de CAEN, France*



ICP

Imperial College Press

*Published by*

Imperial College Press
57 Shelton Street
Covent Garden
London WC2H 9HE

*Distributed by*

World Scientific Publishing Co. Pte. Ltd.
5 Toh Tuck Link, Singapore 596224
*USA office:* 27 Warren Street, Suite 401-402, Hackensack, NJ 07601
*UK office:* 57 Shelton Street, Covent Garden, London WC2H 9HE

**British Library Cataloguing-in-Publication Data**
A catalogue record for this book is available from the British Library.

First published 2002
Reprinted 2005

**ZEOLITES FOR CLEANER TECHNOLOGIES**



ISBN 1-86094-329-2

This book is printed on acid-free paper.

Printed in Singapore by Mainland Press

# PREFACE

Clean Technologies, Environmentally Friendly Processes, Clean Products and Green Chemistry..., are terms encountered with an increased frequency by politicians as well as by scientists. This is hardly surprising. On the one hand, the remarkable innovations witnessed during the 20th century by clever chemistry have led to significant improvements in food supply, health and quality of life. On the other hand however, the manufacture, processing, use and disposal of chemicals have unfortunately resulted in significant damage to human health and to our environment. It is an important objective for the Chemical Industry to break this link between innovation and damage. A minimalist attitude would be to simply comply with environmental regulations. A more responsible approach is necessary in which "*...fundamental knowledge of chemical processes and products is applied to achieve elegant solutions with the ultimate goal of hazard-free, waste-free, energy efficient synthesis of non-toxic products without sacrificing efficacy of function ...*" (*).

The remarkable properties of zeolites allow the development of cleaner and more efficient processes for the production of fuels and chemicals. This was clearly highlighted during the lectures delivered by experts in the Petroleum Refining, the Petrochemicals and the Fine Chemicals Industries at the meeting Zeolites for Cleaner Technologies, a Pre-Conference School held in early July 2001 in Poitiers and organized before the 13th International Zeolite Conference (Montpellier, July 2001). While the emphasis was placed on catalytic processes, the significant role played by separations based on zeolite adsorbents was also demonstrated. Pollution abatement through ion-exchange, adsorption and catalysis over zeolites was also considered. The preparation of catalysts and the chemistry of catalytic processes were outlined in general lectures, both aspects being more specifically developed in the presentation of commercially proven processes or emerging technologies.

The Poitiers School was held in an informal and friendly atmosphere, giving the opportunity to world famous academic and industrial experts, as well as PhD students, to exchange their views. The key role of zeolites in sustainable development and the constant creativity and dynamism of academic and industrial researchers in the synthesis of molecular sieves, and in the design and development of new zeolite catalysts were particularly emphasized.

This book gathers the fifteen lectures presented at the Poitiers School with an introductory chapter; its targeted audience is industrial and academic chemists involved in research, development and education. The aim of the introductory chapter is to provide the basics on zeolites to newcomers to the field. It is, however, a minimum requirement, and the serious reader is encouraged to deepen and broaden his knowledge by going to the more specialized books cited in reference. The other chapters should constitute a valuable source of information for experienced researchers and serious newcomers; they will find the latest fundamental and industrial developments in zeolite science and technology with special emphasis on environmental issues.

It is our secret hope that this book will become rapidly obsolete because of the fast pace of scientific and technological accomplishments in the field of zeolites applied to our environment.

*Anastas P.T., Warner J.C., *Green Chemistry : Theory and Practice*, (Oxford University Press, New York, 1998).

November 2001

Michel Guisnet
Jean-Pierre Gilson

*The organization of the Poitiers School was made possible by the financial support of various institutions and companies; the help of the zeolite team of the Laboratory of Catalysis in Organic Chemistry from the Poitiers University is greatly appreciated. The kind assistance of this team, especially of P. Ayrault and D. Martin, during the preparation of this book is most particularly acknowledged.*

**Support and Sponsorship**

# CONTENTS

CHAPTER 1

# INTRODUCTION TO ZEOLITE SCIENCE AND TECHNOLOGY

M. GUISNET
*Laboratoire de Catalyse en Chimie Organique, Université de Poitiers, 40, Avenue du Recteur Pineau, 86022 Poitiers Cedex, France*

J.-P. GILSON
*Laboratoire de Catalyse et Spectrochimie, ISMRA-CNRS, 6, Bd du Maréchal Juin, 14050 Caen Cedex, France*

## 1.1 Introduction

The history of zeolites began some 250 years ago with the discovery by the Swedish mineralogist Crönsted of a mineral (stilbite) exhibiting intumescence when heated in a flame (1). This new family of minerals (hydrated aluminosilicates) was called zeolite from the Greek words "zeo" and "lithos" meaning "to boil" and "stone". For nearly 200 years, the beautiful and large natural zeolite crystals were only attractions displayed in museum collections. The advent of man-made synthetic zeolites and the realization that they existed not only as minor constituents in volcanic rocks but also as sizable sedimentary deposits led to their large-scale use in many applications. The synthetic zeolite chemists have been (and continue to be) particularly creative: while some 40 different zeolites were discovered in nature, approximately 130 zeolites were synthesized. Moreover, theoretical arguments indicate that a much greater number of zeolite structures is possible (2).

The first synthetic zeolites (X, Y, A) have rapidly found applications in three main areas (Table 1.1):

- **Adsorption:** first in the drying of refrigerant gas and natural gas (1955), followed by the n-/iso-butane separation on the A zeolite (ISOSIV process, 1959)
- **Catalysis:** with the use of X and Y zeolites in isomerization (1959) and cracking (1962)
- **Ion-exchange:** in detergents, A zeolites replacing the polluting polyphosphates (1974)

In the case of these three major areas of applications, the zeolite crystallite size has to be small: generally 1 $\mu$m for adsorption and catalysis, the optimal size for ion exchange in detergents being 3-4 $\mu$m. This illustrates the advantage of synthetic zeolites since the precise engineering of their properties (crystal size, composition, polarity...) is now possible in contrast to their natural counterparts.

**Table 1.1** Milestones in zeolite synthesis and applications

- **Mid 1930s-1940s**: Pioneering work of Barrer in adsorption and synthesis
- **1949-1954**: Discovery and synthesis of zeolites A, X, Y (Milton-Breck)
- **1954**: Commercialization of zeolites A, X, Y (Union Carbide). Applications in
  - drying, n-iso/-alkane separation (Union Carbide, 1959)
  - catalysis: isomerization on Y (Union Carbide, 1959), cracking on X and Y (Mobil, 1962)
  - ion exchange: detergents, A zeolites instead of phosphates (Henkel, 1974)
- **1967-1969**: Synthesis of high silica zeolites MFI and BEA (Mobil)
  Applications of MFI in shape selective processes
  Dewaxing (1981)
  Xylene isomerization and production (1974)
- **1980s**: Secondary synthesis (dealumination, isomorphous substitution)
- **1982-1986**: Synthesis of aluminophosphates, SAPO, MeAPO, ... (Union Carbide)
  Application in Isodewaxing (SAPO11, Chevron, 1997)
  Methanol to olefins (SAPO34, UOP-Norsk Hydro)
- **1983**: Synthesis of titanium silicalites TS1 (Enichem)
  Application in phenol hydroxylation (1986)
- **1992**: MCM41 mesoporous molecular sieves (Mobil)
- **1994, 1998**: Nanocrystalline zeolites, delamination (Corma)

Another key step was the demonstration by P.B. Weisz and coworkers (3-5) of the **shape selectivity** of zeolite catalysts related to **molecular sieving** (1960). This initiated further research in the synthesis of new zeolites as well as industrial applications based on this property. The first commercial shape-selective process, Selectoforming, was developed by Mobil (1968) and allowed the selective cracking of the low octane (n-alkane) components of light gasoline over a natural zeolite (erionite) (6).

Afterwards, the synthesis of various new zeolites, especially ZSM5 (MFI, 1967), the discovery of new shape selective transformations such as the (accidental) discovery of the remarkably stable and selective conversion of methanol into gasoline range hydrocarbons over HZSM5 (7), the development of post-synthesis treatments of zeolites, ... combined to make them the single most important family of catalysts used all other the world.

During the last 20 years, great progress was made in the synthesis of molecular sieves (Table 1.1) and in the understanding of the catalytic transformation of organic molecules on zeolites. This fundamental knowledge was successfully applied to introduce large-scale processes in the fields of oil refining, petrochemicals and, with increasing success, in the clean synthesis of fine chemicals.

In this book, the main catalytic processes in oil refining and petrochemicals are reviewed with special emphasis on environmental issues; they play a historic role in catalysis and illustrate nicely the interplay between chemistry, processes and products. The immense potential (hardly exploited) of zeolites in the clean synthesis of fine chemicals is demonstrated with various examples.

This introductory chapter should serve as a guide for young researchers in the field of zeolites. It highlights briefly what these materials are, what makes them so particular and suited for molecular separations and highly selective catalytic transformations. A list of reference books covering various aspects of zeolite science (structure, synthesis, characterization, applications) is given at the end of this chapter and should be consulted by the serious reader. It is also generally worth paying a visit to the original publications to better grasp the context of many new concepts and technological innovations.

## 1.2 Composition, pore structure and active sites of zeolites

Zeolites are **crystalline aluminosilicates** with the general formula $M_{2/n}O.Al_2O_3.ySiO_2$ where n is the valence of the cation M and y may vary from 2 to infinite. Structurally, zeolites are crystalline polymers based on a three-dimensional arrangement of $TO_4$ tetrahedra ($SiO_4$ or $AlO_4^-$) connected through their oxygen atoms to form subunits and finally large lattices by repeating identical building blocks (unit cells). The structural formula of zeolites (i.e., the chemical composition of the unit cells) is the following:

$$M_{x/n}(AlO_2)_x\,(SiO_2)_y$$

where n is the valence of cation M, x +y the total number of tetrahedra per unit cell and *y/x* the atomic Si/Al ratio varying from a minimal value of 1 (Lowenstein rule) to infinite.

### *1.2.1 Pore structure*

As adsorption and catalytic processes involve diffusion of molecules in the zeolite pores, only those with a minimum of 8 tetrahedral (8T) atoms apertures allowing this diffusion are generally considered. Most of the zeolites can be classified into three categories:

- *small pore zeolites* with eight membered-ring pore apertures (8 T atoms and 8 0) having free diameters of 0.30 – 0.45 nm,
- *medium pore zeolites* with ten membered–ring apertures, 0,45 – 0.60 nm in free diameter

- *large pore zeolites* with 12 membered-ring apertures, 0.6-0.8 nm.

However, zeolite-like materials with ultralarge pores such as Cloverite (20 T, 0.6 x 1.32 nm), VPI-5 (18 T, 1.27 nm), $AlPO_4$–8 (14 T, 0.79 x 0.87 nm) were recently synthesized. A comparison between the pore openings of zeolites and the kinetic diameter of guest molecules shows clearly that zeolites can be used for molecular sieving. It should however be stressed that these values are temperature-dependant: temperature increases both the flexibility of the guest molecules and the breathing motions of the host zeolite pore mouth and framework.

Zeolite structures are designated by a **3 capital letter code** according to rules set by the Commission of the International Zeolite Association (IZA). For instance, FAU stands for the faujasite structure to which the well-known X and Y zeolites belong. The $5^{th}$ edition of the *Atlas of Zeolite Framework Types*, recently published by IZA (8), describes 133 zeolite structures. Regular updates are found on the website of the IZA. A very useful short notation is used for the description of the pore system(s), each porous network is characterized by the channel direction, the number of atoms (in bold type) in the apertures, the crystallographic free diameter of the aperture (in Å), the number of asterisks indicating whether the system is one-, two- or three- dimensional. However, this short notation does not indicate whether the pore system is made of interconnected cages or uniform channels. The IZA coding of the pore systems of some commercially available zeolites is given in Figure 1.1 along with the representation of their framework.

### *1.2.2 Active sites*

Each zeolite type can be easily obtained over a wide range of compositions directly by synthesis and/or after various post synthesis treatments. Moreover, various compounds can be introduced or even synthesized within the zeolite pores (ship in a bottle synthesis). This explains why zeolites can be used as acid, base, acid-base, redox and bifunctional catalysts, most of the applications being however in acid and in bifunctional catalysis.

#### 1.2.2.1 Zeolites as acid catalysts - active sites

The most important process involving a zeolite, Fluid Catalytic Cracking (FCC) uses a catalyst containing an acid FAU zeolite (Chapter 5). Other examples of processes using acid zeolite catalysts will be examined in this book, like Methanol to Olefins (Chapter 12), Acetylation (Chapter 14) etc.

## a.1) Large pore zeolites

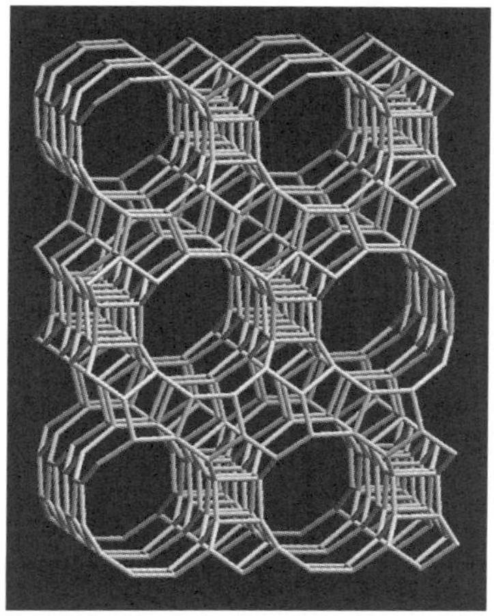

BEA (Beta): <100> **12** 6.6x6.7** ↔ [001] **12** 5.6x5.6*

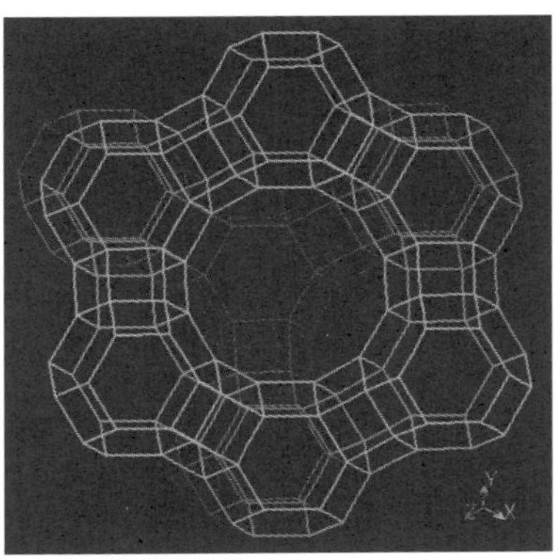

FAU (Faujasite (X,Y)):
<111> **12** 7.4x7.4***
Supercages ∅ :13 Å

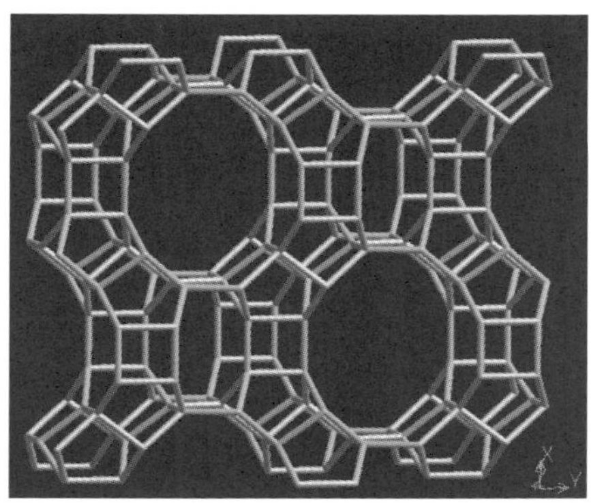

MOR (Mordenite):
[001] **12** 6.5x7.0* ↔ {[010] **8** 3.4x4.8
↔ [001] **8** 2.6x5.7}*

**Fig. 1.1** Porous network of commercially important zeolites

## a.2) Large pore zeolites

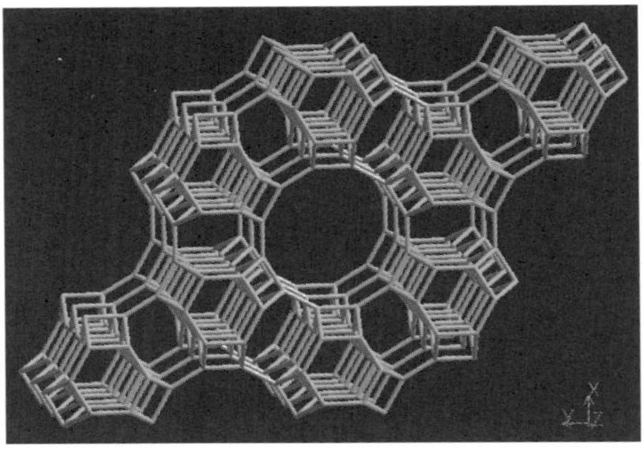

LTL (Linde Type L): [001] **12** 7.1x7.1*

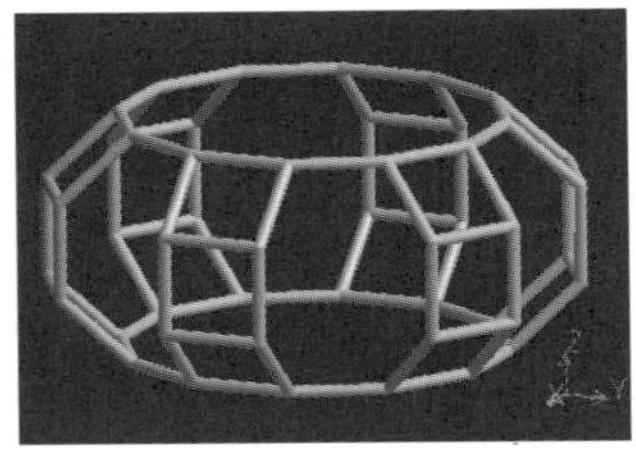

Cages: 4.8x12.4x10.7 Å

## b.1) Medium pore zeolites

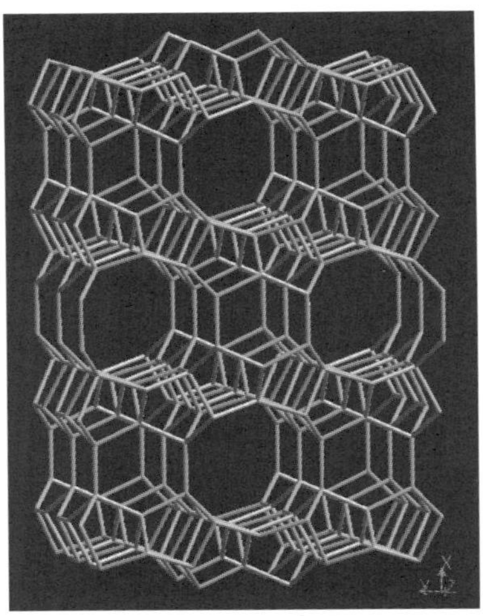

FER (Ferrierite):
[001] **10** 4.2x5.4* ↔ [010] **8** 3.5x4.8*

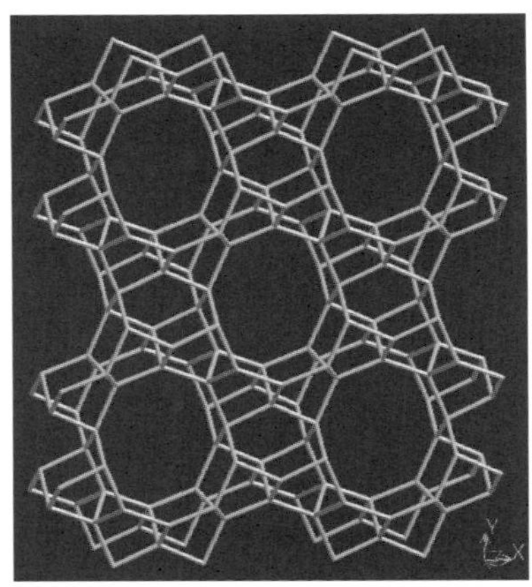

TON (Theta 1,ZSM22) :
[001] **10** 4.6x5.7*

**Fig. 1.1** (*continued*)

## b.2) Medium pore zeolites

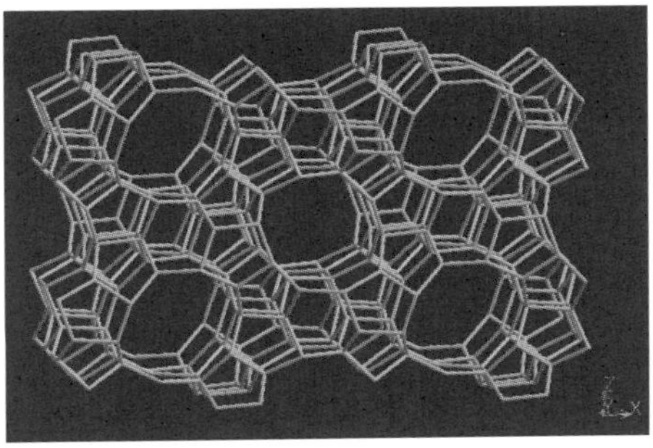

MFI (ZSM5) :
{[100] **10** 5.1x5.5 ↔ [010] **10** 5.3x5.6}***

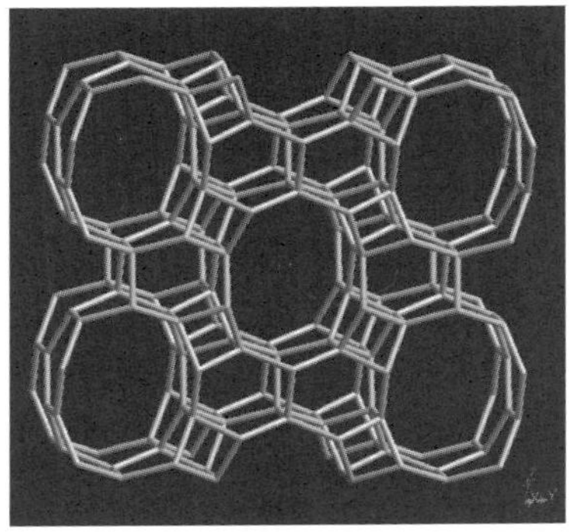

AEL (SAPO11):
[001] **10** 4.0x6.5*

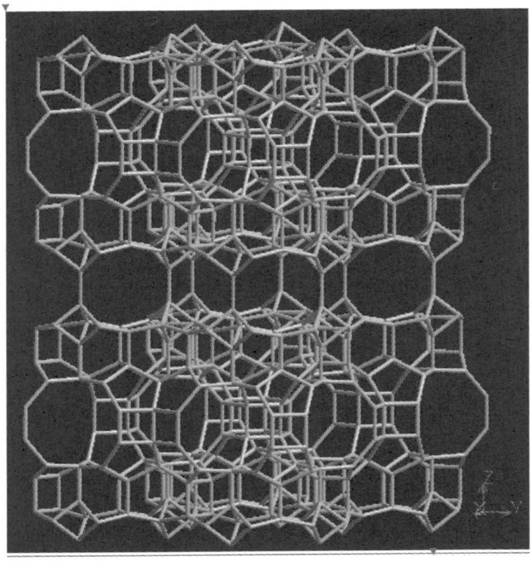

MWW (MCM22):
⊥[001] **10** 4.0x5.5** | ⊥ [001] **10** 4.1x5.1**

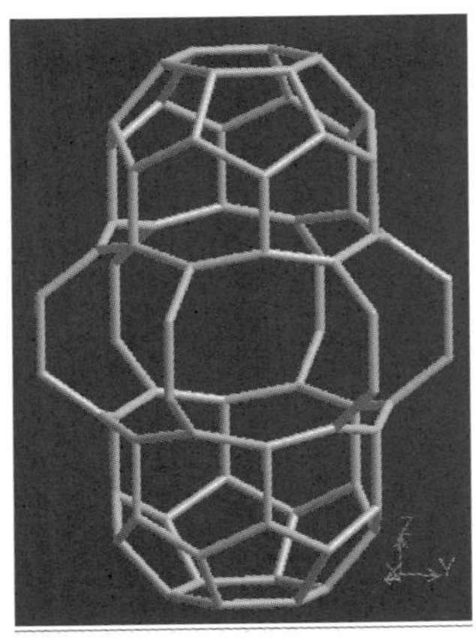

Cages: 7.1x18.4 Å

**Fig. 1.1** (*continued*)

## c) Small pore zeolites

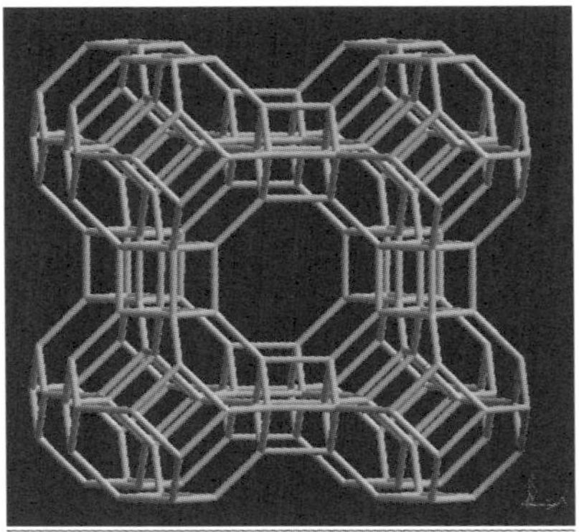

LTA (Linde Type A):
<100> **8** 4.1x4.1***
Cages ∅: 11.4 Å

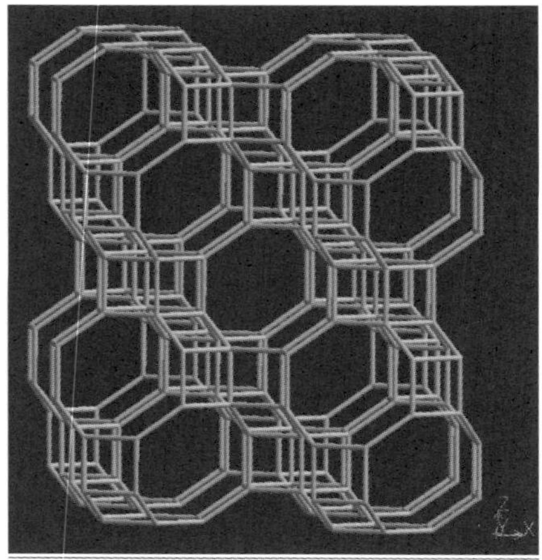

CHA (SAPO34):
⊥ [001] **8** 3.8x3.8***
Cages ∅: 6.5x11 Å

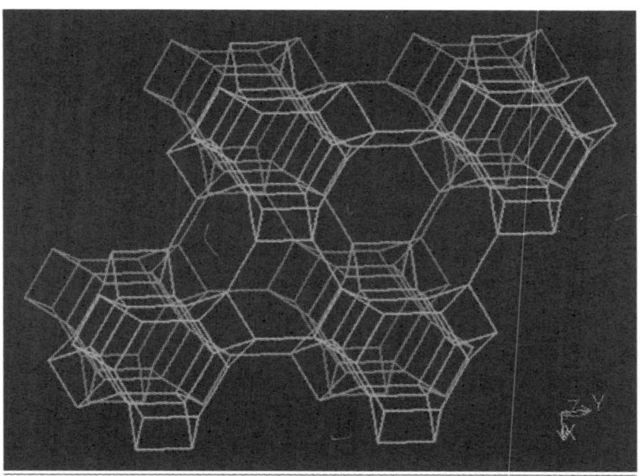

ERI (Erionite):
⊥ [001] **8** 3.6x5.1***

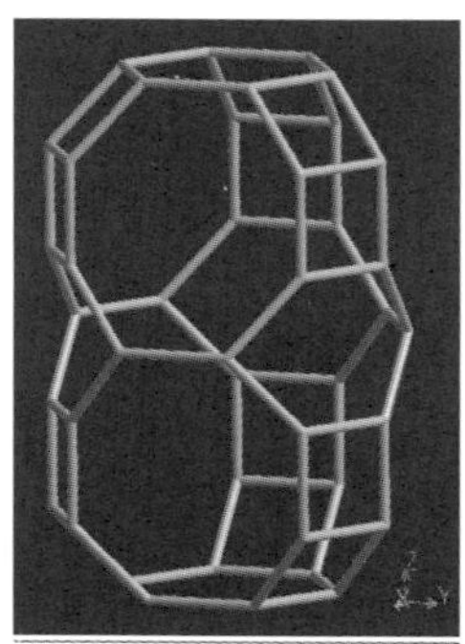

Cages: 6.3x13 Å

**Fig. 1.1** (*continued*)

Most of the hydrocarbon reactions as well as many transformations of functionalized compounds are catalyzed by protonic sites only (9). In zeolites, they are associated with bridging hydroxyl groups attached to framework oxygens linking tetrahedral Si and Al atoms : (Al(OH)Si). **The maximum number of protonic sites is equal to the number of framework aluminum atoms**, the actual number being smaller due to cation exchange, dehydroxylation and dealumination during activation at high temperatures. The number (and density) of protonic sites can therefore be adjusted either during the synthesis or during post synthesis treatments of the zeolite : dealumination, ion-exchange, etc. However, as aluminum atoms cannot be adjacent (Lowenstein's rule), the maximum number of protonic sites is obtained for a framework Si/Al ratio of 1 (8.3 mmol $H^+$ $g^{-1}$ zeolite). Moreover, no purely protonic zeolite is stable with this low framework Si/Al ratio, hence this maximum number can never be achieved.

In order to design zeolite catalysts, the parameters controlling the other features of the protonic sites and particularly their strength (Figure 1.2) have also to be assessed.

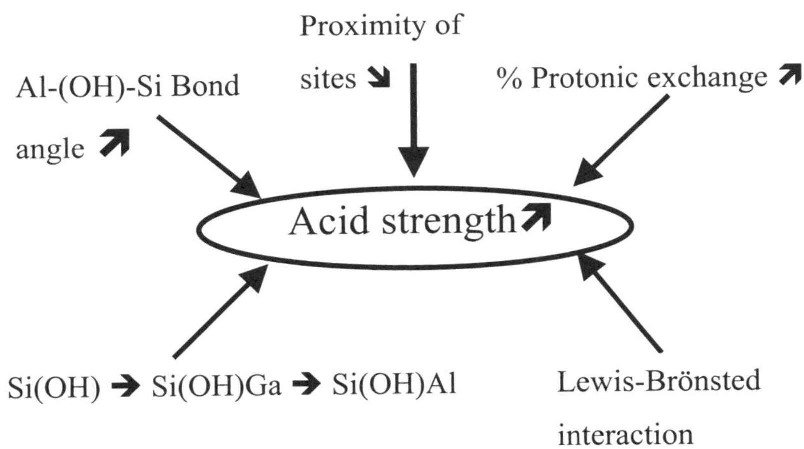


**Fig. 1.2** Parameters determining the acid strength of the zeolite protonic sites

A first feature of zeolites is their **stronger acidity compared to amorphous aluminosilicates**. This is for instance evidenced by the higher heats of adsorption of nitrogen bases. To explain this stronger acidity, Mortier (10) proposed the existence of an enhanced donor-acceptor interaction in zeolites. This interaction was extended by Rabo and Gajda (11) into a resonance model of the (Al(OH)Si) bond structure with bridging hydroxyls (I) and terminal silanols (II) as extreme limits:

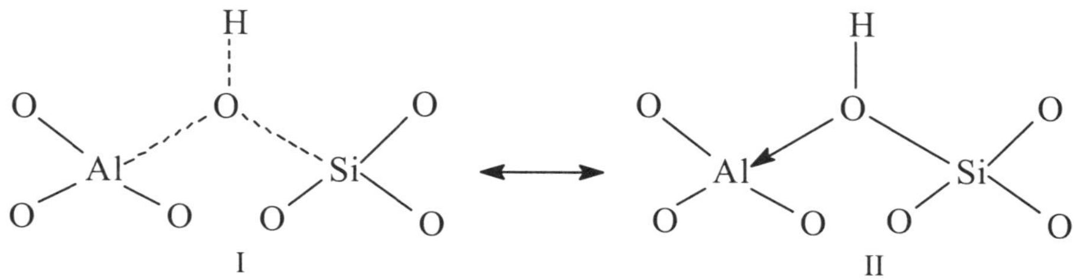

Indeed, NMR data indicates that the OH groups of amorphous aluminosilicates are primarily terminal whereas those of zeolites are primarily bridging, the interaction of O with Al weakening the OH bond and increasing the acid strength (11).

A relation exists between the T-O-T **bond angles and the acid strength of the associated proton in the zeolites** (11). The greater the angle, the stronger the sites. Thus, the protonic sites of HMOR (bond angle range between 143-180°) and HMFI (133-177°) are stronger than those of HFAU(138-147°). This explains why HMOR is and HFAU is not active in n-butane and n-hexane isomerizations at 200-250°C, two reactions requiring very strong acid sites. Again, the protonic sites (bond angles) of the zeolites are influenced by the basicity of the reactants and the temperature will also play a role.

The synthesis of **metallosilicates** containing trivalent elements in the framework other than Al (B, Ga, In, Fe) is of interest in designing the acid strength. The ranking in acid strength drawn from theoretical calculations is in relatively good agreement with acidity measurements (12). Thus, according to IR spectroscopy (wavenumber of the OH groups) and ammonia TPD over MFI samples, the acid strength is in the following order:

$$B(OH)Si < In(OH)Si << Fe(\text{-}OH)Si < Ga(OH)Si < Al(OH)Si$$

The protonic site strength depends on **the degree of exchange of Na cations** in the zeolite. As could be expected, the greater the exchange level, the stronger the protonic sites. However at high exchange degree, there is not only creation of very strong protonic sites but also an increase in the strength of the protonic sites already present in the zeolite (9).

Theoretical and experimental approaches of the **framework Si/Al ratio** effect on the acidity led to the conclusion that the strength of the protonic sites of zeolites is influenced by the presence of neighbors (11, 12). Each framework Al atom has 4 Si atoms (Lowenstein's rule) in the first surrounding layer (nearest neighbors) and, depending on the zeolite topology, 9-12 Al or Si atoms in the second layer (Next-

Nearest-Neighbors or NNN). According to the NNN concept, the acid strength of a protonic site depends on the number of Al atoms in the NNN position (13) ; the strength is maximum at 0 Al as NNN and minimum at full occupancy of the NNN sites with Al. Using statistical calculations, Wachter (14) determined that the Si/Al value for which all Al atoms were isolated was 7 for all zeolites with 9 Al or Si NNN (e.g FAU, LTA). Barthomeuf (15) improved this idea by using topological densities to include the effects of layers 1 to 5 surrounding the Al atom. The values of Si/Al for which **the protonic sites are isolated, hence have the maximum strength**, were calculated for 33 different zeolites and found between 5.8 (FAU) and 10.5 (Bikitaite-BIK).

**Extraframework aluminum species** created by mild steaming were shown to increase the catalytic activity of zeolites. This increase in activity was ascribed to the creation of sites exhibiting "enhanced" acidity through interaction of bridging hydroxyl groups (Brönsted sites) and neighboring small extraframework aluminum species (Lewis acid sites) (16). Their exact nature is still a matter of debate, but is schematically represented as:

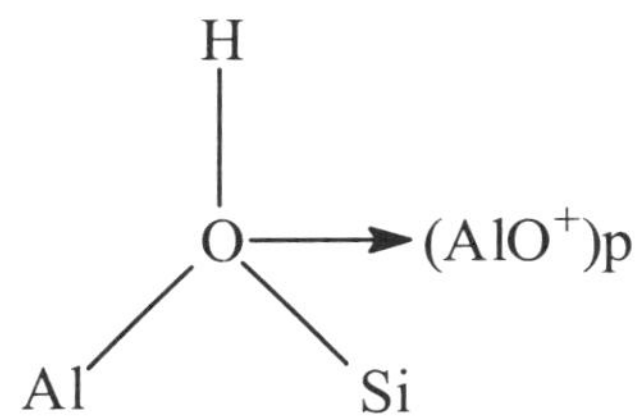


In accord with this proposal, the removal of these EFAL species by acid treatment (17) or with ammonium hexafluorosilicate causes a significant decrease in acidity and catalytic activity.

The creation by mild steaming of "enhanced acidity" sites resulting from the interaction of protonic sites and EFAL species was demonstrated in the case of HFAU (18) by the appearance in their IR spectra of two additional bands at 3600 and 3525 $cm^{-1}$. These bands result from a bathochromic shift of the high and low frequency framework hydroxyl bands of the parent zeolite (OH located in supercages and in hexagonal prisms, respectively). This bathochromic shift is in agreement with an electron withdrawal from the bridging hydroxyls by EFAL species, leading to a weakening of the OH bond, hence an increase in acid strength.

The **accessibility** of the protonic sites also plays a significant role in the catalytic activity of zeolites. Obviously this accessibility depends both on the location of the OH in the zeolite and on the size of the reactant molecules. Thus, the portion of protonic sites of HFAU zeolites located in the supercages is accessible to many organic molecules whereas the others, located in the hexagonal prisms, are inaccessible to all the organic molecules. HMOR also has protonic sites accessible (in the large channels) to many organic molecules and less accessible sites (in the side pockets). With HMFI, all the protonic sites being located at the channel intersections are equally accessible (or inaccessible) to reactant molecules. The same can be said for HERI, the protonic sites of which are located in large cages with small apertures, hence only accessible to linear organic molecules.

Finally, it is increasingly clear that molecules confined in the zeolitic nanocavities see their electronic properties modified. It has been shown for instance (19) that the dipole moment of acetonitrile increases significantly upon its introduction in the side pockets of MOR compared to the linear channels of the same zeolite. The guest molecule is made more basic and is easily protonated in such a confined environment. Zeolites also act as solid solvents and the anionic framework acts as the conjugate base of the proton thereby stabilizing some charged intermediates along concerted catalytic pathways.

### 1.2.2.2 Zeolites as redox or as bifunctional catalysts

Redox sites can be introduced in zeolites enabling them to catalyze redox reactions, or bifunctional processes when the zeolites have also acidic sites. Elements with redox character can be:

- either incorporated in the framework during hydrothermal synthesis or through post-synthesis treatments, for instance in the case of titanium zeolites
- introduced in the micropores after the synthesis by ion-exchange ($Pt(NH_3)_4^{2+}$ for instance) followed by calcination and reduction (creation of small metallic Pt clusters for example)
- associated with zeolites in mechanical mixtures (for instance an acidic zeolite mixed with $Pt/Al_2O_3$ during an extrusion process)

**Redox catalysis**

The discovery of titanium silicalite-1 (TS1) by Enichem scientists (20-22) and its commercial use as a catalyst for a variety of selective oxidations with aqueous hydrogen peroxide under mild conditions (Figure 1.3) constituted a major breakthrough in oxidation catalysis.

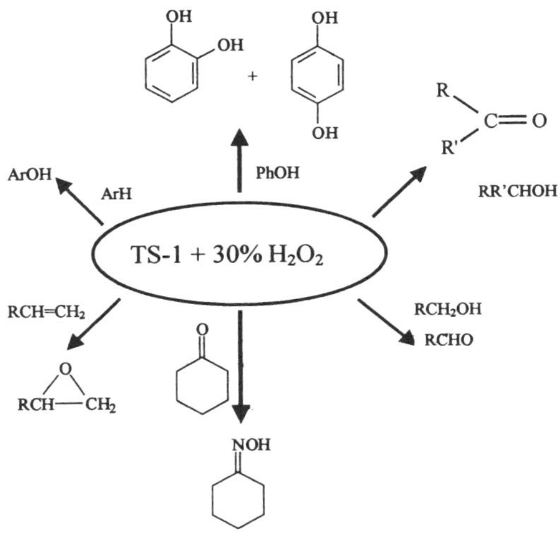


**Fig. 1.3** Oxidation with dilute aqueous hydrogen peroxide catalyzed by TS1

TS1 is a crystalline molecular sieve with the MFI structure in which titanium (IV) is isomorphously substituted for silicon in the framework. The location of Ti in the framework has been demonstrated by using several techniques: XRD, UV-Visible spectroscopy, EXAFS-XANES (23). Its remarkable efficiency for selective oxidation of various functional groups with dilute aqueous hydrogen peroxide can be attributed (24) to:

- the isolation of titanium sites preventing the undesired decomposition of $H_2O_2$ requiring two adjacent sites for its catalysis;
- the hydrophobicity of the lattice enabling the preferential adsorption of the hydrophobic substrate in the zeolite micropores in the presence of water.

The discovery of TS1 led rapidly to the development of a process for phenol hydroxylation (25). This process has numerous advantages over the previous processes using peracid or $Co^{2+}$, $Fe^{2+}$ as catalysts: higher conversion of phenol (30% instead of 5-9%) requiring less phenol separation/recycle steps, comparable or higher yields relative to both hydrogen peroxide and phenol, wider range of catechol/hydroquinone ratio (0.5-1.3 instead of 1.2-1.5 or 2.0-2.3) (24, 26).

Unfortunately, the use of TS1 (as well as TS2 discovered in 1990 by the group of Ratnasamy (27)) in catalytic oxidations is restricted to the relatively small substrates able to enter the pores of these zeolites (apertures  0.55 nm). Therefore, many research groups attempted to incorporate titanium in large pore molecular sieves: BEA zeolites, mesoporous molecular sieves: MCM41 and MCM48. Other transition metal zeolites were also synthesized and tested in oxidation; one of the main problems of these systems is the release of redox cations in liquid phase (24). Progress remains to be made to develop molecular sieves catalyzing the oxidation (olefin epoxidation, aromatic hydroxylation, etc.) of bulky molecules. A more detailed description of titanium zeolite catalysts and their applications can be found in ref. 24 and 26.

Noble metal zeolite catalysts are used in various processes, most of them occurring through bifunctional hydrogenating/acid catalysis. One exception, however, is the selective aromatization of n-alkanes (e.g. n-hexane into benzene) proceeding through monofunctional metal catalysis. Indeed the PtLTL catalyst used commercially does not present any protonic sites.

### Hydrogenating/acid bifunctional catalysis

**Bifunctional zeolite catalysts** are used in various commercial processes: light alkane hydroisomerization (chapter 7), hydrocracking (chapter 6),hydrodewaxing (chapter 8), light alkane aromatization and hydroisomerization of the $C_8$ aromatic cut (chapter 9). The hydrogenation/dehydrogenation components included in zeolite catalysts can be very different and located in different positions:

*i) Highly dispersed noble metals (Pt, Pd...)* located in the zeolite cages (e.g. PdHFAU hydrocracking catalysts) or deposited on alumina (e.g. Pt/$Al_2O_3$- HMOR catalysts for the isomerization of the $C_8$ aromatic cut)

*ii) Metal sulfides (e.g. NiMoS)* deposited on alumina associated with HFAU zeolites in hydrocracking catalysts

*iii) Metal oxides (e.g $Ga_2O_3$)* associated with a HMFI zeolite in aromatization catalysts, the hydrogenating sites being formed in a series of complex steps early in the reaction (28)

Bifunctional catalytic reactions involve a series of catalytic steps over acidic and hydrogenating-dehydrogenating sites with formation of intermediate compounds. Thus n-hexane (hydro)isomerization involves successively n-hexane dehydrogenation in n-hexenes (metal catalyzed), skeletal isomerization of n-hexenes into isohexenes over protonic acid sites followed by the (metal catalyzed) hydrogenation of isohexenes into isohexanes (Figure 1.4).

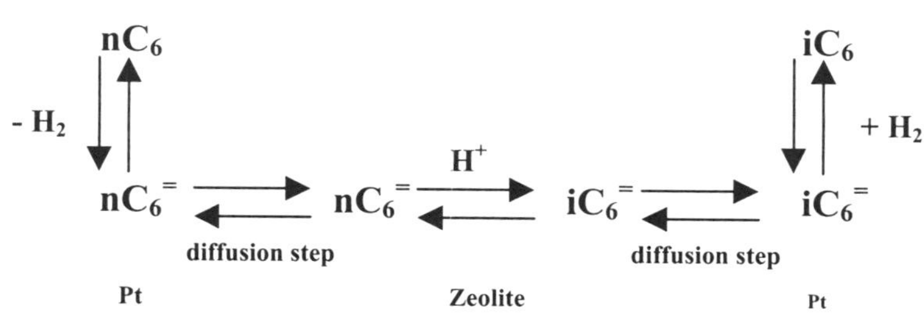


**Fig. 1.4** Bifunctional mechanism of n-hexane isomerization over Pt H zeolites

Under the operating conditions (hydrogen pressure, temperature, ...), hexenes and isohexenes are not thermodynamically favored and appear only as traces in the products. The activity, stability and selectivity of bifunctional zeolite catalysts depend mainly on two parameters: the balance between hydrogenating and acid functions (29) and the zeolite pore structure (30, 31). In most of the commercial processes, the balance is in favor of the hydrogenating function, hence the reaction is limited by the acid function.

## 1.3 Shape selective catalysis

### *1.3.1 Introduction (32, 33)*

The term **Shape Selective Catalysis** was coined 40 years ago by Weisz and Frilette (3) to describe the unexpected catalytic behavior of an LTA zeolite exchanged with $Ca^{2+}$ (CaA). They found that at 260°C, CaA dehydrated 60% of 1-butanol but did not convert isobutanol. CaA also cracked n-hexane (but not 3-methyhexane) exclusively into linear products (4). These results demonstrated that reactions occurred inside the pores of CaA (0.5 nm pore openings) making it impossible for branched molecules to enter or exit these pores. The first shape selective catalytic process **Selectoforming**, was based on this simple and elegant concept of shape selectivity by sieving (6). Table 1.2 shows the significant highlights of the history of shape selective catalysis.

**Table 1.2** History of shape selective catalysis

| | |
|---|---|
| 1925 | First molecular sieve effect (Adsorption on Chabazite, Weigel, Steinhoff) |
| 1932 | "Molecular sieve": porous material that acts as sieve on a molecular scale (Mac Bain) |
| 1960 | Shape selective catalysis through molecular sieving (Weisz-Frilette) |
| 1966 | Shape selective catalysis in Encyclopedia of Chemistry (Reynolds Publ. Co) |
| 1968 | First commercial process selectoforming (natural erionite, Weisz, Chen) |
| 1967 | Synthesis of ZSM5 (Mobil) |
| | → Many shape selective commercial processes: methanol to gasoline; dewaxing, xylene isomerization, etc. |
| 1968 | Transition state selectivity (Csicsery) |
| 1970s | Concentration Effect (Venuto, Rabo, Poutsma etc.) |
| 1984 | Nest effect→confinement effect (Derouane, 1986) |
| 1991 | Pore mouth catalysis (Martens et al.) |
| 1995 | Key-lock mechanism (Martens et al.) |

The concept of **transition state selectivity** was first proposed by Csicsery (1971) to explain why symmetrical trialkylbenzenes were not formed during disproportionation of dialkylbenzene over H mordenite although their diffusion in mordenite channels was not hindered (32). However, it was the discovery of the synthesis of intermediate pore size zeolites and especially ZSM5 that led to an explosive development of research in shape selective catalysis and to an extraordinary expansion of the catalytic applications of zeolites (34).

It has been concluded that, in most cases, catalytic reactions over zeolites occur within their intracrystalline cages and channels. Zeolite catalysts can therefore be considered as a succession of **nano** or **molecular reactors**. The consequence is that the activity, selectivity, but also the stability of all the reactions carried out over zeolite catalysts, depend (slightly or significantly) on the shape and size of cages, channels and of their apertures, hence that shape selectivity is a general characteristic of zeolite catalyzed reactions.

It should be emphasized that the active sites located on the external surface, often in small amounts compared to the inner sites (<1% for crystallites of 1 $\mu$m), play a catalytic role. Generally, this leads to a selectivity decrease, the external surface lacking the shape selective properties of the inner pores.. However, recent results show that reactions which can occur only on the external surface of zeolites or just within the pore mouth are very selective, suggesting a shape selective influence of external surface depending on the nature of the substrate (Table 1.2).

### *1.3.2 Shape selectivity due to molecular sieving*

Separation of molecules with different sizes can be achieved by a proper choice of zeolites (nature of the zeolite and adjustment of the pore architecture, especially the pore size). The simplest forms of shape selectivity come from the impossibility of certain molecules in a reactant mixture entering the zeolite pores (**reactant shape selectivity**) or of certain product molecules (formed inside the pore network) exiting from these pores (**product shape selectivity**). In practice, reactant and product shape selectivities are observed not only when the size of molecules is larger than the size of the pore apertures (**size exclusion**) but also when their diffusion rate is significantly lower than that of the other molecules. Differences of diffusivities by 2 orders of magnitude are required to produce significant selectivities between reactant species (35).

At this juncture several remarks are worth making:

i) As is the case for all heterogeneous catalysis reactions, the selectivity is not only determined by differences in diffusivities between reactant molecules, but also by differences between their reactivities (i.e., by the relative rates of diffusion and reaction - interplay between physical and chemical phenomena).

ii) Weisz (36) showed that a new regime of diffusion arises in zeolites when the dimensions of the catalyst pores approach those of molecules. In this diffusion regime, called **configurational diffusion**, even subtle differences in the molecule or pore dimensions can result in large changes in diffusivities. This regime is fundamentally different from the Knudsen and Fickian diffusions. Thus in CaA zeolite, the diffusivity of trans 2-butene was found to be at least 200 times larger than that of the cis isomer even though the two molecules differ in size by only 0.2Å (37).

iii) The nature of the intracrystalline surface of zeolites may differ significantly, ranging from highly hydrophilic for low Si/Al ratios to hydrophilic for high Si/Al ratios (38). This will significantly affect the relative adsorption of polar and non polar molecules (39).

**Reactant shape selectivity** was the basis of the Selectoforming process previously mentioned. The n-alkanes of light gasoline (essentially n-pentane, n-hexane) enter the pores of the erionite catalysts and are transformed into propane and n-butane, whereas the branched alkanes are excluded from the pores and do not react (Figure 1.5a).

**Product shape selectivity** plays a key role in various processes developed by Mobil for the selective synthesis of para dialkylbenzenes (32): selective toluene disproportionation, (STDP) toluene alkylation by methanol (TAM) or by ethylene (PET) etc. In all these processes, the bulkier ortho and meta isomers are also formed in the pores, but their exit is hampered by their slower diffusion; this selective removal of a product (para isomer) allows the bulkier isomers (ortho and meta) to be isomerized into the less bulky species (para). Above thermodynamic yields are thus possible by the simple application of LeChatelier's principle (Figure 1.5b). Thus, the diffusion coefficient of paraxylene in ZSM5 zeolite modified by coking at high temperature in toluene disproportionation, is of several orders of magnitude higher than those of ortho and metaxylene. It should be emphasized that in the case of dialkyl-benzene synthesis, the situation is ideal. Indeed, the ortho and meta isomers trapped in the pores transform rapidly into the para isomer. In other reactions, the molecules trapped in the pores transform only very slowly into desorbable molecules, their accumulation and their transformation into larger molecules ('coke') leading to a fast deactivation of the zeolite catalyst.

Venuto et al. (40) proposed that coke formation in a Y zeolite at a high temperature was an example of **reverse shape selectivity**: the bulky molecules formed in the supercages (1.3 nm) cannot escape, the pore apertures (0.7 nm) being too narrow. This reverse shape selectivity, related to molecular sieving, has, of course, a negative effect; coke molecules limiting or inhibiting the access of reactant molecules to the active sites hence causing catalyst deactivation.

As stated above, shape selectivity due to molecular sieving depends on the relative rates of diffusion and reaction, hence on the respective sizes and shapes of molecules and pores and on the characteristics of active sites (e.g. concentration, nature and strength of acid sites). Obviously the diffusion rate, hence the selectivity depend also on the length of the diffusion path (i.e., on the size of the zeolite crystallites). The selectivity of a zeolite catalyst can be optimized by an adequate choice of operating conditions (temperature especially has a large effect on the reaction rates) and by adjustment of the zeolite characteristics. For instance, in toluene disproportionation the selectivity to paraxylene of the MFI zeolite is significantly increased by deposits of silica, magnesia, coke etc, on the outer surface which limit the desorption of ortho and meta isomers (41). However, this positive effect is also partly due to the selective deactivation of the (non selective) external surface sites.

### *1.3.3 Spatioselectivity or transition state selectivity*

**Transition state shape selectivity** (or spatioselectivity) occurs when the formation of reaction intermediates (and/or transition states) is sterically limited by the space available near the active sites. This spatioselectivity depends on the size and shape of cages, channels and channel intersections. This type of selectivity was first proposed by Csicsery (34) to explain the absence of 1, 3, 5- trialkylbenzenes in the disproportionation products of dialkyl-benzenes transformation over H-mordenite although these trialkylbenzenes could diffuse in the zeolite channels. The space available in these channels was not sufficient to accommodate the diphenylmethane intermediates involved in the formation of 1, 3, 5-trialkyl benzenes; they are bulkier than those involved in the formation of 1, 2, 3 and 1, 2, 4 trialkylbenzenes (Figure 1.5 c).

Contrary to molecular sieving, spatioselectivity does not depend on the relative rates of diffusion and reaction, hence both can be easily distinguished: changing the crystallite size has no effect on spatioselectivity whereas it increases selectivity by molecular sieving. However, the two types of selectivity may act simultaneously as was shown by Song et al. (42) for naphthalene isopropylation over H-mordenite.

As many organic compounds may transform simultaneously through mono molecular (intramolecular) and bimolecular (intermolecular) processes, it is easy to understand that the shape and size of the space available near the active sites often determine the selectivity of their transformation. Indeed the transition state of a bimolecular reaction is always bulkier than that of a monomolecular reaction, hence the first type of reaction will be much more sensitive to steric constraints than the second one. This explains the key role played by the pore structure of zeolites on the selectivity of many reactions. A typical example is the selective isomerization of xylenes over HMFI: the intermediates leading to disproportionation, the main secondary reaction over non-spatioselective catalysts, cannot be accommodated at its channel intersections (32). Furthermore, if a reaction can occur through mono and bimolecular mechanisms, the significance of the bimolecular path will decrease with the size of the space available near the active sites (41).

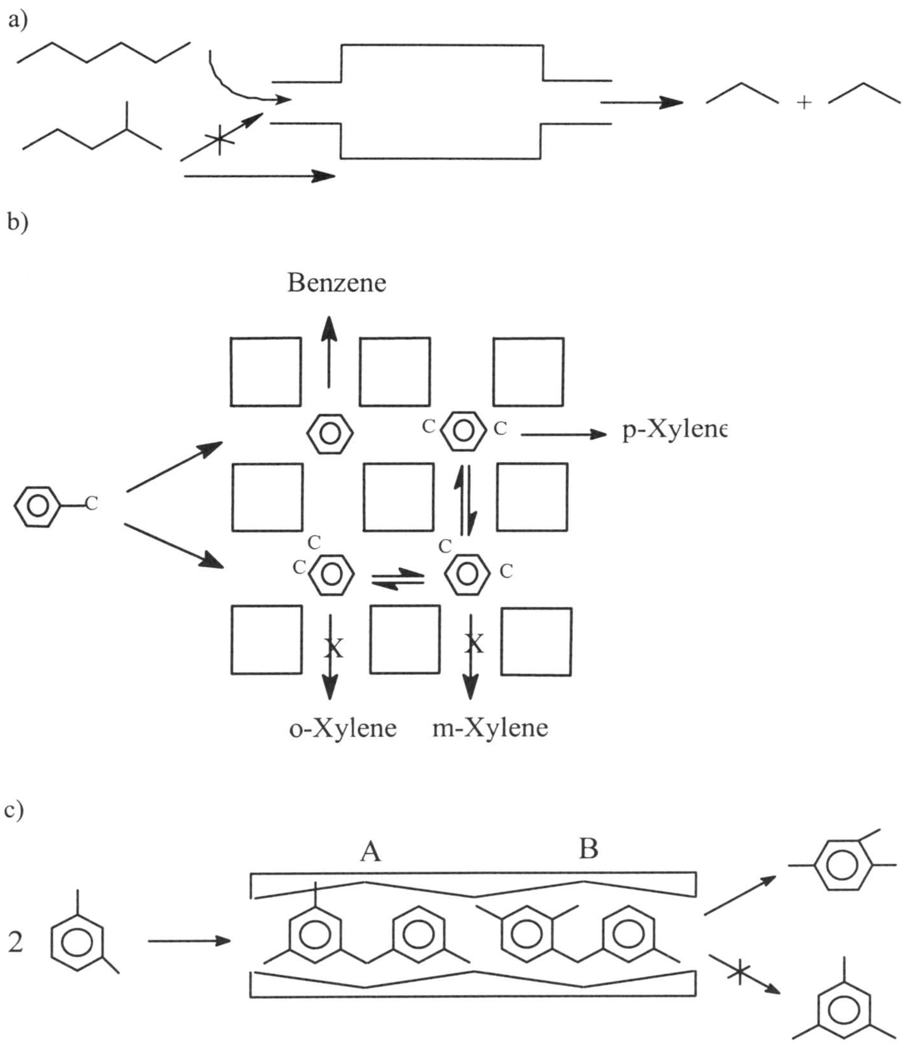


**Fig. 1.5** Schematic representation of shape selective effects: a) Reactant selectivity: Cracking of an n-iso $C_6$ mixture. b) Product selectivity: Disproportionation of toluene into para-xylene over a modified HFMI zeolite. c) Spatioselectivity: Disproportionation of meta-xylene over HMOR. The diphenylmethane intermediate A in formation of 1,3,5 trimethylbenzene is too bulky to be accommodated in the pores, which is not the case for B

Spatioselectivity plays a significant role in the formation of the heavy secondary reaction products responsible for deactivation ('coke'). Indeed, coke formation involves various bimolecular steps (condensation, hydrogen transfer) that, as indicated above, are very sensitive to steric constraints. Therefore, the rate of coke formation will greatly depend on the size and shape of cages, channels and their intersections. However, as discussed earlier in 1.3.2, coke formation involves not only chemical steps (involving spatioselectivity) but also physical retention in the zeolite pores due to steric blockage (reverse shape selectivity), at least at high reaction temperatures (43, 44).

### *1.3.4 Shape selectivity related to molecular concentration in zeolite micropores*

The interactions between organic molecules and the pore walls of similar size are very strong (Type I adsorption isotherms) and zeolites may be considered as **solid solvents** (11, 15, 45). The reactants' concentration in zeolite micropores is therefore considerably higher than in the gas phase with a significant positive effect on the reaction rates. This effect is all the more pronounced as the reaction order is greater, favoring more the bimolecular over the monomolecular reactions.

The concentration of reactant molecules in the zeolite micropores is largely responsible for the observation that zeolite activities of zeolites are much higher than that of more conventional catalysts (11, 46, 47). It is the case in catalytic cracking (FCC) where REHY zeolites were found, depending on the hydrocarbon reactant, 10 to 10,000 times more active than amorphous silica alumina. Moreover, the selectivity was completely different, the gasoline fraction in the cracked product being richer in aromatics and alkanes (at the expense of naphthenes and alkenes) on zeolites than on silica alumina. This drastic change in selectivity is due to different ratios between the rates of hydrogen transfer (bimolecular reaction) and cracking (monomolecular reaction), much higher on zeolites than on silica alumina (46). The effect is dramatic as far as the gasoline yield is concerned: paraffins and aromatics being more 'refractory' towards cracking, the 'zeolitic' gasoline is less prone to secondary cracking and yields (but not octane) are much higher on zeolites than on amorphous silica aluminas. The use of zeolites spread so rapidly in FCC for this very reason.

In other cases however, the high concentration of some reactants leads to undesired side reactions. In the alkylation of isobutane with butenes, zeolites are very efficient catalysts but lack stability because olefins are more strongly adsorbed than the paraffin and over a matter of minutes, their oligomerization takes over the alkylation reaction and deactivates the catalyst by pore blocking.

### *1.3.5 Other types of shape selectivity*

Various other types of shape selectivity have been proposed:

The **cage** or **window effect** was proposed by Gorring (48) to explain the non-linear effect of chain length observed in hydrocracking of various n alkanes over T zeolite, chabazite (CHA) and erionite (ERI). Thus, when a $nC_{22}$ alkane is cracked over erionite, there are two maxima in the size distribution of the product molecules at carbon numbers of 4 and 11 and a minimum at carbon number of 8. The diffusivities of n-alkanes also change in a similar periodic manner by over two orders of magnitude between the minimum at $C_8$ and the maxima. This shows that for diffusion, and hence for shape selective effects, not only the size but also the structure of the reactant and product molecules need to be considered.

The concept of **molecular traffic control** was proposed by Derouane and Gabelica (49) to explain the unexpected absence of counter-diffusion effects during methanol conversion over zeolites such as MFI presenting interconnected channels with different sizes and tortuosity: the smallest molecules (e.g. methanol) would diffuse through the sinusoidal channels while the bulkiest (e.g. aromatics) exit through the slightly larger linear channels.

The diffusion of molecules in the channels of monodimensional zeolites may present peculiarities, accounting for shape selective effects. A first type of shape selectivity may be influenced by the diffusion in single file (**single file diffusion**) (50) preventing bimolecular transformations (51, 52). Another type of shape selectivity called '**tunnel shape selectivity**' has been recently proposed (53) to explain the high selectivity to orthoxylene during metaxylene isomerization over MCM 41 mesoporous molecular sieves. The xylene molecules entering the non interconnected monodimensional channels of these molecular sieves undergo successive disproportionation and transalkylation reactions with consequently a purely bimolecular mode of xylene isomerization.

Fraenkel et al. (54) were the first to propose that the external surface of zeolites could be responsible for shape selective catalysis. Acid sites located in the half cavities on the external surface of HMFI would be responsible for the selective formation of 2,6– and 2,7–dimethylnaphthalene during naphthalene methylation (nest effect). This explanation was afterwards rejected on the basis of adsorption experiments. However, a nest effect was recently proposed to be responsible for the shape selective properties of the MCM-22 (MWW) zeolite and its delaminated analog (ITQ-4) in aromatics alkylation (55).

The possibility of selective reactions on the external surface of zeolites, more exactly at the **pore mouth,** was recently addressed by Martens et al. (56, 57) to explain the unusual selectivity of several intermediate pore size zeolites and especially of ZSM-22 (TON) in long chain n-alkanes isomerization. Over PtHTON, this isomerization is very selective towards monobranched isomers even though they cannot desorb from the narrow channels of this monodimensional zeolite. **Pore mouth catalysis** could also be responsible for the selective isomerization of n-butene into isobutene observed over aged HFER samples. However, in this case carbonaceous compounds trapped in the pores in the vicinity of the external surface were proposed to be the active species (58). Furthermore, the second branching of n-alkanes over PtHTON was shown to occur only in positions determined by the distance between pore apertures at the surface of the zeolite crystallites. This type of selectivity was considered to be due to a **key-lock catalysis** (57) i.e. similar to what occurs in enzymatic catalysis.

This concept of zeolites as enzyme mimics was used by Derouane and Vanderveken (59) to explain the selective aromatization of n-hexane on Pt/LTL catalysts: confinement effects combined with the unique pore structure of LTL zeolite would be responsible for the fast and selective conversion of n-hexane to benzene.

## 1.4 Conclusion: Toward a scientific design of zeolite catalyst

The design of solid catalysts for a specific reaction is a difficult task still largely based on trial and error, including of course the new sophisticated techniques of 'High Throughput Experimentation'. In the case of zeolite catalysts, the approach can be based on more scientific concepts available (e.g. shape selectivity) and on well-established relationships between active sites properties and zeolite composition. Moreover, various well-known methods exist for tailoring the pores and the active sites of zeolites: ion exchange, dealumination, grafting or deposition of elements on the external surface of crystallites, etc. (60). There are already many examples of commercial zeolite catalysts developed through an essentially 'ab-initio' approach:

- erionite catalysts used in selectoforming, a process based on the simple concept of reactant shape selectivity
- MFI catalysts selectivated by coking at high temperature and used for selective disproportionation of toluene into paraxylene (product shape selectivity)
- Pt-mordenite catalyst for isomerizing $nC_5$-$C_6$ alkanes with an optimal acidity (presence of very strong protonic sites due to interaction of framework hydroxyl groups with extraframework aluminum species)

The high degree of flexibility in the tailoring of zeolites properties partly explains why only a relatively small number of zeolites are used commercially today: only a dozen of zeolites (and their modifications) are presently in use while more than 130 structures are known today (61).

Even emerging concepts such as those related to the shape selectivity of the external surface of the crystallites (nest effect, pore mouth catalysis, ...) have already been applied to develop new catalytic processes: isodewaxing, selective cumene synthesis, etc. The development of methods for obtaining large external surfaces such as the synthesis of nanocrystallites (62, 63), delamination (64, 65), mesoporous zeolites (66), ... should lead to new applications of this concept. Of course, molecular modeling will play an ever increasing role in the rational design of zeolite based catalysts: zeolites are indeed prime candidates for thought or virtual experiments as well as computer screening of their properties for specific reactions.

Whereas zeolite catalysts have led to the development of many environmentally friendly processes in the refining and petrochemical industries, only few are used commercially in the field of fine chemicals (Table 1.3). Many research reports have however demonstrated that acid zeolite catalysts (and mesoporous molecular sieves), were active and selective in the synthesis of organic compounds and could substitute the highly polluting and corrosive acids ($AlCl_3$, $H_2SO_4$, $H_3PO_4$) still often in use. As is the case with the other solid catalysts, one of the main limitations of using zeolites catalysts in this field is related to the difficulty of substituting rapidly well established processes (non-catalytic or homogeneous catalysis) with completely different processes (heterogeneous catalysis); the relatively small volume of fine chemicals also makes large capital expenditures for new plants and costly R&D programs difficult. The relatively fast deactivation of zeolite catalysts during the synthesis of organic compounds is another important reason. Their deactivation is often due to heterogeneity in the acid strength of the active sites as well as to severe diffusion limitations of bulky and polar reactants or products. The consequence is that secondary transformations into bulkier products ending up trapped in the zeolite micropores, block the access to incoming reactant molecules ('coke' formation). Therefore, the choice of operating conditions, reactor type and zeolite characteristics (smaller crystallites, mesopores, delamination, etc.) favoring product desorption is crucial.

All these aspects were thoroughly discussed by lecturers and participants during the round table organized during the Poitiers School on "The Future Trends in Zeolite Applications". Special emphasis was placed on the role played by the sites at the external surface (pockets, etc.) or at the pore mouth, by mesopores, extraframework aluminum species, as well as by the polarity of reactant and product molecules. Other important topics dealt with the remarkable catalytic properties of BEA zeolites for fine chemical synthesis, the potential of mesoporous molecular sieves, zeolitic membranes and the role of combinatorial catalysis in the development of zeolite catalysts. It is our hope that the fruits of these discussions will appear in the literature or even better as new and environmentally friendly products or processes.

**Table 1.3** Commercial processes using zeolite catalysts

A. Refining:

| Process | Feed | Products or goal | Catalyst |
|---|---|---|---|
| Cracking(FCC) | Vacuum distillates (and residues) | Gasoline<br>Light olefins | 15-40%HFAU (USHY-REY)+MFI |
| Selectoforming | Light gasoline | Octane number ↗ | Erionite doped with Ni |
| Hydrocracking<br><br>Dewaxing | Vacuum gasoils<br><br>Middle distillates and lubricants | Gasoline Middle distillates<br>Cold flow<br>Properties ↗ | Pd/HFAU<br>NiMOS.<br>$/Al_2O_3$/HFAU<br>Ni/MFI-MOR |
| Hydroisomerization * | $nC_4$<br>Light gasoline ($nC_5$,$nC_6$) | $iC_4^=$<br>$iC_5$-$iC_6$ | Pt HMOR<br>Pt HMOR |
| Isomerization | $nC_4^=$<br>$nC_4^=$-$nC_5^=$ | $iC_4^=$<br>$iC_4^=$-$iC_5^=$ | HFER<br>HMFI, HFER |
| Oligomerization | $C_3^=$ | Diesel oil | HMFI |

* with separation of n-isoalkanes on 5A zeolites (LTA)

B. Petrochemicals:

| | | | |
|---|---|---|---|
| Aromatization | $C_6$-$C_7$ | Benzene-toluene | Pt KLTL |
| Aromatization | $C_3$-$C_4$ | Benzene-toluene xylenes (BTX) | Ga MFI |
| Xylene isomerization* | $C_8$ aromatics (xylenes + ethylbenzene) | Paraxylene<br>Paraxylene-benzene | Pt HMOR<br>Pt HMFI |
| Toluene disproportionation<br>Selective toluene disproportionation | Toluene | Xylene-benzene<br><br>Paraxylene-benzene | HMOR<br><br>HMFI |
| Transalkylation | Toluene-trimethylbenzenes<br>Benzene-diisopropylbenzenes<br>Benzene-diethylbenzenes | Xylenes<br>Cumene<br><br>Ethylbenzene | HMOR<br>HMOR<br><br>HMOR |
| Benzene alkylation | Ethylene (ethanol)<br>Propene<br>Long-chain alkenes | Ethylbenzene<br>cumene<br>Linear alkylbenzenes(LAB) | HMFI/HBEA<br>HBEA/HMOR/HMWW<br>? |
| Methanol to olefins | Methanol<br>Methanol | $C_3^=$,$C_4^=$<br>$C_2^=$, $C_3^=$ | HMFI<br>SAPO34 |

* with separation of paraxylene on BaX zeolites (FAU)

C. Specialty and Fine Chemicals:

| | | | |
|---|---|---|---|
| Cracking of MTBE | Methyl tertiobutylether | $iC_4^=$ | BMFI |
| Oxidation | Phenol ($H_2O_2$) | Catechol, hydroquinone | Ti silicalite (TS1) |
| Hydration | Cyclohexene | Cyclohexanol | HMFI |
| Amination with $NH_3$ | Methanol<br>Ethyleneoxide<br>Acetaldehyde<br>Formaldehyde<br>Acetaldehyde<br>Isobutene | Methylamines, dimethylamine<br>Ethanolamines<br>Methylpyridines<br>Pyridine + methylpyridines<br><br>Tertiobutylamine | RHO, CHA<br>?<br>MFI<br>MFI<br><br>MFI |
| Isomerization | 3-phenylpropanal<br>Chlorotoluene<br>Dichlorotoluene | Phenylacetone<br>Metachlorotoluene<br>Metachlorotoluene | ?<br>?<br>? |
| Acetylation with acetic anhydride | Anisole<br>Veratrole | Paramethoxyacetophenone<br>Dimethoxyacetophenone | HBEA<br>HFAU |

*A more detailed presentation of commercial processes using zeolite catalysts can be found in Ref. 67*

## General books on zeolites


- Breck D.W., *Zeolite Molecular Sieves: Structure, Chemistry and Use*, (Wiley John and Sons, 1974).
- Rabo J., ed., *Zeolite Chemistry and Catalysis* ACS Monograph 171, 1976.
- Barrer R.M., *Hydrothermal Chemistry of Zeolites*, (Academic Press, 1982).
- Jacobs P.A. and Martens J.A., "Synthesis of High-silica Aluminosilicate Zeolites", *Stud. Surf. Sci. Catal.* **33** (1987).
- Chen N.Y., Garwood W.E. and Dwyer F.G., "Shape Selective Catalysis in Industrial Applications", *Chemical Industries* **36** (Dekker Marcel, 1989).
- van Bekkum H. et al., eds., "Introduction to Zeolite Science and Practice", *Stud. Surf. Sci. Catal.* **58** (1991).
- Derouane E.G. et al., eds., *Zeolite Microporous Solids: Synthesis, Structure and Reactivity*, NATO ASI Series, C, **352** (Kluwer, 1992).
- Barthomeuf D. et al., eds., *Guidelines for Mastering the Properties of Molecular Sieves*, NATO ASI Series, B **221**, (Kluwer).
- Nagy J.B. et al., *Synthesis, Characterization and Use of Zeolite Microporous Materials*, (Deca Gen Ltd., 1988).
- Song C. et al., eds. *Shape Selective Catalysis* ACS Symposium Series **738** (2000).

**From the International Zeolite Association**

- Baerlocher Ch., Meier W.M. and Olson D.H. *Atlas of Zeolite Framework Types*. 5th Revised Edition (Elsevier, 2001).
- Treacy M.M.J. and Higgins J.B., *Collection of Simulated XRD Powder Patterns for Zeolites*, 4th Revised edition (Elsevier, 2001).
- Robson H. Ed., Lillerund K.P., *Verified Syntheses of Zeolite Materials*, 2nd revised Edition (Elsevier, 2001).

**References**


1. Cronstedt A.F., *Akad. Handl. Stockholm,* **18** (1756) 120.
2. Dyer A. *An Introduction to Zeolite Molecular Sieves*, (Wiley, Chichester, 1988).
3. Weisz P.B. and Frilette V.J., *J. Phys. Chem.* **64** (1960) 382.
4. Weisz P.B., Frilette V.J., Maatman R.W. and Mower E.B., *J. Catal.* **1** (1962) 307.
5. Weisz P.B., *Erdoel und Kohle* **18** (1965) 525.
6. Chen N.Y., Mazuik J., Schwartz A.B. and Weisz P.B., *Oil and Gas J.* **66** (1968) 154.
7. Chang C.D. and Silvestri A.J., *J. Catal.* **47** (1977) 249.
8. Baerlocher Ch., Meier W.M. and Olson D.H. , eds.,, *Atlas of Zeolite Framework Types*, 5th Revised Edition (Elsevier, Amsterdam, 2001).
9. Guisnet M., and Edimbourg in *Supported Catalysts and their Applications,* Ed. Sherrington D.C. and Kybett A.P., (The Royal Society of Chemistry, Cambridge, 2001) 55.
10. Mortier W.J., *Proceeding 6th Int. Zeolite Conference*, Ed. Olson D. and Bisio A., (Butterworth, Guildford, 1984) 734.
11. Rabo J. and Gajda G.J., in *Guidelines for Mastering the Properties of Molecular Sieves*, Ed. Barthomeuf D. et al., **221** (NATO ASI Series B: Physics, Plenum Press, New York, 1990) 273.
12. Martens J.A., Souverijns W., van Rhyn W. and Jacobs P.A., in *Handbook of Heterogeneous Catalysis,* Ed. G. Ertl et al., **1** (Wiley, 1997) 324.
13. Pines L.A., Maher P.J., Wachter W.A., *J. Catal.* **85** (1984) 466.
14. Wachter W.A., *Proceedings 6th Int. Zeolite Conference*, Ed. Olson D. and Bisio A., (Butterworth, Guildford, 1984) 141.
15. Barthomeuf D., *Materials Chemistry and Physics* **17** (1987) 49.
16. Mirodatos C. and Barthomeuf D., *J. Chem. Soc. Chem. Commun.* **181**, 38.
17. Wang Q.L., Giannetto G. and Guisnet M., *J. Catal.* **130** (1991) 471.
18. Khabtou S., Chevreau T. and Lavalley J.C., *Microp. Mater.* **3** (1994) 133.
19. K. Smirnov and Thibault-Starzyk F., *J.Phys.Chem B.* **103**, (1999) 8595.
20. Taramasso M., Perego G. and Notari B., US Patent **4410501** (1983).

21. Taramasso M., Manara G., Fattore V.and Notari B., US Patent **4666692** (1987).
22. Notari B., *Stud. Surf. Sci. Catal.* **37** (1988) 413.
23. Vayssilov G.N., *Catal. Rev. Sci. Eng.* **39** (1997) 209.
24. Clerici M.G., in *Fine Chemicals through Heterogeneous Catalysis,* Ed. Sheldon R.A. and van Bekkum H., (Wiley-VCH, Weinheim, 2001) 538.
25. Esposito A., Taramasso M. and Neri C., US Patent **4396783** (1983).
26. Bellussi G. and Perego C. in *Handbook of Heterogeneous Catalysis,* Ed. Ertl G., Knözinger H. and Weitkamp J., **5** (Wiley, 1997) 2334.
27. Reddy J.S., Kumar R. and Ratnasamy P., *Appl. Catal.* **582** (1990) L1.
28. Guisnet M. and Gnep N.S., *Catalysis Today*, **31** (1996) 275.
29. Weitkamp J., *Ind. Eng. Chem. Prod. Res. Dev.* **21** (1992) 550.
30. Guisnet M., Alvarez F., Giannetto G. and Perot G., *Catalysis Today* **1** (1987) 415.
31. Martens J.A., Thielen M. and Jacobs P.A., *Stud. Surf. Sci. Catal.* **46** (1989) 49.
32. Chen N.Y., Garwood W.E., Dwyer F.G. , eds., "Shape Selective Catalysis in Industrial Applications", *Chemical Industries*, **36** (1989).
33. Song C., Garces J.M. and Sugi Y. , eds., "Shape Selective Catalysis", *ACS Symposium Series* **738** (1999).
34. Csicsery S.M., *J. Catal.* **23** (1971) 124.
35. Weisz P.B. in ref. 32, p.18.
36. Weisz P B., *Chemtech* **3** (1973) 498.
37. Chen N.Y. and Weisz P.B. *Chem. Eng. Prog. Symp.* **73** (1967) 86.
38. Chen N.Y., *J. Phys. Chem.* **80** (1976) 60.
39. Chen N.Y. and Miale J.N., US Patent **4420561** (1983).
40. Venuto P.B.and Hamilton L.A., *Ind. Eng. Chem. Prod. Res. Develop.* **6** (1967) 190.
41. Olson D.H. and Haag W.O., *ACS Symposium Series* **248** (1984) 275.
42. Song C., Garces J. and Sugi Y., in ref. 32 p.1.
43. Guisnet M., Magnoux P. and Martin D., *Stud. Surf. Sci. Catal.* **111** (1997) 1.
44. Guisnet M. and Magnoux P., *Stud. Surf. Sci. Catal.* **88** (1994) 53
45. Derouane E.G., *J. Mol. Catal. A: Chemical* **134** (1998) 29.
46. Gates B.C., Katzer J.R. and Schuit G.L.A. , eds., *Chemistry of Catalytic Processes*, (Chemical Engineering Series, McGraw Hill Book Company, New York, 1979) Ch.1.
47. Fraissard J., *Stud. Surf. Sci. Catal.* **5** (1980) 343.
48. Gorring R.L., *J. Catal.* **31** (1973) 13; Gorring R.L. and Danills R.H., *ACS Symposium Series* **248** (1984) 51.
49. Derouane E.G., Gabelica Z., *J. Catal.* **65** (1980) 486.
50. Kärger J. et al., *J. Catal.* **136** (1992) 283.
51. Kopelmann R., *Science* **241** (1988) 1620.

52. Adeeva V. et al., *J. Catal.* **151** (1995) 364.
53. Guisnet M., Morin S. and Gnep N.S., in ref. 32, p.334.
54. Fraenkel R., Cherniavsky M. and Levy M., *Proc. 8$^{th}$ Intern. Congress Catal.*, **4** (Dechema, Frankfurt am Main, 1984) 545.
55. Corma A., Martinez-Soria V. and Schnoeveld E., *J. Catal.* **192** (2000) 163.
56. Martens J.A., Parton R., Uytterhoeven L., Jacobs P.A., Froment G.F., *Appl. Catal.* **76** (1991) 95.
57. Martens J.A. et al., *Angew. Chem.* **34** (1995) 252.
58. Andy P. et al., *J. Catal.* **173** (1998) 322.
59. Derouane E.G. and Vanderveken D., *Appl. Catal.* **45** (1988) L15.
60. Stostak R., *Stud. Surf. Sci. Catal.* **58** (1991) 153.
61. Marcilly, Ch., *Stud. Sruf. Sci. Catal* **135** (2001) 37.
62. Schoeman B.J., Sterte J. and Otterstedt J.E., *Zeolites* **14** (1994) 110.
63. Camblor M.A. et al., *Stud. Surf. Sci. Catal.* **105** (1997) 341.
64. Corma A. et al., *Nature* **396** (1998) 353.
65. Corma A., Diaz V., Domine M. and Fornés V., *J. Am. Chem. Soc.* **122** (2000) 2804.
66. Jacobsen C.J.H. et al., *J. Am. Chem. Soc.* **122** (2000) 7116.
67. Tanabe K. and Höelderich W. F., *Appl. Catal. A : General* **181** (1999) 399.

CHAPTER 2

# THE CHEMISTRY OF CATALYTIC PROCESSES

A. CORMA, A. MARTÍNEZ
*Instituto de Tecnología Química, UPV-CSIC*
*Avda. de los Naranjos s/n, 46022 Valencia, Spain*

## 2.1 Introduction

Zeolites have found wide application as catalysts in the oil refining and petrochemical industry, where they have been gradually replacing amorphous catalysts. The superior catalytic performance of zeolites is related to some important properties, namely:

a) high concentration of active sites, whose number and strength can be modified in a wide range,
b) high thermal and hydrothermal stability, which is important in processes occuring at high reaction temperatures and when high temperature regeneration is required,
c) shape selectivity, which allows for directing the reaction pathway towards the formation of desired products by means of reactant/product size exclusion or transition state selectivity.

At present, the oil refining industry is faced with important challenges, such as the processing of heavier and more contaminated crudes, the increasing demand for higher quality transportation fuels with reduced emissions of contaminants, and the need for more petrochemical feedstocks (e.g. olefins, aromatics). In this context, there is no doubt that zeolites (and related molecular sieves) can help refiners to achieve the new goals. Recent advances in zeolite synthesis and post-synthesis modifications are expected to contribute to the development of improved catalysts and processes.

In most of the processes, zeolites are involved in acid-catalyzed reactions that proceed through the formation of carbocation-like intermediates. Therefore, the chemistry of the catalytic processes on zeolite catalysts is closely related to the chemistry of the carbocations with the particularities imposed by the restricted microporous environment. The development of new zeolite catalysts with improved activity and/or selectivity necessarily requires knowledge of the nature of the active sites and the chemistry involved in a given catalytic process.

Here we will describe the main aspects of the chemistry involved in selected zeolite-catalyzed processes in the field of oil refining and petrochemistry, such as short paraffin aromatization, skeletal isomerization of n-paraffins and n-olefins, isoparaffin/olefin alkylation, and catalytic cracking.

## 2.2 Aromatization of short alkanes (LPG)

Aromatics are important raw materials for the petrochemical industry and they are also used as high octane blending components of gasoline. On the other hand, the availability of liquified petroleum gas (LPG) in the refinery is expected to increase as the FCC units are operated at higher severities in order to maximize the production of butenes, isobutane and propylene. Therefore, there is a clear incentive in converting the low value LPG into aromatics.

Aromatization of short alkanes can be carried out on a purely acidic HZSM-5 zeolite. In this case, activation of the alkane is thought to occur by protonation on the zeolite acid sites, with formation of a penta-coordinated carbonium ion (1). In the case of propane, protonation at a C-H bond will lead to the formation of $H_2$ and a *sec*-propyl cation (dehydrogenation), while protonation at a C-C bond will produce methane and a primary ethyl cation (cracking):

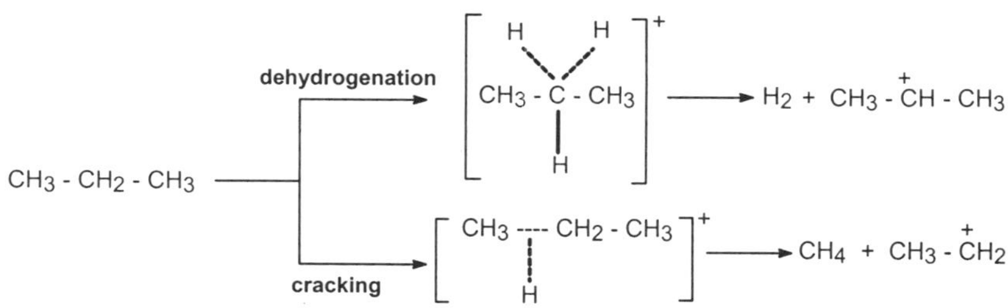


The carbocations can be desorbed as olefins by giving back a proton to the zeolite. The ethylene and propylene formed can then undergo oligomerization, cyclization, and dehydrogenation by hydride transfer to give aromatics, as it is observed when the reaction is carried out on H-ZSM5 at temperatures above 500°C and long contact times. However, aromatization on purely acidic HZSM-5 produces large amounts of methane and ethane, thus limiting the formation of aromatics.

The rate of alkane dehydrogenation and the yield of aromatics are significantly enhanced when a metal, such as Ga, is present in close proximity to the zeolite acid sites, as shown in Table 2.1. In fact, Ga/H-ZSM-5 is the catalyst used for LPG aromatization in the Cyclar process developed by BP/UOP (2). The catalyst typically contains gallium in an amount between 1 and 5 wt%, and the H-ZSM-5 has a Si/Al ratio in the range of 15 to 30 to achieve optimum performance. A significant advantage of the Cyclar process that makes it economically attractive is that significant amounts of hydrogen, a refinery deficitary and high valuable product, are also produced during the aromatization reaction. Yield of aromatics and hydrogen of ca. 65% and 5% are typically obtained in the Cyclar process. These yields are little affected by the relative proportions of propane and butane in the feed stream. The catalyst becomes deactivated after a relatively short time on stream, and the process is carried out in a series of adiabatic stacked reactors with continuous regeneration (3).

**Table 2.1** Influence of the proximity of the Ga and $H^+$ sites on propane conversion and product selectivity (TOS= 4 h)

| Catalyst | MFI | Two beds $Ga/\gamma\text{-}Al_2O_3$ + MFI | Ga/MFI |
|---|---|---|---|
| Conversion, % | 29 | 23 | 20 |
| Selectivity, % | | | |
| $C_1 - C_3$ | 52 | 51 | 25 |
| $C_4 - C_6$ | 8 | 15 | 3 |
| BTX* | 39 | 33 | 65 |
| $C_{9+}$ | 1 | 1 | 7 |

* B= benzene, T= toluene, X= xylenes.

Gallium can be introduced by impregnation, ion exchange or directly incorporated in framework positions during the synthesis of the ZSM-5 zeolite (4, 5). In the latter case, most of the Ga is removed from the framework during calcination, resulting in a more active and selective catalyst than the original sample. These findings suggest that the gallium species active in dehydrogenation are extraframework species, probably of the type $Ga_2O_3$ and/or GaO(OH) (6), which are also present in samples prepared by impregnation or ion exchange. In fact, when Ga-H-ZSM-5 is prepared by calcination of a physical mixture of $Ga_2O_3$ and HZSM-5 zeolite, especially if the mixture is molturated, the activity and selectivity of the final catalyst is very similar to the one prepared by impregnation (7, 8, 9). It is, however, difficult to determine the exact nature of the active extraframework Ga species. In a recent study using *in-situ* X-ray adsorption at the Ga *K*-edge, Meitzner et al. (10) have suggested the presence of highly dispersed $GaH_x$ hydride species, formed upon reduction of $Ga^{3+}$ during hydrogen pretreatment or during propane reaction, coordinated to basic zeolite oxygens.

Besides Ga, other metals such as Zn (11, 12) and Pt (13) have also been used in combination with ZSM-5 zeolite for $C_2$-$C_4$ aromatization. However, besides aromatization, Pt also catalyzes other undesired reactions, such as hydrogenolysis, hydrogenation and dealkylation that leads to excessive formation of methane and ethane, and limits the selectivity to aromatics. Therefore, Ga- and Zn-ZSM-5 catalysts are preferred over Pt-ZSM-5 except, perhaps, in the case of the more refractory ethane, in where a higher dehydrogenating function is needed to activate the reactant. The catalytic performance of Ga and Zn/ZSM-5 for propane aromatization is compared in Table 2.2. The results obtained on the purely acidic H-ZSM-5 are also included in the table. As observed, a higher conversion and yield of aromatics is obtained for the Ga/ZSM-5 catalyst.

**Table 2.2** Conversion of propane on H-ZSM-5 (Si/Al= 23), 1.8 wt%Zn/H-ZSM-5, and 1.2 wt%Ga/H-ZSM-5 zeolites at 550°C, 101 kPa $C_3$, and W/F= 10.8 g h $mol^{-1}$

| | Catalyst | | |
|---|---|---|---|
| | H-ZSM-5 | Zn/H-ZSM-5 | Ga/ZSM-5 |
| Conversion, % | 56.3 | 72.5 | 85.5 |
| Yield of aromatics, % | 16.9 | 36.1 | 61.7 |
| Product distribution, % | | | |
| $CH_4 + C_2H_6$ | 43.0 | 42.3 | 22.7 |
| $C_2H_4 + C_3H_6 + C_4H_8$ | 21.8 | 7.0 | 4.4 |
| $C_4H_{10}$ | 4.1 | 0.7 | 0.4 |
| $C_{5+}$ | 1.1 | 0.1 | 0.4 |
| Aromatics | 30.0 | 49.9 | 72.1 |
| Aromatic distribution, % | | | |
| Benzene | 25.4 | 42.2 | 40.2 |
| Toluene | 46.2 | 36.8 | 40.7 |
| $C_8$ alkylaromatics | 26.0 | 14.1 | 12.6 |
| $C_{9+}$ | 2.4 | 6.9 | 6.5 |

As stated above, the aromatization of short alkanes is carried out in presence of bifunctional catalysts, in where the dehydrogenating function is given by the metal component (Ga, Zn, Pt) and the H-ZSM-5 zeolite carries the acid sites. Although there is still some uncertainty concerning the initial activation of the alkane, probably both the metal and the zeolite acid sites are involved in this step. Metal sites can dehydrogenate the alkane to give the corresponding alkene, which can then be protonated on the Brönsted acid sites of the H-ZSM-5 zeolite to produce the carbocation.

Experimentally it is observed that the rate of propane dehydrogenation is faster when both $H^+$ and Ga are present in Ga/ZSM-5 catalysts (14). In order to explain this promoting effect of protons, Meriaudeau and Naccache proposed a bifunctional mechanism involving the following steps (15):

a) Dissociative adsorption of the alkane with formation of gallium hydride and gallium alkoxide species:

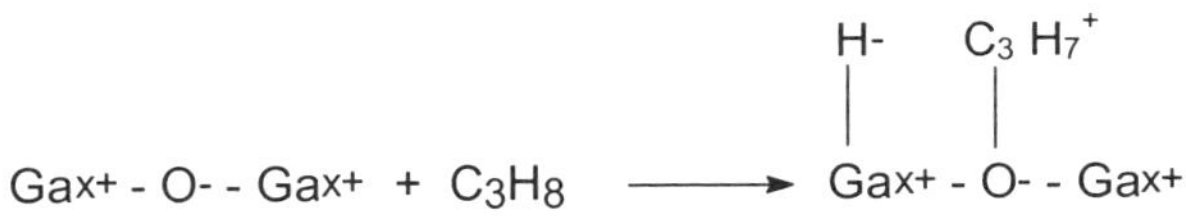


b) Rapid exchange of the alkyl carbenium ions with zeolite protons:

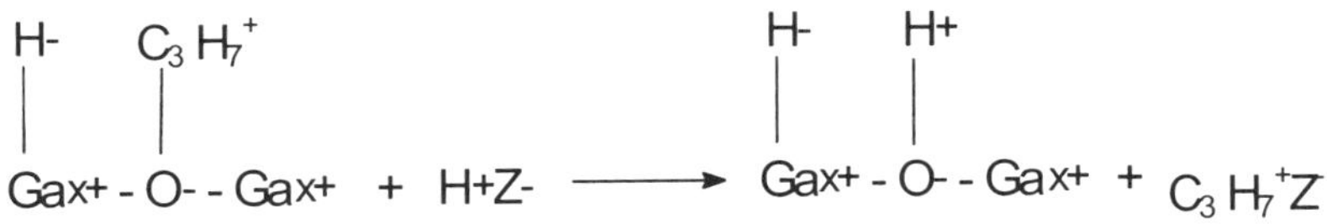


c) A rapid desorption of propene from the zeolite:

$$C_3H_7^+Z^- \longrightarrow C_3H_6 + H^+Z^-$$

Once the alkane has been dehydrogenated, the alkene formed oligomerizes on the zeolite acid sites. Some of the oligomers produced can also crack on the protonic sites to form lighter olefins. The oligomers are further dehydrogenated on the metal sites and the dienes formed undergo cyclization on the acid sites. Finally, dehydrogenation of the cyclic olefins will lead to the formation of aromatics. The alkylaromatics formed can also undergo dealkylation, transalkylation, and isomerization reactions on the acid sites of the HZSM-5 zeolite. A general reaction pathway for the aromatization of short alkanes on monofunctional H-ZSM-5 and bifunctional Ga/H-ZSM-5 catalysts is given in Fig. 2.1.

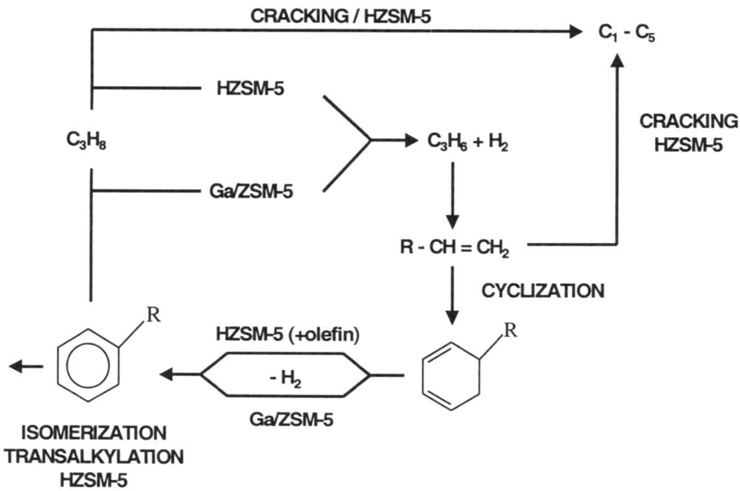


**Fig. 2.1** Simplified reaction mechanism for the aromatization of propane on HZSM-5 and bifunctional (Ga, Zn, Pt)/HZSM-5 catalysts

## 2.3 Skeletal isomerization of short n-olefins

Skeletal isomerization of $C_4$-$C_5$ n-olefins has been recognized as an attractive alternative for increasing the availability of isobutene and isopentenes (16), which are used as raw materials for the production of high octane blending MTBE and TAME oxygenates and other chemical products.

The skeletal isomerization of $C_4$ and $C_5$ n-olefins is an acid-catalyzed reaction requiring relatively strong acid sites that proceeds via carbenium ion intermediates formed upon protonation of the double bond (17). Double bond *cis-trans* isomerization usually occurs on the acid sites before skeletal isomerization. The general reaction mechanism for branching isomerization is depicted in Fig. 2.2. Protonation of the double bond leads to a secondary carbenium ion, which then rearranges into a protonated cyclopropane (PCP) structure. In the case of n-butenes, skeletal isomerization would involve the formation of a less stable primary carbenium ion.

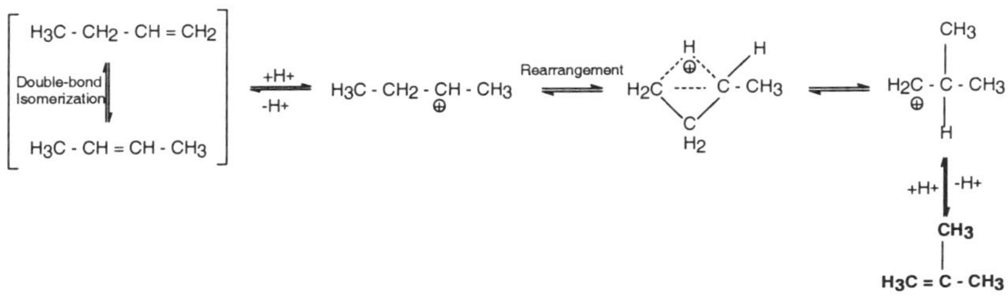


**Fig. 2.2** Reaction pathway for the monomolecular isomerization of n-butenes on acid catalysts

From a practical point of view, the skeletal isomerization should be carried out at temperatures as low as possible, since at low temperatures the equilibrium is shifted toward the formation of branched products. Unfortunately, at low temperatures the selectivity to isoolefins decreases due to the competing olefin oligomerization reactions. The extent of oligomerization can be decreased by using high reaction temperatures and low olefin partial pressures. However, at higher temperatures other undesired reactions, such as cracking of the dimers (producing mainly $C_3$ and $C_5$ fragments from octenes), hydrogen transfer and coking, leading to catalyst deactivation, also occur (18).

Zeolites have been studied for this reaction with promising results. Medium pore are preferred over large pore zeolites because bimolecular dimerization-oligomerization and hydrogen transfer reactions leading to coke precursors are diminished within the restricted space in the narrower channels of the former. Among the medium pore zeolites, better results were obtained for those having a monodirectional channel structure, such as ZSM-22, Theta-1, and ZSM-23. The selectivity to isobutene has been shown to improve by decreasing the density of Brönsted acid sites (increasing Si/Al ratio) (19, 20, 21, 22) and by decreasing their acid strength by replacing framework $Al^{3+}$ by other trivalent cations, such as $Ga^{3+}$ and $Fe^{3+}$ (23, 24, 25, 26).

Recently, a catalyst based on ferrierite zeolite has been reported in a Shell patent (27). In contrast to other medium pore zeolites, ferrierite produced high yields of isobutene with excellent stability with time on stream at low temperatures (350ºC) and without any feed dilution. This was ascribed to the particular structure of ferrierite having intersecting 10MR (4.2 x 5.4 Å) and 8MR (3.5 x 4.8 Å) channels which, according to the authors (28), induced the selective formation of isobutene by a bimolecular mechanism involving the formation of trimethylpentene dimers and their cracking into $C_4$ fragments (including isobutene) before they can escape from the zeolite pores. Such a mechanism is illustrated in Fig. 2.1. The dimerization-cracking mechanism as the prevailing mechanism for isobutene formation has been recently questioned on the basis of isotopic experiments using $^{13}$C-labelled molecules (29) and kinetic considerations (opposite effect of reaction temperature and n-butene partial pressure on isobutene and by-products formation) (30, 31).

In order to explain the beneficial effects of coke deposits on the isobutene selectivity of Ferrierite, Guisnet et al. (32, 33) proposed a pseudo-monomolecular pathway that avoids the involvement of a highly unstable primary carbocation. In this mechanism the active sites are carbenium ions attached to the coke molecules at the pore mouth of the zeolite (Fig. 2.2). n-Butene reacts with those carbenium ions to form a secondary carbenium ion that isomerizes into a tertiary carbenium ion via methyl and hydride shifts, and then suffers β-scission to produce isobutene while regenerating the active site.

## 2.4 Skeletal isomerization of n-paraffins

### *2.4.1 Isomerization of gasoline-range n-paraffins*

Isomerization of n-paraffins in the $C_5$-$C_6$ range, such as those present in the LSR (light straight run) fraction, is industrially carried out to improve the octane number of the gasoline. Skeletal isomerization of n-paraffins is an acid-catalyzed reaction that is thermodynamically favored at lower temperatures. Therefore, acid catalysts with strong acidity have to be used in order to perform the reaction at temperatures as low as possible. The process is carried out in the presence of hydrogen and a bifunctional catalyst, which typically consists of a noble metal (Pt) supported on an acidic carrier.

According to the accepted bifunctional mechanism (Fig. 2.3), the n-paraffin is first dehydrogenated on a metal site to give the corresponding olefin, which then diffuses to the acid site where it is protonated to form a carbenium ion. Then the carbenium ion can suffer branching isomerization through the formation of a protonated cyclopropane (PCP) intermediate (34, 35), or it can undergo cracking by β-scission to produce an olefin and a smaller carbenium ion.

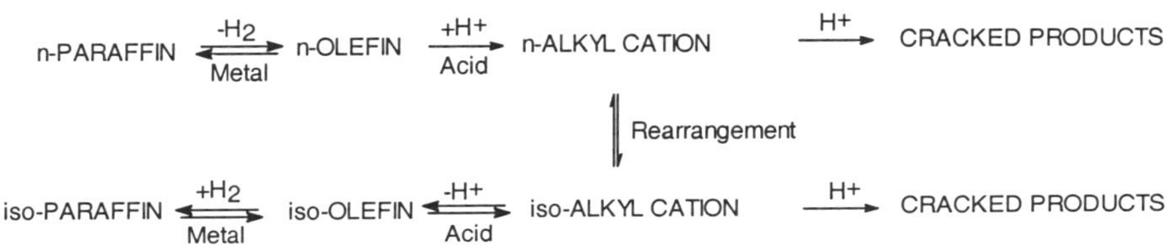


**Fig. 2.3** Bifunctional mechanism for n-paraffin isomerization

Recently, it has also been proposed (36) that desorption of the branched carbenium ion can also occur by $H^-$ species formed by dissociation of the hydrogen reactant on the Pt centers and its diffusion to the acid sites by a hydrogen spillover mechanism:

$$\text{iso-}C_n^{+} + H^{-} \cdots\cdots H^{+} \leftrightarrow \text{iso-}C_n + H^{+}$$

In order to obtain high yields to the branched isomers, an adequate balance between the metal and the acid function is required. According to Guisnet et al. (37), this is achieved for $Pt/H^+$ ratios above 0.15. Under these conditions, the rearrangement of the carbenium ion becomes the controlling step of the process. Too high metal/acid ratios produce an increase in the formation of $C_1$-$C_2$ hydrocarbons by hydrogenolysis, whereas too low metal/acid ratios favor the cracking of the branched paraffins into lighter products.

The first generation of LSR isomerization catalysts consisted of Pt supported on chlorinated or fluorinated alumina. These catalysts are used at temperatures below 200ºC. However, they suffer from three major drawbacks:

a) Deactivation by water
b) Low sulfur resistance
c) Halide has to be continuosly added into the stream, with the corresponding corrosion problems

Pt supported on zeolites, and particularly on modenite, have also been successfully commercialized for LSR isomerization. Pt/mordenite catalysts are less active than Pt/Cl-$Al_2O_3$, and are used at higher reaction temperatures (250ºC), but they have the advantage of being more resistant to water and sulfur poisons present in the feed.

A maximum catalytic activity was obtained for Pt/mordenite with a Si/Al ratio of about 10, for which all the framework Al atoms are isolated, and consequently, present the highest acid strength (38, 39). However, if the only factor controlling the zeolite acidity is the framework Si/Al ratio, one should obtain the same activity for a given Si/Al ratio, irrespective of the method of dealumination used. It is found experimentally that the maximum isomerization activity for Pt/Mordenite catalysts occurs at different Si/Al ratios depending on the method of dealumination used (40, 41, 42, 43, 44). This indicates that other factors, besides the framework Si/Al ratio, play an important role. One of these factors is the presence of extraframework Al (EFAL) species. A higher activity was obtained for the zeolite with Si/Al= 15 dealuminated by acid treatment, followed by a further dealumination by mild steaming. In this zeolite small and controlled amounts of EFAL are generated, which can have a synergetic effect on the Brönsted acid sites of the zeolite associated with framework Al (FAL), increasing their acid strength (45). When the amount of EFAL in the zeolite was optimized, a maximum activity was found for a FAL/EFAL ratio of about 3 (Fig. 2.3). Due to the effect of EFAL on the acid strength of the zeolite Brönsted sites, the maximum activity is obtained at a higher framework Si/Al ratio, thus decreasing the coking tendency of the zeolite and improving catalyst life (46, 47).

It would be highly desirable to extend the isomerization reaction to n-paraffins larger than $C_6$. In this case, high isomerization yields, especially of higher octane multibranched isomers, has to be achieved. It has to be considered, however, that the cracking tendency of the branched paraffins increases with an increase in the degree of branching and the length of the hydrocarbon chain. In fact, Pt/mordenite produces low yields of isomers during n-heptane isomerization because of extensive recracking of the isoheptanes formed. As observed in Fig. 2.4, much better catalytic performance for n-heptane isomerization is obtained when Pt is supported on a nanocrystalline Beta zeolite (48). This is explained by the combination of Brönsted acid sites of a lower acid strength than mordenite and a faster diffusion of the branched isomers through the small crystallites (10-20 nm) of the nanocrystallite Beta sample, thus decreasing the probability of recracking.

*2.4.2 Isomerization of n-butane to isobutane*

According to the Brouwer's mechanism (35), non-branching (scrambling) and branching rearrangement of carbenium ions occur through a common PCP intermediate, as depicted in Fig. 2.4 for the $C_4H_9^+$ cation. In this case, opening of the cyclic intermediate at side **a** leads to a n-butyl cation in which a terminal and a non-terminal carbon atom have exchanged positions (C scrambling), while opening of the intermdiate at side **b** would lead to the tert-butyl cation (branching isomerization) through a highly unstable primary carbenium ion.

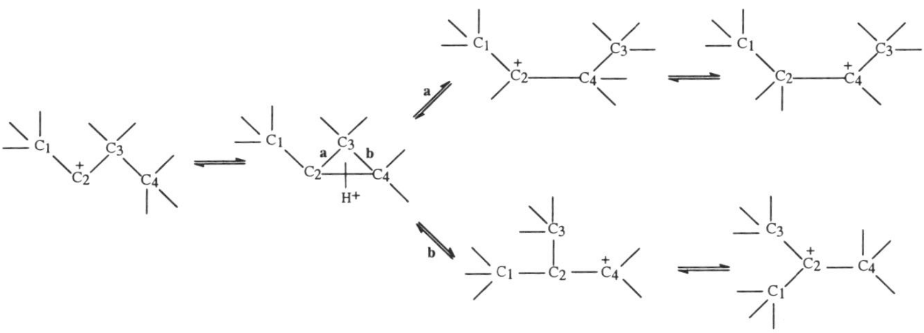


**Fig. 2.4** Isomerization of *sec*-butyl cations via PCP intermediates

However, in a theoretical study using ab initio methods which include electron correlation and extended basis sets, Corma et al. (49) found that the protonated cyclopropane (PCP) species is not a common intermediate for these reactions, but a transition state. These authors proposed a new mechanism wherein the protonated methyl-cyclopropane ring is the transition state for the carbon scrambling reaction, and the isomerization of the linear n-butyl cation into the branched tert-butyl cation occurs through a primary cation, as shown in Figure 2.5. The activation energies calculated assuming this mechanism are in very good agreement with those obtained experimentally.

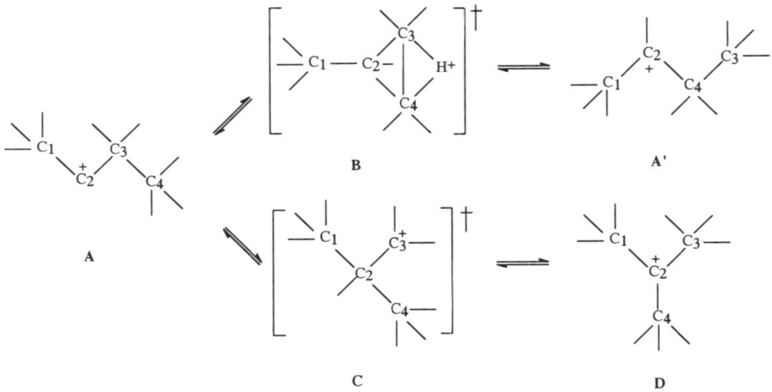


**Fig. 2.5** Alternative mechanism proposed for the skeletal isomerization of *sec*-$C_4^+$

Isobutane can also be formed from n-butane by a bimolecular dimerization-cracking mechanism (Fig. 2.6). In this case, a $C_8^+$ intermediate is formed by reaction of n-butene (produced by dehydrogenation of n-butane or by deprotonation of a butyl cation) to give a dimethylhexyl cation. This can be rearranged into a trimethylpentyl cation (for example 2,2,4-trimethylpentyl cation), which can then be easily cracked giving isobutene and a highly stable *tert*-butyl cation. Isobutene can then be hydrogenated to form isobutane, or protonated to give a *tert*-butyl cation, which can react with isobutene to continue the chain mechanism, or converted to isobutane by hydride transfer.

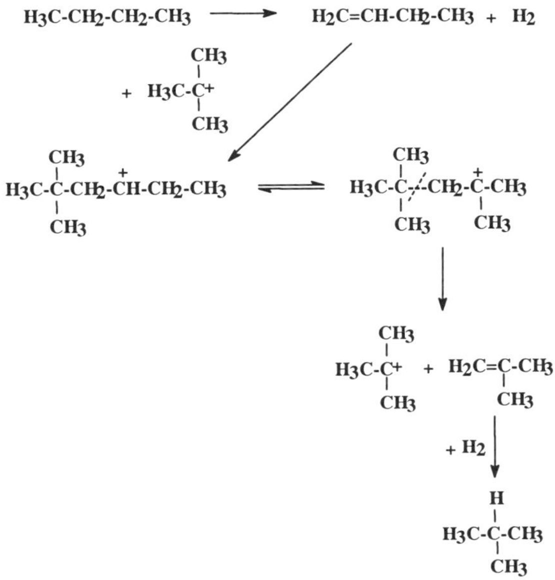


**Fig. 2.6** Formation of isobutane from n-butane by a dimerization-cracking mechanism involving branched $C_8^+$ intermediates

Isomerization of n-butane is even more thermodynamically limited than isomerization of $C_{5+}$ n-paraffins. In order to carry out the reaction at low temperatures, very strong acid sites are required. In fact, isomerization of n-butane hardly takes place on Pt/mordenite catalyst under conditions where isomerization of n-pentane and n-hexane occurs. At higher reaction temperatures isobutane starts to be formed on Pt/mordenite but with a low selectivity, suggesting a different mechanism operating for n-$C_4$ and higher n-paraffins isomerization on this type of catalysts. In fact, the relatively large amounts of $C_3$ and $C_5$ products observed besides isobutane on Pt/zeolite catalysts (Table 2.3) strongly suggests that isobutane is preferentially formed by the bimolecular dimerization-isomerization-cracking mechanism described above.

**Table 2.3** Isomerization of n-butane on mordenite zeolite (250°C, 1 atm)

| Time on Stream (min) | 5 | 20 | 25 |
|---|---|---|---|
| n-$C_4$ conversion (%) | 60 | 22 | 15 |
| Selectivity | | | |
| iso-$C_4$ | 49 | 65 | 74 |
| $C_3$ | 35 | 22 | 21 |
| $C_5$ | 16 | 13 | 5 |

Solid 'superacids' of the sulfated zirconia type were found active for n-butane isomerization at low reaction temperatures (50). These catalysts, however, were rapidly deactivated with time on stream. The isomerization selectivity and the stability of sulfated zirconia catalysts can be incerased by the introduction of Pt and by carrying out the reaction in the presence of $H_2$. Higher catalytic activities were obtained when Pt was impregnated after the impregnation of zirconia gel with 0.5 M $H_2SO_4$ (51). Sulfated zirconia promoted with Fe or Mn showed an even higher activity than unpromoted SZ for the low temperature isomerization of n-butane (52).

## 2.5 Isobutane/butene alkylation

Environmental concerns are driving forces to find suitable solids acids which can be used instead of the hazardous HF and $H_2SO_4$ commercial catalysts for the production of high octane alkylation gasoline (53). Zeolites could be, in principle, good candidates for this reaction, and thus zeolite Y, either in the protonic form or exchanged with di- or trivalent cations, has been widely studied (54, 55, 56, 57, 58, 59). On the basis of the nature of the products formed on both crystalline aluminosilicates and $H_2SO_4$ (Table 2.4), it appears that the general reaction mechanism operating on zeolites is the same as that on liquid acids (53).

**Table 2.4** Comparison of product composition obtained during isobutane/2-butene alkylation on $H_2SO_4$ and a solid silico-aluminate catalyst

| Compound | Al-Si (wt%)[a] | | $H_2SO_4$ (vol%) |
|---|---|---|---|
| iso-P | 3.40 | 5.78 | 4.16 |
| 2,3-DMB + MP | 3.59 | 3.46 | 4.58 |
| 2,4-DMP | 4.14 | 3.58 | 2.37 |
| 2,2,3-TMB | 0.21 | 0.10 | |
| All MH | 0.35 | 0.32 | |
| 2,3-DMP | 1.67 | 1.26 | 1.38 |
| 2,2,4-TMP | 16.22 | 17.70 | 30.64 |
| All DMH | 8.59 | 8.89 | 9.02 |
| 2,2,3-TMP | 3.54 | 3.97 | |
| 2,3,4-TMP + 2,3,3-TMP | 47.34 | 48.16 | 41.55 |
| 2,2,5-TMH | 0.78 | 0.71 | 1.88 |
| Other $C_{9+}$ | 9.21 | 6.06 | 4.41 |

[a] Different catalysts and conditions used.

This involves a chain mechanism in which the initiation step is the protonation of the olefin, followed by a hydride transfer from a molecule of isobutane to the *sec*-butyl carbocation to give n-butane and a *tert*-butyl cation:

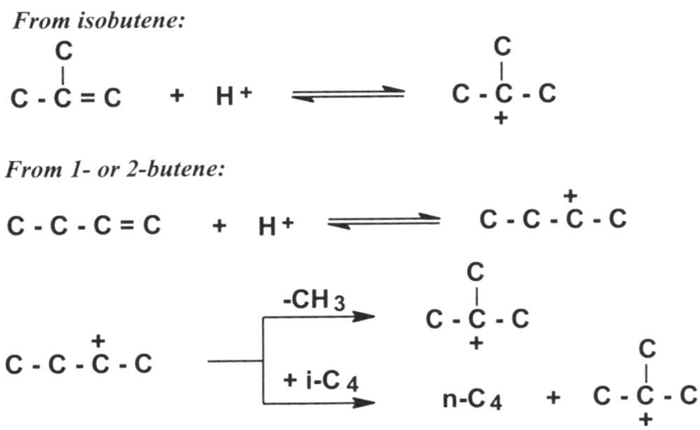

Then, the *tert*-butyl cation intermediate can attack a molecule of butene to give the corresponding $C_8$ carbocation. Depending on the particular butene isomer that is alkylated, a different $C_8$ carbocation will be formed (2,2,4-TMP$^+$ from isobutene, 2,2,3-TMP$^+$ from 2-butene, and 2,2-DMH$^+$ from 1-butene):

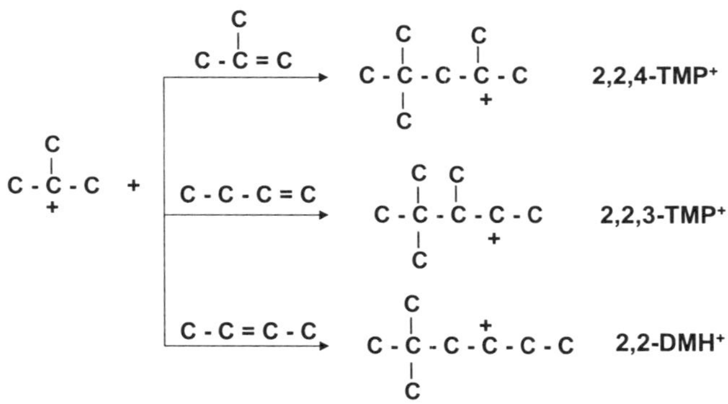


The $C_8$ carbenium ions formed may isomerize via hydride transfer and methyl shifts to form more stable carbenium ions:

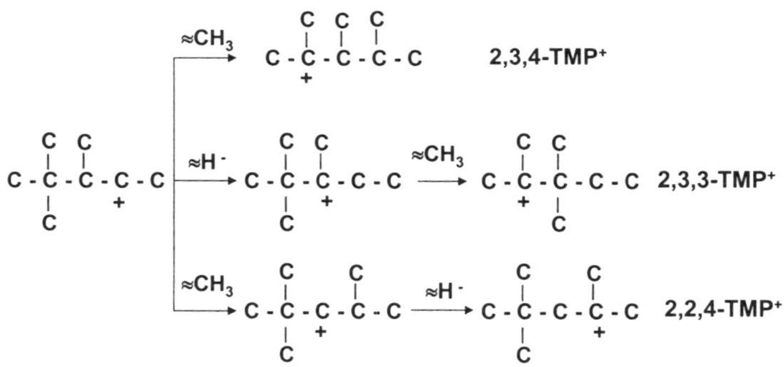


Finally, these carbenium ions suffer rapid hydride transfer from isobutane, leading to the different octane isomers and regenerating the *tert*-butyl cation to perpetuate the chain sequence:

This simplified mechanism cannot explain, however, the full spectrum of products obtained during isobutane/butene alkylation. The main reactions competing with alkylation are the oligomerization of the olefins, and particularly the dimerization of the butenes, and the cracking of larger isoalkylcations that leads to a mixture of isoparaffins with less than 8 carbon numbers. A simplified scheme of the main alkylation and competing reactions occurring during isobutane/2-butene alkylation is given in Fig. 2.5. Dimerization of 2-butene leads to the formation of dimethylhexanes (DMH) which have a lower octane number than trimethylpentanes (TMP).

In the case of zeolites, an additional pathway has been proposed involving the formation of a non-classical carbonium ion by protonation of isobutane, which can alkylate the olefin with formation of a protonated cyclopropane intermediate (60, 61):

Then, cracking of any the C-C bonds in the protonated cyclopropane ring will lead to the different trimethylpentyl cations:

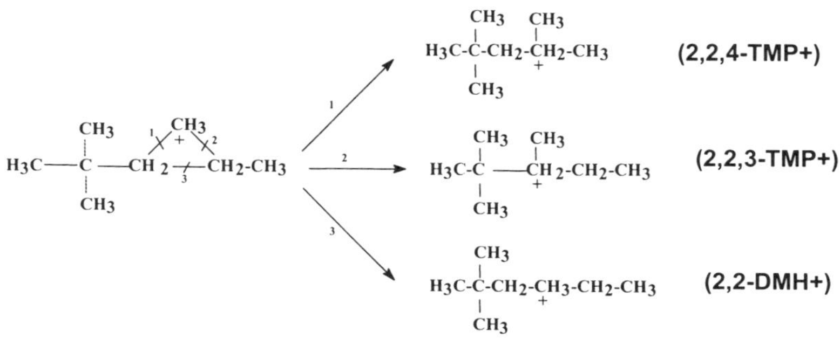


When zeolites are used as alkylation catalysts, a high initial activity and selectivity to the desired TMP is obtained during the initial stages of the reaction. However, after a few minutes on stream the activity of the zeolite for hydrogen transfer, which is the reaction responsible for chain propagation, is lost, and then the reaction enters a critical stage in which the product distribution is better explained in terms of olefin oligomerization instead of isobutane/butene alkylation (57, 62). This can be clearly seen in Table 2.5 presenting the butene conversion and product selectivities obtained for a USY zeolite at different times on stream.

**Table 2.5** Conversion and product distribution obtained at different TOS on a USY zeolite ($a_0$= 24.35 Å) during the alkylation of isobutane with 2-butene in a fixed-bed reactor (T= 80°C, P= 20 bar, WHSV= 2 $h^{-1}$, I/O= 15)

| | Time on Stream (min) | | |
|---|---|---|---|
| | 1 | 10 | 15 |
| 2-$C_4$= Conversion (%) | 99 | 63 | 47 |
| Product distribution (wt%) | | | |
| $C_5 - C_7$ | 33 | 8 | 3 |
| $C_{8's}$ | 41 | 55 | 57 |
| $C_{9+}$ | 26 | 37 | 40 |
| $C_8$'s distribution ( wt%) | | | |
| TMP | 74 | 23 | 1 |
| DMH | 26 | 30 | 15 |
| Octenes | 0 | 47 | 84 |

The zeolite composition and structure, which can affect hydrogen transfer activity, are important parameters determining the activity, selectivity, and stability of the zeolite during isobutane alkylation. In the case of USY zeolites, a maximum initial 2-butene conversion was observed for a framework Si/Al ratio of about 6 (63). However, the TMP/DMH ratio, which can be taken as a measure of the alkylation/oligomerization ratio, continuously increased when decreasing the framework Si/Al ratio. On the other hand, the amount and nature of extraframework Al (EFAL) species also affected the alkylation properties of USY zeolites (64).

Zeolite Beta has also been studied for isobutane/butene alkylation (65, 66), but it was less selective to the desired TMP than USY, suggesting some diffusional limitations for these highly branched products at the relatively low reaction temperatures used. In fact, an increase of activity was observed when decreasing the crystal size of the Beta zeolite (66). As for USY zeolites, the activity, selectivity and deactivation rate of Beta zeolite were influenced by the presence of EFAL species (67). Medium pore zeolites, such as ZSM-5 and ZSM-11 were also found active for alkylation, but at temperatures above 100°C (68, 69). Moreover, the product obtained on ZSM-5 and ZSM-11 contained more light compounds ($C_5$-$C_7$), and the $C_8$ fraction was almost free of trimethylpentanes, indicating serious pore restrictions for the formation of the desired alkylation products.

## 2.6 Catalytic cracking

Catalytic cracking is probably the most important conversion unit in modern refineries. Essentially, catalytic cracking involves the C-C bond rupture of hydrocarbons contained in the feedstock (typically a vacuum gasoil) to produce more valuable low molecular weight hydrocarbons including light olefins for petrochemistry, gasoline, and diesel.

The actual technology involves the formulation of multifunctional cracking catalysts which are composed of different amorphous and crystalline acid functions, and a series of additives for metal passivation, $SO_x$ removal, promotors for total combustion, and octane enhancing additives. Among them, zeolite Y is the main component controlling the activity and selectivity of the cracking catalysts.

From the point of view of the mechanism, the acid sites of the zeolite are believed to be the catalytic active sites. By looking to the products distribution obtained during the cracking of the different hydrocarbons contained in typical FCC feedstocks (paraffins, naphthenes, and alkylaromatics), it appears that the reaction must involve the formation of carbenium and carbonium-like intermediates (70).

When the cracking of paraffins is considered, there is still some controversy regarding the intiation step and the nature of the acid sites involved. The following possibilities have been suggested:

a) abstraction of a hydride ion from the paraffin by a Brönsted acid site, giving $H_2$ as a product (71, 72, 73)
b) abstraction of hydride by a Lewis acid site from the catalyst (74, 75, 76)
c) protonation of olefins that are either present in the feed or formed by thermal cracking on the zeolite Brönsted sites (77)

More recently, it has been proposed (78, 79) that on zeolite catalysts the reaction can also start by protonation of a C-C bond by the acid site of the zeolite forming a pentacoordinated carbonium ion transition state. This can then eliminate $H_2$ or a short alkane molecule leaving an adsorbed carbenium ion on the zeolite (protolytic cracking), as shown below:

$$R_1-CH_2-CH_2-R_2 + H^+ \rightleftharpoons R_1-CH_2-CH_3^+-R_2 \begin{cases} \nearrow R_1-CH_2-CH^+-R_2 + H_2 \\ \searrow R_1 + CH_3-CH_2-R_2 \end{cases}$$

Once a carbenium ion is formed by either of the mechanisms described above, it can desorb as an olefin and restoring the zeolite Brönsted acid site:

$$[C_mH_{2m+1}]^+ \Longleftrightarrow [C_mH_{2m}] + H^+$$

or isomerize via hydride or methyl shifts into a more stable carbenium ion before it cracks via β-scission:

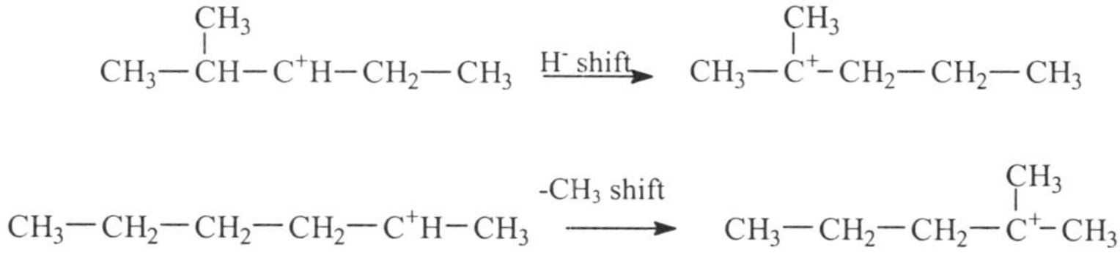

Then, the branched carbenium ion can either desorb giving a branched olefin in the gas phase and restoring the Brönsted site, or abstracting a hydride ion from a neutral molecule, desorbing as a branched paraffin and leaving an adsorbed carbenium ion, or may crack by β-scission giving an olefin in the gas phase and a shorter carbenium ion:

$$CH_3-C(CH_3)_2-CH_2-C^{+}H-CH_3 \xrightarrow{\text{ß scission}} CH_3-CH=CH_2 + CH_3-C^{+}(CH_3)_2$$

Chain propagation then occurs by hydride transfer from a reactant molecule to the adsorbed carbenium ion:

$$CH_3-C^{+}H-CH_2-CH_2-CH_3 + R' \longrightarrow CH_3-CH_2-CH_2-CH_2-CH_3 + R'^{+}$$

Chain termination occurs when the adsorbed carbenium ion is desorbed as an olefin, restoring the Brönsted site:

$$CH_3-C^{+}H-CH_2-CH_2-CH_3 \longrightarrow CH_3-CH=CH-CH_2-CH_3 + H^{+}$$

or by protonation of the adsorbed carbenium ion which desorbs as a paraffin while regenerating a Lewis acid site

$$CH_3-C^{+}H-CH_2-CH_2-CH_3 + H^{-} \longrightarrow CH_3-CH_2-CH_2-CH_2-CH_3$$

A simplified reaction network for the cracking of alkanes on zeolite catalysts is presented in Fig. 2.7.

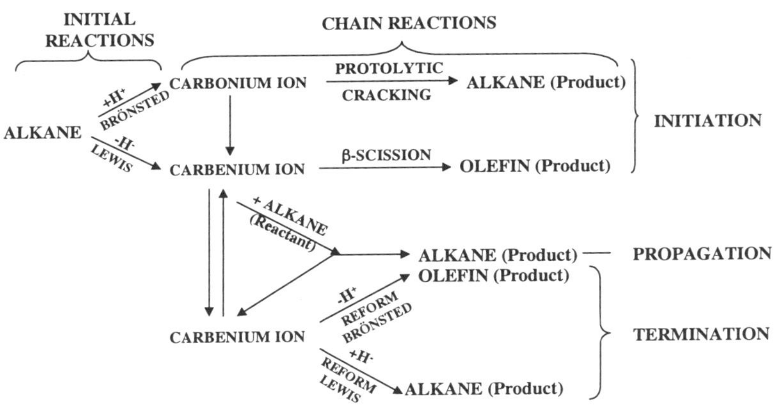


**Fig. 2.7** Reaction pathway for alkane cracking on zeolite catalysts

In summary, it could be said that, in the case of n-alkanes, the first molecules of reactant are cracked by protolytic cracking. Then, once a smaller carbenium ion is left on the catalyst surface, cracking can continue by one of the following routes:

a) protolytic cracking, if the carbenium ion desorbs and regenerates the original Brönsted site, and
b) β-scission, if hydride transfer from a reactant molecule occurs.

The relative extension of protolytic to β-scission cracking has a clear impact on the final product distribution observed during catalytic cracking. Indeed, protolytic cracking favors the formation of undesired $C_1$ and $C_2$ products as compared to β-scission. On the other hand, hydride transfer is believed to be responsible for the saturation of olefins, with the corresponding increase in gasoline stability and selectivity, but producing a decrease of the research octane number (RON) of the gasoline.

The ratio of protolytic (monomolecular) to β-scission (bimolecular) cracking depends on reaction conditions (temperature, hydrocarbon partial pressure), as well as on the characteristics of the zeolite (chemical composition and pore dimensions). It has been shown (80) that the ratio of protolytic cracking to β-scission increases with increasing the framework Si/Al ratio of the zeolite, meanwhile hydrogen transfer and coke formation decrease. Thus, during gasoil cracking an increase in the framework Si/Al ratio of the USY zeolite produces a decrease in the selectivity to gasoline owing to a larger recracking, which gives $C_1$-$C_4$ gases, while selectivity to coke decreases and the olefin/paraffin ratio of the gasoline increases due to a lower hydrogen transfer, with the corresponding increase in RON.

The medium pore ZSM-5 zeolite is also used as a cracking additive in small amounts (0.5-3 wt%) to enhance the gasoline octane. Due to the pore dimensions, ZSM-5 increases the octane rating of the gasoline by selectively upgrading low octane gasoline components into lower molecular weight compounds with a higher octane. It mainly acts by cracking low-octane linear and monobranched paraffins and olefins in the gasoline range, which can access to the acid sites. Experiments in a riser pilot plant (81) showed that ZSM-5 alters the distribution of olefins in the gasoline range by decreasing linear and monobranched $C_{7+}$ olefins and increasing $C_{6-}$ branched olefins with a high RON. Linear and branched olefins in the upper gasoline range are thus removed before they are converted into low octane paraffins and aromatics by hydrogen transfer reactions on the Y zeolite component of the cracking catalyst. Similarly, $C_{7+}$ paraffins are also selectively cracked on the ZSM-5 zeolite, increasing the yield of $C_5$-$C_6$ isoparaffins.

An aromatics concentration effect is also responsible for the increase of octane produced by ZSM-5 addition. Gasoline yield slightly decreases while $C_3$-$C_4$ gases, mainly propylene, increase (82). Increasing the Si/Al ratio of the ZSM-5 zeolite, less relative light gasoline loss is experienced as the cracking of the intermediate olefins is reduced relative to isomerization (83). Moreover, the $C_4$=/$C_3$= and MON/RON ratios are also increased with the Si/Al ratio of ZSM-5. It was also shown that the gain in RON and MON depends on the ZSM-5 content and the severity of the FCC unit (84).

Production of light olefins (propylene, n-butenes and isobutene) will be one of the main targets of FCC untis in the near future. These olefins can be fed to alkylation and etherification units to produce additional high octane environmentally acceptable gasoline components, or used as petrochemical feedstock. Johnson and Avidan (85) used higher amounts of ZSM-5 (10-20%) to increase the production of light olefins, mainly propylene.

The large pore tridirectional zeolite Beta has also been evaluated for catalytic cracking (86, 87). While zeolite Beta was found more active than USY for cracking n-heptane, it was less active for gasoil cracking, indicating some problems of accessibility for the more bulky molecules present in the gasoil. However, zeolite Beta was less active for hydrogen transfer reactions, which should be reflected in a higher gasoline octane and more olefinic $C_3$-$C_4$ products. Moreover, when zeolite Beta is used as an additive instead of the medium pore ZSM-5, it produced higher yields of desirable $C_4$ olefins (88).

## References


1. Haag W.O.and Dessau R.M., *Proc. 8th Int. Congr. Catal.*, **2** (Berlin, 1984) 305.
2. Doolan P.C. and Pujado P., *Hydrocarbon Process.* **68** (1989) 72.
3. Martindale D.C., Kuchar P.J. and Kolson R., *AIChE Meet.*, (Denver, CO, March, 21-24, 1988).
4. Gricus-Kofke T.J., Gorte J. and Kokotailo G.T., *Appl. Catal.* **54** (1989) 179.
5. Simmons D.K., Szostak R., Agrawal P.K. and Thomas T.L., *J. Catal.* **106** (1987) 287.
6. Giannetto G., Monque R. and Galiasso R., *Catal. Rev.-Sci. Eng.* **36** (1994) 271.
7. Price G.L. and Kanazirev U., *J. Catal.* **126** (1990) 267.
8. Price G.L., *J. Catal.* **130** (1991) 611.
9. Dooley K.M., Chang C. and Price G.L., *Appl. Catal.* 84 (1992) 117.
10. Meitzner G.D., Iglesia E., Baumgartner J.E. and Huang E.S., *J. Catal.* **140** (1993) 209.
11. Mole T., Anderson J.R. and Creer G., *Appl. Catal.* **17** (1985) 141.
12. Kanai J. and Kawata N., *J. Catal.* **114** (1988) 284.
13. Mériaudeau P. and Naccache C., *Catal. Rev.-Sci. Eng.* **39** (1997) 5.
14. Meriaudeau P. and Naccache C., *J. Mol. Catal.* **59** (1990) L31.
15. Meriaudeau P. and Naccache C., *J. Mol. Catal.* **59** (1990) L31.
16. Butler A.C. and Nicolaides C.P., *Catal. Today* **18** (1993) 443.
17. Basini L., Aragno A. and Raffaelli A., *J. Phys. Chem.* **95** (1991) 211.
18. Szabo J. et al., *J. Mol. Catal.* **67** (1991) 79.
19. Houzvicka J., Nienhuis J.G., Hansildaar S. and Ponec V., *Appl. Catal. A* **165** (1997) 443.
20. Seo G., Jeong H.S., Hong S.B. and Uh Y.S., *Catal. Lett.* **36** (1996) 249.
21. Asensi M.A., Corma A. and Martínez A., *J. Catal.* **158** (1996) 561.
22. Asensi M.A. and Martinez A., *Appl. Catal. A* **183** 155 (1999).
23. Meriaudeau P., Vu Anh T., Le Ngoc H. and Naccache C., *Stud. Surf. Sci. Catal.* **105** (1997) 1373.
24. Asensi M.A. et al., *Appl. Catal. A* **174** (1998) 163.
25. Houzvicka J., Nienhuis J.G., Hansildaar S. and Ponec V., *Appl. Catal. A* **165** (1997) 443.

26. Meriaudeau P. et al., C., *J. Chem. Soc., Faraday Trans.* **94** (1998) 467.
27. Grandvallet P. et al., EP Patent **501 577** (1992), assigned to Shell Int. Res.
28. Mooiweer H.H. et al., *Stud. Surf. Sci. Catal.* **84** (1994) 2327.
29. Meriaudeau P., Bacaud R., Hung L.N.and Vu T.A., *J. Mol. Catal. A* **110** (1996) 177.
30. Houzvicka J., Diefenbach O. and Ponec V., *J. Catal.* **164** (1996) 288.
31. Houzvicka J. and Ponec V., *Catal. Rev.-Sci. Eng.* **39** (1997) 319.
32. Guisnet M.et al., *J. Chem. Soc., Chem. Commun.* (1995) 1685.
33. Andy P. et al., *J. Catal.* **173** (1998) 322.
34. Condon F.E. in *Catalysis,* P.H. Emmet, ed., (Reinhold, New York, 1958) Vol. VI, p.121.
35. Brouwer D.M., *Rec. Trav. Chim.* Pays Bas **87** (1968) 1435.
36. Zhang A., Nakamura I., Aimoto K. and Fujimoto K., *Ind. Eng. Chem. Res.* **34** (1995) 1074.
37. Guisnet M., Alvarez F., Gianetto G. and Perot G., *Catal. Today* **1** (1987) 415.
38. Koradia P.B., Kiouski J.R. and Asim M.Y., *J. Catal.* **66** (1980) 290.
39. Guisnet M. et al., *Appl. Catal.* 71 (1991) 283.
40. Koradia P.B., Kiovski J.R. and Asim M.Y., *J. Catal.* **66** (1980) 290.
41. Guisnet M. et al., *Appl. Catal.* **71** (1991) 283.
42. Voorhies A. and Brian M., *AIChE J.*, (1968) 482
43. US Pat. **4,018,711** (1979)
44. Eberly P.E., Kimberlin C.N. and Voorhies A., *J. Catal.* **22** (1971) 419.
45. Corma A., Frontela J., Lázaro J. and Pérez M., *Prep. ACS Petrol. Div.* **36** (4) (1991) 833.
46. Derouane E.G., *Stud. Surf. Sci. Catal.* **20** (1985) 221.
47. Alvarez F. et al., *Stud. Surf. Sci. Catal.* **49** (1989) 1339.
48. Chica A. and Corma A., *J. Catal.* **187** (1999) 167.
49. Boronat M., Viruela P. and Corma A., *Appl. Catal. A* **146** (1996) 207.
50. Arata K., *Adv. Catal.* **37** (1990) 165.
51. Hino M. and Arata K., *J. Chem. Soc., Chem. Commun.* (1995) 789.
52. Arata K., *Appl. Catal. A* **146** (1996) 3.
53. Corma A. and Martínez A., *Catal. Rev.-Sci. Eng.* **35** (1993) 483.
54. Garwood W.E. and Venuto P.B., *J. Catal.* **11** (1968) 175.
55. Minachev Kh.M. et al., *ACS Symp. Ser.* **55** (1977) 89.
56. Kirsch F.W., Potts J.D. and Barmby D.S., *J. Catal.* **27** (1972) 142.
57. Weitkamp J., *Stud. Surf. Sci. Catal.* **5** (1980) 65.

58. Weitkamp J. and Ernst S., in *Proc. 13th World Petroleum Congress*, **3** (1992) 315..
59. Corma A., Martínez A. and Martínez C., *J. Catal.* **146** (1994) 185.
60. Olah G.A., *Chem. Brit.* **8** (1972) 281.
61. Khadzhiev S.N. and Gerzeliev I.M., *Prep. ACS Petrol. Div., NY Meeting*, (August 1991) 799.
62. Weitkamp J., *Proc. Int. Zeol. Conf.* **5** (1980) 858.
63. Corma A., Martínez A. and Martínez C., *J. Catal.* **146** (1994) 185.
64. Corma A., Martínez A. and Martínez C., *Appl. Catal. A* **134** (1996) 169.
65. Unverricht S., Ernst S. and Weitkamp J., *Stud. Surf. Sci. Catal.* **84(C)** (1994) 1693.
66. Corma A., Gomez V. and Martínez A., *Appl. Catal. A* **119** (1994) 83.
67. Corma A. et al., *Appl. Catal. A* **142** (1996) 139.
68. Chu Y.F. and Chester A.W., *Zeolites* **6** (1986) 195.
69. Weitkamp J. and Jacobs P.A., in *New Frontiers in Catalysis*, Ed. Guczi L., Solymosi F., and Tétényi P., (Proc. 10th Inter. Congr. Catal., Part B, 1993) 1735.
70. Wojciechowski B.W. and Corma A., *Catalytic Cracking Catalysts, Chemistry and Kinetics*, (M. Dekker, New York, 1986).
71. Planelles J., Sanchez-Marin F., Tomás F. and Corma A., *J. Mol. Catal.* **32** (1985) 365.
72. Abbot J. and Head I.D., *J. Catal.* **125** (1990) 187.
73. Poustma M.L., in *Zeolite Chemistry and Catalysis* Ed. Rabo J.A., (ACS Monograph, Washington DC, 1976, 117), 505.
74. Tung S.E. and McIninch E., *J. Catal.* **10** (1968) 166.
75. Nace D.M., *Ind. Eng. Chem., Prod. Res. Dev.* **8** (1969) 31.
76. Borodzinski A., Corma A. and Wojciechowski B.W., *Can. J. Chem. Eng.* **58** (1980) 219.
77. Greensfelder B.S., Voge H.H. and Good G.M., *Ind. Eng. Chem.* **41** (1949) 2573.
78. Haag W.O. and Dessau R.M., *Proc. 8th Int. Congr. Catal.*, (Berlin, 1984, Vol. II) 305.
79. Corma A., Planelles J., Sanchez-Marin J. and Tomás F., *J. Catal.* **92** (1985) 284.
80. Gianetto G., Guisnet M. and Perot G., *J. Chem. Soc., Chem. Commun.*, **3225** (1989)

81. Pappal D.A. and Schipper P.H., *ACS Symp. Ser.* **452** (1991) 45.
82. Madon R.J., *J. Catal.* **129** (1991) 275.
83. Huehler C.W., in *Fluid Cracking Catalysts*, Eds. Occelli M.L. and O'Coonor P., (Marcel Dekker, New York, 1998) 31.
84. Elia M.F., Iglesias E., Martínez A. and Pérez M.A., *Appl. Catal.* **73** (1991) 195.
85. Johnson T.E. and Avidan A.A., *NPRA Annual Meeting, San Antonio, TX, Mar. 21-23*, (1993) paper AM-93-51.
86. Corma A., Fornés V., Montón J.B. and Orchillés A.V., *J. Catal.* **107** (1987) 288.
87. Corma A., Fornés V., Melo F.V. and Perez-Pariente J., *ACS Symp. Ser.* **375** (1988) 49.
88. Bonetto L., Camblor M.A., Corma A. and Pérez-Pariente J., *Appl. Catal.* **82** (1992) 37.

CHAPTER 3

# PREPARATION OF ZEOLITE CATALYSTS

T.G. ROBERIE, D. HILDEBRANDT, J. CREIGHTON
*Grace Davison, 7500 Grace Drive, 21044, Columbia, Md., USA*

J.-P. GILSON
*Laboratoire de Catalyse et Spectrochimie, ISMRA-CNRS,*
*6, Bd du Maréchal Juin, 14050 Caen Cedex, France*

## 3.1 Introduction

Industrial catalysts are produced in a large variety of shapes ranging from powders, tablets, granules, beads, extrudates, etc. In general, the physical form of the catalyst is governed by the needs (kinetics, mass transport, heat transfer and hydrodynamics) of the particular industrial process (1a). Catalyst deactivation (1b) plays, for instance, a dominant role in the catalyst and reactor designs. The very stable (lifetime measured in months or weeks) $C_{5-6}$ paraffins isomerization catalysts (Pt/H-MOR) are marketed in the form of $Al_2O_3$ bound extrudates ; they enter the fixed bed reactor and leave it only for ex-situ regeneration or disposal. The slowly deactivating (lifetime measured in days or weeks) naphtha reforming catalysts are often supplied as millimeter sized beads (spheres produced by the oil-drop technique) to allow a smooth transport between the reaction and regeneration/rejuvenation zones. The fast deactivating (lifetime measured in seconds) FCC catalysts are shaped as micron sized spheres to allow a short contact time during the reaction and a prompt transfer to and from the regenerator. The morphology of the catalyst can make a difference not only on the pressure drop within the reactor but also on the diffusional speed of the reagents and products. The latter can have a significant impact on the selectivity of certain reactions.

In petrochemical and oil refining operations, the zeolite is primarily responsible for the catalyst's activity, selectivity and stability (catalytic, thermal and hydrothermal). The fluid catalytic cracking process (FCC) is the most widely used of the oil refining process and is characterized by the use of a finely divided catalyst, which is moved through the processing unit. The catalyst particles are of such a size (about 70 μm) that when aerated with air or hydrocarbon vapor, the catalyst behaves like a liquid and can be moved easily through pipes.

A general review of the manufacture of common industrial zeolites will be presented, with particular emphasis on faujasite type materials and catalysts in the case study of FCC.

Spray drying is the most widely used industrial process involving particle formation and drying. It is highly suited for the continuous production of dry solids from pumpable suspensions. Operating conditions and dryer design are selected according to the drying characteristics of the product and powder specifications. The discussion will be centered on spray drying principles as well as the different types of equipment commonly used in commercial applications. Other catalyst forming techniques will be discussed more briefly. The reader will notice that contrary to other aspects of zeolite science and technology, the source of open information is scant. This feature underlines the central economic role played by catalyst preparation and may give the impression this aspect of zeolite science and technology is very empirical. Advances in process control and quality programs have, however, made catalyst preparation remarkably efficient and a source of ever diminishing waste streams.

## 3.2 Zeolites: Their nature, properties and synthesis

### *3.2.1 A short refresher on zeolites*

The object of this paragraph is not to give a comprehensive overview of zeolites but mainly to highlight definitions and properties critical to the forthcoming discussion. A useful introduction is the subject of Chapter 1 of this book and more detailed information can be found in the references therein.

Zeolites are crystalline microporous aluminosilicates possessing an anionic framework; the electroneutrality of the crystals requires the presence of cations in extra-framework positions, such as $Na^+$, $K^+$, $H^+$, ½ $Ca^{++}$, ½ $Mg^{++}$, 1/3 $La^{+++}$, 1/3 $Ce^{+++}$, etc. By extension, molecular sieves are defined as microporous crystalline structures with a variable elemental composition leading to frameworks with variable charge as outlined in Table 1.

**Table 1** Elemental composition of molecular sieves

| Framework: (Composition)$^{CHARGE}$ | Examples |
|---|---|
| $(M^{+2}O_2)^{-2}$ | Be, Mg, Zn, Co, Fe, Mn |
| $(M^{+3}O_2)^{-1}$ | Al, B, Ga, Fe, Cr |
| $(M^{+4}O_2)^{0}$ | Si, Ge, Mn, Ti |
| $(M^{+5}O_2)^{+1}$ | P |

It is then straightforward to classify the molecular sieves as a function of their composition as is done below:

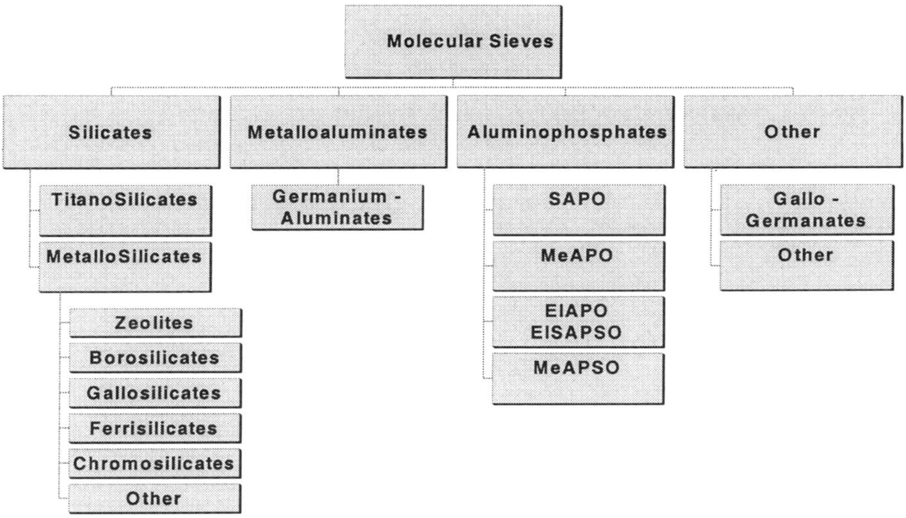


**Fig. 3.1** Classification of molecular sieve materials

It should be reminded at this juncture (see Chapter 1) that only a fraction of the zeolite structures are used commercially and that with few exceptions (SAPO-11 for instance), almost no metallosilicates have found a large scale use. As far as the catalytic and adsorption properties are concerned, zeolite molecular sieves are further classified according to their pore openings, defined by the number of $O^{2-}$ anions delineating their pore mouths. The most useful zeolites are those containing 12, 10 or 8 of these $O^{2-}$ and are commonly referred to as 12-, 10-, and 8- ring structures. The zeolites FAU and MFI are the major components of the FCC catalysts.

### *3.2.2 Zeolite synthesis*

During their use and especially in FCC, zeolites will have to withstand harsh thermal and hydrothermal conditions. Three key parameters will influence their stability: their Al content (the lower the more stable the zeolite), crystal size (the larger the more stable the zeolite) and Na level (the lower the more stable the zeolite). These parameters are set during the synthesis or the consecutive processing steps. All zeolites are produced by hydrothermal synthesis (2) under alkaline conditions (OH- mineralizing ion).

### 3.2.2.1 FAU zeolite synthesis and modifications

The FAU synthesis is a well mastered process taking place at atmospheric pressure and yielding materials whose $SiO_2/Al_2O_3$ molar ratio ranges from 4.5 to 6.0; the crystal size ranges between 0.3 – 2.0 μm and the BET surface area is typically 850-900 $m^2.g^{-1}$. Quite often however, the $SiO_2/Al_2O_3$ ratio needs to be adjusted (increased) outside the range achievable via direct classical synthesis and other means have to be used to produce Silica-rich FAU. The most commonly encountered processes to increase the $SiO_2/Al_2O_3$ ratio are:

- *Direct Synthesis*: the crystallization is done in the presence of organic templates (2) such as TBAOH, crown ethers, ...These syntheses unfortunately suffer from two economic drawbacks, i.e.; the long synthesis times (between 3 to 8 days) compared to the classical synthesis (typically 16 hours) and the use of costly, sometimes toxic, and not always recyclable organic molecules
- *Hydrothermal Treatment*: a post-synthesis treatment widely practiced in the industry because it also modifies the texture (creation of desired mesoporosity) of the zeolite. Such a treatment also takes place in the regenerator of every FCC unit
- *Extraction of Al from the framework*: chelating agents such as EDTA remove Al and leave defects behind according to the following overall mechanism:

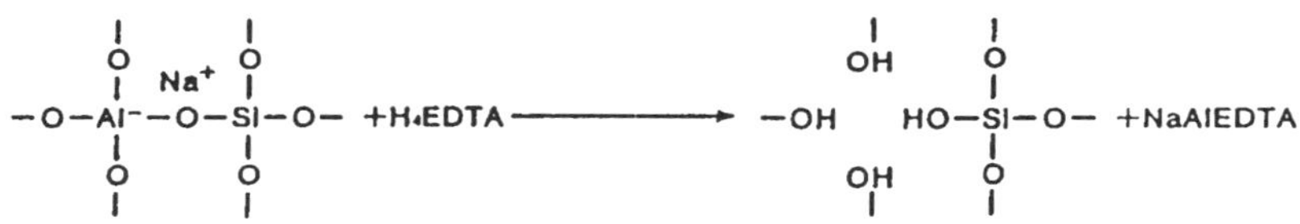


- *Chemical substitution*: a variant of the process above where Si is inserted in the framework in the place of the Al, therefore minimizing the level of defects. The two most discussed processes are based on the (gas phase) $SiCl_4$ or the (liquid phase) $(NH_4)_2SiF_6$; they suffer however from the environmental penalty associated with the use of halogenated reactants. The overall chemistry is as follows:

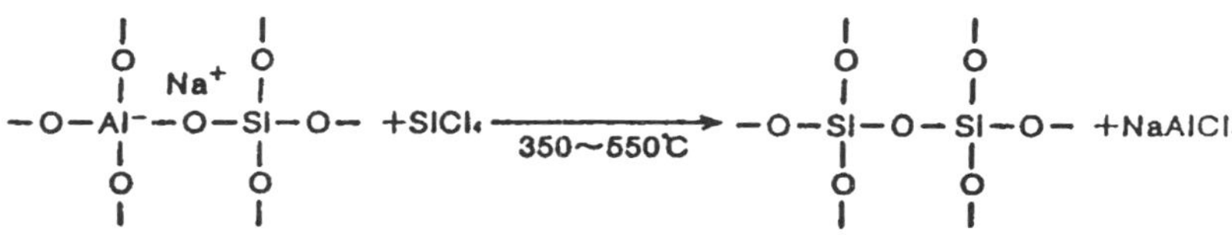

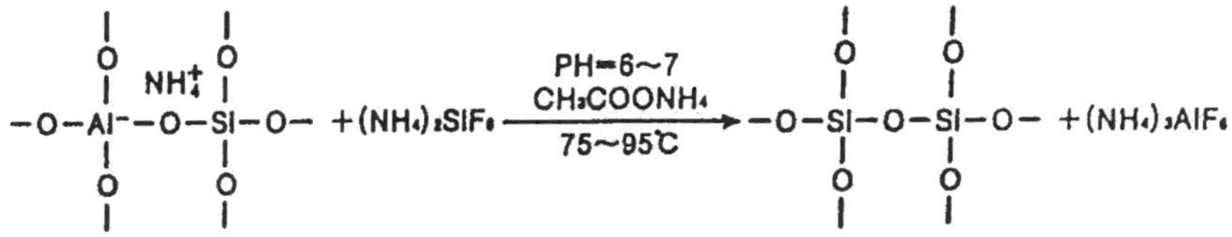


A closer look at the mechanism of hydrothermal dealumination (hydrolysis of Si-O-Al bonds) illustrates that a healing process also takes place during this transformation (3), but contrary to the chemical substitution highlighted above, the source of Si originates from the zeolite, and results in a partial destruction of the zeolites and Si migration. The mesoporosity created during the process is beneficial to the diffusion of the large molecules of oil.

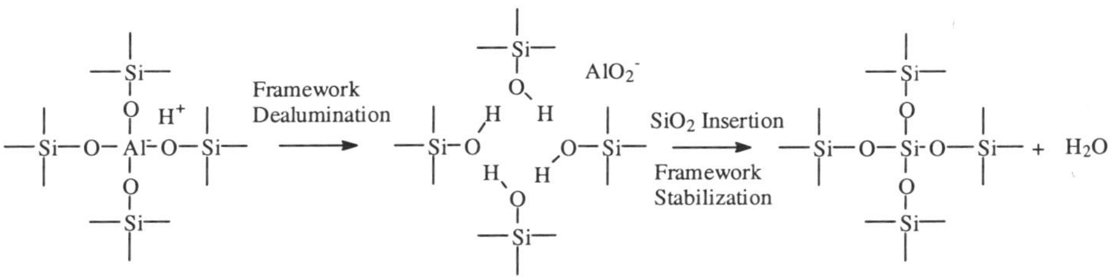


**Fig. 3.2** Schematic hydrothermal dealumination mechanism

The most common way to track down the fate of the Al is by measuring the (cubic) unit cell content of the FAU; a direct measure of the framework Al content is obtained by the so-called Breck curve, Figure 3.3. The FAU zeolite can then be fine-tuned to cover most of its applications in FCC (gasoline yield with high unit cell size materials, octane yield with low unit cell size materials) and Hydrocracking (high middle distillate yields with low unit cell size materials).

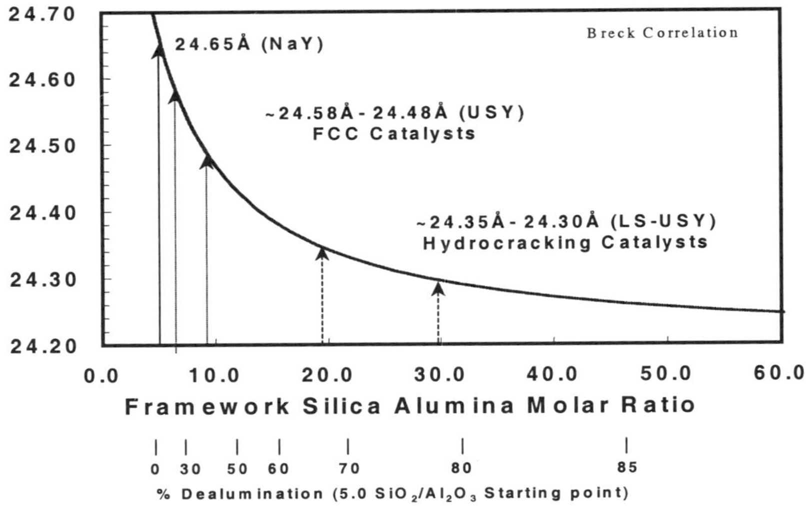


**Fig. 3.3** Relationship between unit cell size and framework al content

3.2.2.2 MFI and BEA zeolite synthesis

Contrary to the FAU zeolite, MFI can be produced already at the synthesis level over a wide range of $SiO_2/Al_2O_3$ ratios (from about 20 to infinity) and crystal size (< 0.1 μm to > 200 μm) by a high pressure synthesis. As far as zeolite BEA is concerned, high pressure synthesis also yields a wide range of compositions ($SiO_2/Al_2O_3$ ratios between 30 and 300+) and a narrower crystal size range (< 0.1 μm to > 2 μm).

## 3.3 FCC catalysts

### *3.3.1 FCC Catalyst composition*

As stated above and discussed in greater detail by Habib (Chapter 5 of this book), FCC is the major catalytic user of zeolites; the process conditions impose on the catalyst manufacturer a tight control of the preparation conditions. The catalyst components are:

- Zeolite *(15 – 45%)*
    - ReY/CREY - HY/ReHY
    - USY/ReUSY
    - Chemically Modified Sieves
- Active matrix components *(0 – 30%)*
- Clay *(10 – 60%)*
- Binder *(10 – 30%)*

### *3.3.2 Overview of the manufacturing processes*

Because the requirements of each FCC unit are very specific (feedstock variability, product slate, type of unit), FCC catalysts are tailor-made and the list above highlights the degree of flexibility (and complexity) imparted on the modern FCC catalysts. Most of the unit operations encountered in catalyst manufacturing are gathered in the FCC catalyst preparation: powder preparation, wetting, mixing and blending of the additives, particle forming, drying, ion-exchanging, calcining, impregnating with active metals, further drying and calcining. FCC catalyst preparation is therefore an excellent case study. The very fast deactivation of the catalyst, the need to supply the endothermic reaction section with heath produced during the exothermic regeneration imposes the use of micron sized particles (~ 60 µm diameter); spray-drying is the method of choice to prepare such materials.

The zeolite is at the core of the catalyst, but the role of the matrix is also crucial and parameters such as its surface area, porosity, composition and acidity need to be controlled very precisely (4). A general scheme for preparation of FCC catalysts is depicted in Figure 3.4.

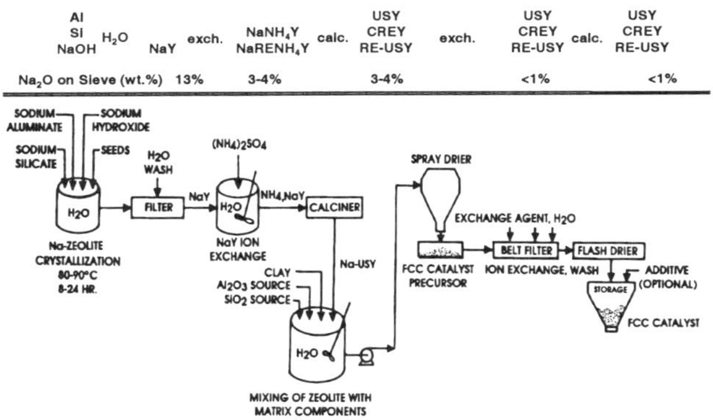


**Fig. 3.4** Generic manufacturing scheme for FCC catalysts

Two features need to be highlighted. First, along the manufacturing scheme (left to right in Figure 3.4), the Na level in the zeolite and the catalyst needs to be drastically reduced in order to give the desired (hydro)thermal stability to the catalyst. This requirement implies repeated washing steps. However, it should be noted that since the early industrial preparations of zeolite containing FCC catalysts, the effluents have been reduced by 99% in order to comply with ever stricter environmental rules. Figure 3.5 illustrates schematically four commonly encountered FCC catalyst preparations. The major differences are found in the first two steps, namely the process of synthesizing the zeolite and the mixing with the matrix. It is seen that the zeolite can be produced separately from the matrix and then blended with the other constituents of the composite catalyst (Grace and Akzo method). Another method is the in-situ crystallization of the zeolite in the matrix (Engelhard process). Both techniques are used commercially. It is of note that the oldest technology, using $SiO_2$ as a binder, is the one leaving the most Na on the catalyst (5).

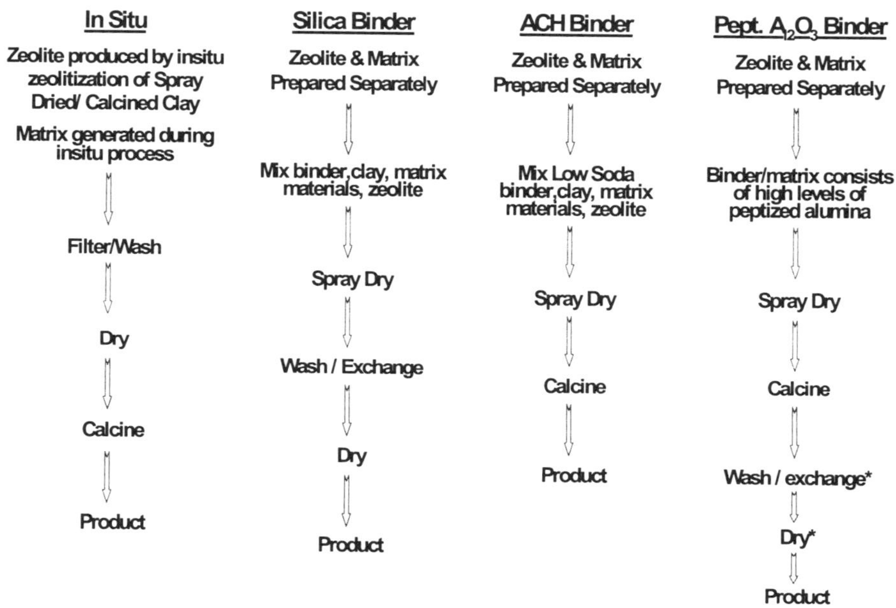


**Fig. 3.5** Four commercial types of FCC preparation

The crucial step in FCC catalyst production is the so called spray drying, where both the shaping and the drying of the composite catalyst are effected. This step is often overlooked in the open literature on zeolite preparation and few authors describe details on this operation.

### *3.3.3 Spray Drying*

#### 3.3.3.1 General features of spray drying

The spray drying process described below has many appealing features for the production of large quantities of catalysts:

- It is highly suited for the continuous production of powder from pumpable suspensions as outlined in Figure 3.4 (slurries in the FCC jargon)
- It forms spherical particles (around 60 μm) that provide for fluid-like flow properties in downstream processing
- The control of key quality parameters is made easier: the particle size distribution, the bulk density and the residual moisture content
- It produces the most homogeneous product for multicomponent systems and the catalyst particles have the same chemical composition as the feed, therefore affording a very high recovery ($99^{+}\%$) and a minimal level of pollution

#### 3.3.3.2 Spray dryer principles

The principle of spray drying is relatively simple and is best understood by studying Figure 3.6.

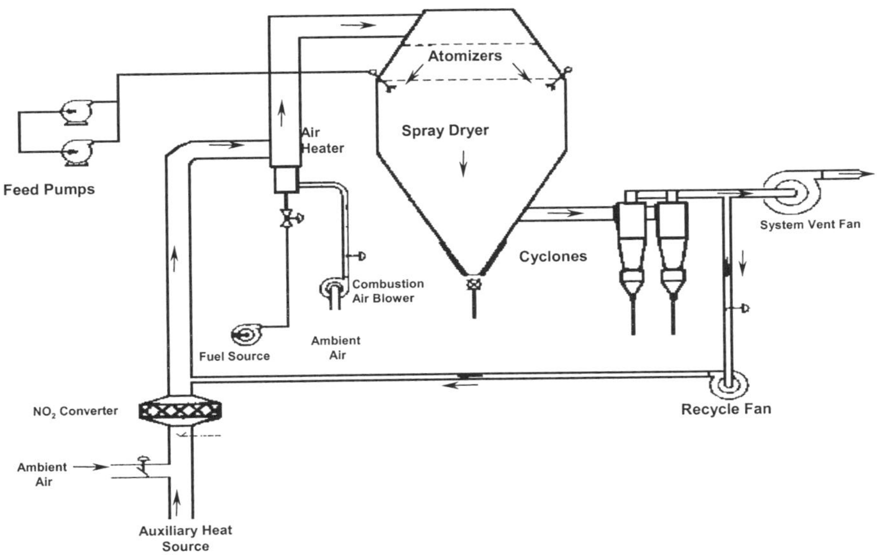


**Fig. 3.6** Typical spray dryer schematic

The catalyst slurry is fed to the top of the spray dryer where it is atomized in the form of small micron sized droplets. These droplets fall through gravity in a flow of hot air following a spiral downward movement. The particles with the desired size are collected at the bottom of the spray dryer and the fines are recovered with the aid of cyclones and recycled. This contained process explains why recoveries of 99% are achieved. The height of a commercial spray dryer is measured in meters and such a process is difficult to downsize, therefore explaining partly the lack of academic research in the field of zeolite catalysts spray drying (6).

**Atomization**

The heart of the spray dryer is the atomizer where the spherical droplets are initially produced. The atomizers can be classified into three different categories:

- *Two Fluid Nozzles*: the spray is created by contacting two fluids, namely the feed slurry and compressed air. The compressed air provides the atomization energy. The device has the advantage of being relatively simple, uses a simple feed pump and affords a wide particle size distribution. On the minus side however, the consumption of compressed air is relatively high (0.25 – 1.0 lb air/lb feed) and this nozzle is typically limited to small plants. Figure 3.7 illustrates such a device
- *Pressure or Single Fluid Nozzle*: the principle is illustrated in Figure 3.8. Such a device produces a narrow particle size distribution of FCC catalyst. The feed pump is more complex (high pressure is required) than in the nozzle described above and is susceptible to plugging; the materials of construction have to be abrasion resistant and the capacity is generally limited to 2 gallons per minute, but multiple atomizing devices can be used in the spray dryer
- *Rotary Atomizers*: typical devices and their principles are illustrated in Figure 3.9. A narrow particle size distribution is obtained, very close to the one observed with the Pressure Nozzles. There is, however, only a single atomizing device fed by a single pump. The atomizing machine is more complex because it requires high rotation rates to produce droplets of the desired size. It is, however, not easily plugged, does not require construction materials as abrasion resistant as the pressure nozzles and has a virtually unlimited capacity.

Large differences in particles size distribution are observed between these types of nozzles as outlined in Figure 3.10.

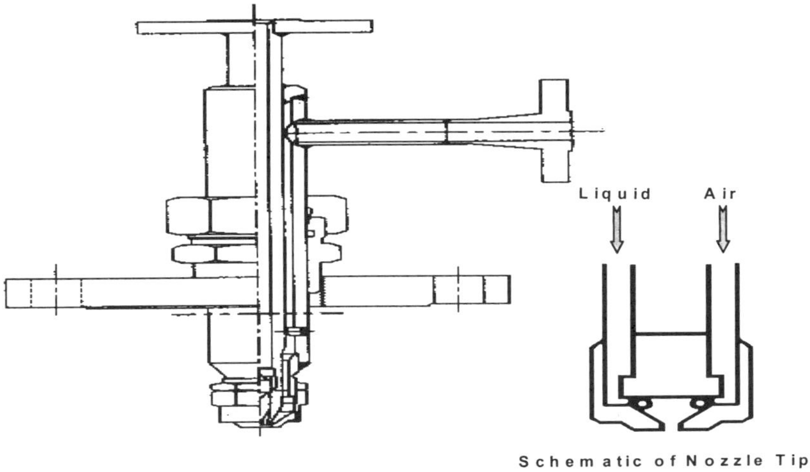


**Fig. 3.7** Two fluid nozzle atomizer

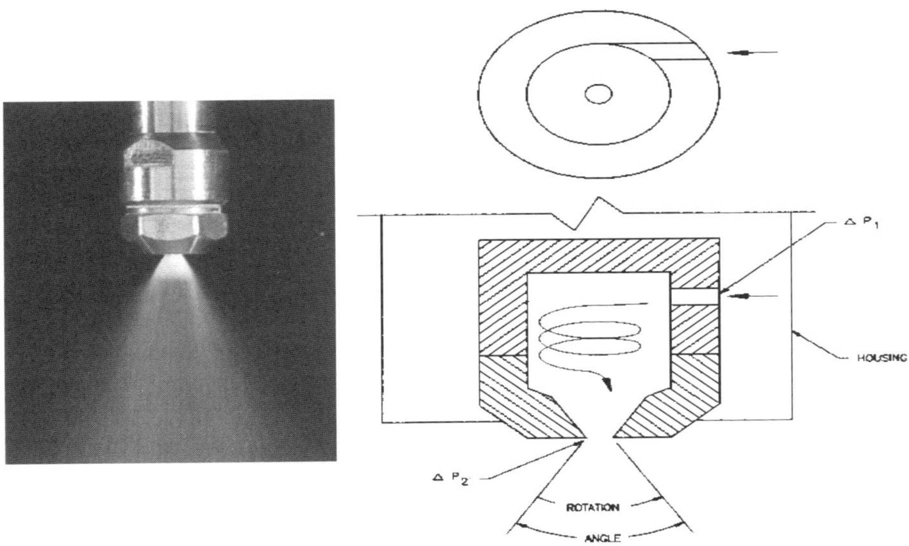


**Fig. 3.8** Pressure or single fluid nozzle atomizer

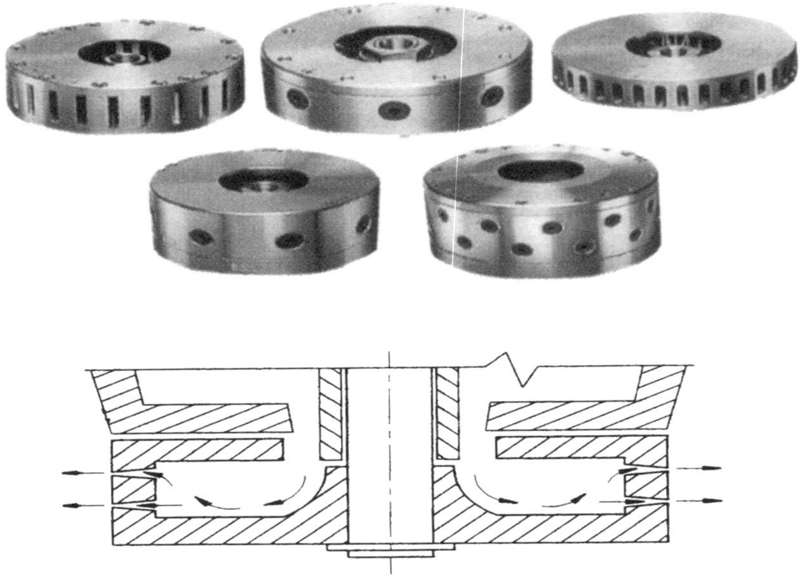

**Fig. 3.9** Rotary atomizer

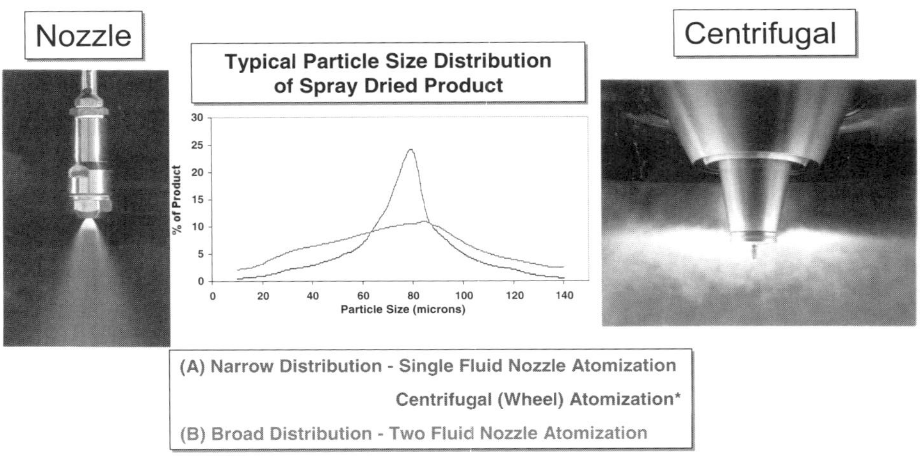


**Fig. 3.10** Comparison of particle size distributions between types of atomizers

## Particle Drying

Once produced by atomization, the droplets have to be dried. The initial contact between the spray droplets and the drying air will control the evaporation rates and the product temperature. In order to have a smooth operation, the largest particles must be dry before hitting the chamber wall to avoid sticking to them. Three drying modes are currently encountered and illustrated in Figure 3.11:

- *Co-current air flow*: the drying air and the particles move (downwards) through the chamber in the same direction
- *Counter-current air flow*: the drying air and the particles move (downwards) through the chamber in opposite directions
- *Mixed flow*: the particles experience both co- and counter-current flows

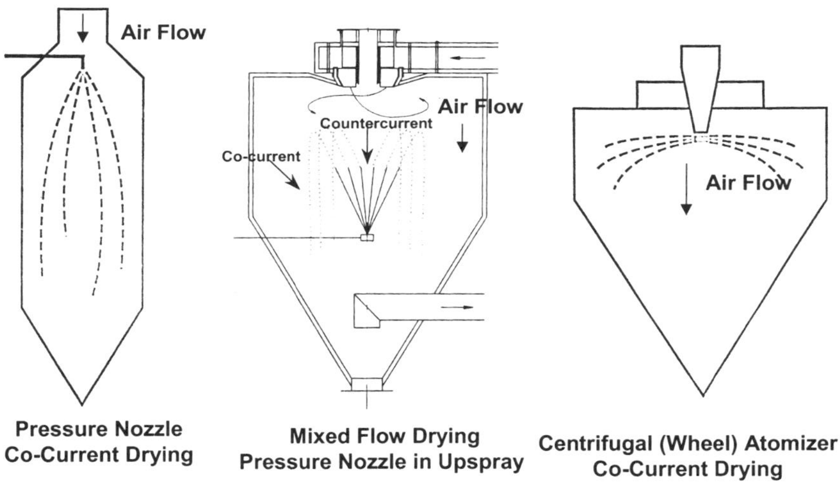


**Fig. 3.11** Drying chamber configurations

As far as the dimensions of the chambers are concerned, they are imposed by the hydrodynamics of the drying process as illustrated in Figure 3.12. It is seen that in order to achieve the particle size (around 70 μm) of interest in FCC catalysts, the drying chamber diameter is between 3-4 meters. Such dimensions make the spray drying process difficult to downscale for a detailed study and modeling of this unit operation.

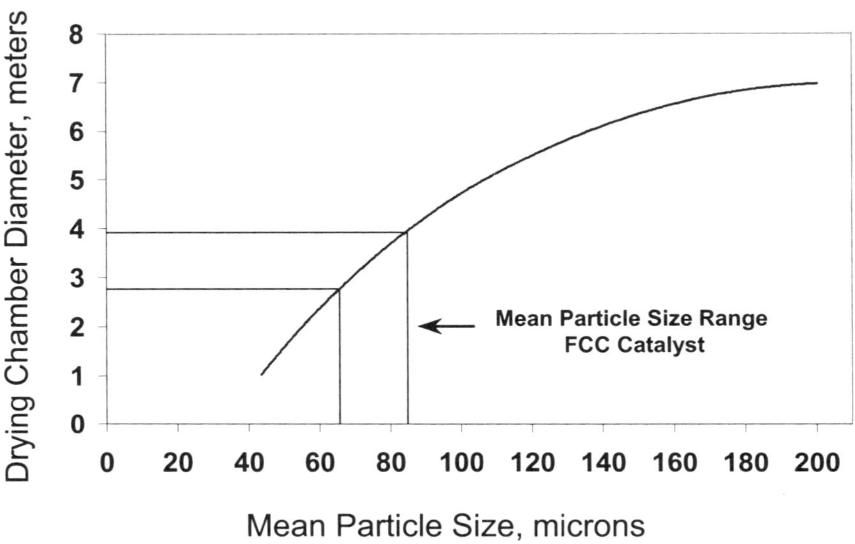


**Fig. 3.12** Mean particle diameter vs. chamber diameter for rotary (wheel) atomizers

## 3.4 Other catalyst forming techniques

Among the other shaping techniques described in the literature (7-10), none is as specific to zeolite catalyst preparation as Spray Drying. Extrusion and Oil-Drop are the most sophisticated of the other techniques commonly used in zeolite catalysts shaping (7). Extrusion is by far responsible for the largest volume of other zeolite catalysts (paraffins and aromatics isomerization, hydroprocessing, ...). Depending on the type of extruders used, the process can be viewed as operating under the following conditions:

- Single screw auger have a plug-flow character
- Double screw auger behave more like back-or cross-mix 'reactors'
- Piston extruders are working under high pressure and extreme plug flow

Oil-Drop is used in the case of slowly deactivating catalysts such as the BP/UOP Cyclar process where $C_{3\text{-}4}$ paraffins are aromatized on a Ga/H-MFI under reforming conditions. In the oil drop technique, a precursor of the binder (an acidified alumina sol for instance in the case of an $Al_2O_3$ binder) is mixed with the active ingredient (the zeolite in this case) in the presence of an organic compound releasing ammonia upon thermal decomposition (urea for instance); the aqueous suspension is then fed as millimeter sized droplets at the top of a long column filled with oil maintained at a temperature around 100°C. During their fall in the hot oil, the $NH_3$ released by the decomposition of, say urea, changes the pH of the alumina sol and provokes its precipitation and the formation of the alumina binder. The spheres are then collected, washed, aged, dried and further processed. It is also possible to oil drop with other binders such as $SiO_2$, $ZrO_2$, ... and eventually use this technique to generate in-situ zeolites (such as MFI) by synthesizing them during the wet spheres aging by addition of suitable templates and seeding agents. One example of such a process is described in (11); the zeolite produced by such an in-situ technique is generally well mixed and distributed homogeneously throughout the binder.

The other shaping techniques are used when specific requirements (hydrodynamics, pressure drop, ...) have to be met on the the reactors used or to be constructed (12). Among those described in the literature, the following can be found described rather qualitatively:

- Granulating
- Briquetting
- Balling/Marumerizing
- Prilling
- Dripping
- Pilling, Tabletting or Pellet Milling

In these agglomeration techniques, many operating parameters are used to improve the quality of the process; a non-exhaustive list reads like:

- Milling (ball, colloid, ...) in order to increase the surface contact of the components of the composite catalyst
- Use of binders (catalytically active or inert) to 'stick' the particles together
- Deaeration of the pastes to remove voids present between particles
- Severe mechanical mixing to promote intimate contact of the catalyst ingredients or to alter pore size distribution
- Peptization by acids or bases
- Use of tabletting/plasticizing agents or die lubricants to facilitate these shaping processes

As far as commercial applications of zeolites are concerned, the following list can be made:

- FCC: use of microspheres with the exception of units (TCC) still using millimeter sized spheres.
- Hydroprocessing: extensive use of extrudates and occasionally of spheres
- Reforming: use of extrudates (semi-regenerative units) and spheres (CCR types of units)
- Chemical applications: use of extrudates, pills and pellets

- Pollution control: monoliths and pellets are the preferred shapes for both stationary and mobile sources

## 3.5 Conclusions and perspectives

It is obvious that the area of catalyst preparation and shaping, although important technologically, is scarcely studied by academic researchers. Their industrial colleagues report their results in often difficult-to-read patent examples or keep them as trade secrets. The problem is made unappealing to the academic researcher because of the empirical image of catalyst preparation and (even more) shaping. In addition, some shaping processes (spray drying) are difficult to downscale to allow proper laboratory study and modeling.

This is however an area for a fruitful cooperation between industry and academia: catalyst shaping is more than physically mixing of an inactive 'binder' and an active zeolite. It has been demonstrated (14, 15) that important chemical changes (leading to more stable zeolites for instance) take place during these processes and also during subsequent treatments (steaming for instance) of finished particles: matrix bound zeolites are for instance more steam stable than the corresponding bare powders (16, 17).

This need for more science and engineering in catalyst shaping is unfortunately counterbalanced by the confidentiality requirements of a core aspect of the catalyst manufacturers' business. If academia and industry find a suitable cooperation framework in this area, the common benefit will be an even more rational design of catalysts.

## References

1. a) Sie S.T. and Krishna R., *Rev. Chem. Eng.* **14** (1998) 159.
   b) Rostrup-Nielsen J.R., *Catal. Today*, **37** (1997) 225.
2. Gilson J.-P., *Zeolite Microporous Solids: Synthesis, Structure and Reactivity*, (Kluwer, Dorderecht, 1992) 19.
3. Szostak R., *Stud.Surf. Sci. Catal.* **58** (1991) 153.
4. Von Ballmoos R., Harris D.H. and Magee J.S., in *Handbook of Heterogeneous Catalysis* Eds. G. Ertl et al. (Wiley, 1997) 1955.
5. Woltermann G.M., Magee J.S. and Griffith S.D., *Stud.Surf. Sci. Catal.* **76** (1993) 105.
6. Masters K., *Spray Drying*, (Wiley, New York, 1985).
7. Doesburg E.B.M., de Jong K.P. and van Hooff J.H.C., *Stud.Surf. Sci. Catal.* **123** (1999) 433.

8. Richardson J.T., *Principles of Catalyst Development*, (Plenum, New York, 1992).
9. LePage J.F., *Applied Heterogeneous Catalysis*, (Technip, Paris, 1987) 75.
10. Stiles A.B., *Catalyst Manufacture*, (Marcel Dekker, New York, 1983).
11. Gilson J.-P., *Method for Preparing Silicate Composition*, US. Patent **4,537,866**, (assigned to UOP, 1985).
12. Moulijn J.A., Makkee M. and van Diepen A., *Chemical Process Technology*, (Wiley, New York, 2001) Chapter 9 and references therein.
13. Farrauto R.J. and Bartholomew C.H., *Fundamentals of Industrial Catalytic Processes*, (Blackie Academic & Professional, London, 1999) Chapter 2 and references therein.
14. Gélin P. and Des Courrières T., *Appl. Catal.* **72** (1991) 179.
15. Corma A. et al., *Appl. Catal.* **66** (1990) 45.
16. Gélin P. and Gueguen C., *Appl. Catal.* **38** (1988) 225.
17. Kubicek N. et al., *Appl. Catal. A* **175** (1998) 159.

CHAPTER 4

# REFINING PROCESSES: SETTING THE SCENE

R.H. JENSEN
*UOP LLC, Des Plaines, IL, USA*

## 4.1 Introduction

Zeolites have become fundamental components of catalysts and adsorbents in refinery processes that convert petroleum into fuels and petrochemicals (1-3). Zeolite-enabled processes are clearly cleaner technologies than their predecessors from the viewpoint of environmental impact, particularly with regard to air quality. A major environmental objective of the 21st century is to further improve air quality by reducing the generation of greenhouse gases and the vehicle emissions that cause air pollution. In refinery operations, improving the efficiency of processes that transform petroleum into useful products will reduce greenhouse gases because less energy is consumed. Requiring refined fuels to meet pertinent quality standards will minimize harmful emissions from vehicles.

## 4.2 Fuel parameters and vehicle emissions

Emissions from gasoline and diesel engines are commonly classified into three categories: volatile organic compounds, toxics and nitrogen oxides. The composition of exhaust gas from a vehicle is dependent on its fuel properties and the vehicle's emission control technology. Table 4.1 summarizes the primary fuel properties of interest and their qualitative effects on emissions.

**Table 4.1** Fuel properties and vehicle emissions

| Fuel Property | Most Significant Emission Impact | |
|---|---|---|
| | Emission Type | Effect |
| RVP | VOC | High vapor pressure causes evaporative emissions |
| Benzene | Toxic | A known human carcinogen |
| Oxygenate | Toxic | Reduces toxic combustion products but contaminates groundwater |
| Sulfur | Toxic, NOx | Inhibits exhaust catalysts, $O_2$ sensors, NOx adsorbers, & particulate traps |
| Distillation | VOC | Affects driveability; and emissions during vehicle start and warm-up |
| Aromatics | Toxic | Form toxic combustion products |
| Olefins | Toxic | Contribute to ozone formation; combustion products form toxic dienes |

Emissions of volatile organic compounds during the transfer and storage of gasoline are a function of the vapor pressure of the gasoline, as determined by the Reid Vapor Pressure method, and by the initial boiling point and shape of the distillation curve. The distillation curve also affects VOC emissions during starting and warming of the engine.

Toxic emissions are formed from the combustion of benzene, other aromatics and olefins. Sulfur indirectly contributes to toxic emissions by interfering with the emission control components of a vehicle.

NOx emissions can be curtailed by equipping vehicles with modern catalytic converter technology. However, sulfur oxides in the exhaust reduce the efficiency of NOx conversion. Sulfur oxides also impede diesel particulate traps. Both SOx and NOx contribute to acid rain, causing other adverse environmental consequences.

## 4.3 Fuel specifications

Specifications for the composition of gasoline and diesel fuel have been steadily evolving, with the objective of reducing emissions of VOCs, toxics and NOx. For gasoline, the trend is to maintain high octane, while meeting stricter requirements for RVP, benzene, sulfur, distillation, aromatics and olefins. For diesel, a high cetane number must be achieved, along with low levels of sulfur and aromatics.

Gasoline specifications derive from studies that correlate reductions in emissions with changes in gasoline properties. Emission reductions calculated from the U.S. EPA Phase II Complex Model are presented as a bar graph in Figure 4.1. The most significant variable for VOC emissions is RVP. A reduction of RVP from 8.7 to 6.6 psi (60 to 45.5 kPa) reduces VOC emissions by 24%. Less significant are E200 and E300, which are the volume percentages distilled over at 200°F and 300°F, respectively. Sulfur and olefins mostly affect NOx emissions, with sulfur obstructing the performance of catalytic converters and oxygen sensors. Toxic emissions are primarily reduced by decreasing RVP, benzene, sulfur and aromatics and by adding oxygenates to gasoline.

Other benefits resulting from adding oxygenates to gasoline include high blending octanes and clean combustion products. Methyl-tertiary-butyl-ether was added to gasoline in various areas of the U.S. for many years. However, it became a contaminant in groundwater and was detectable by taste and odor at a level of 2–5 parts per billion. Consequently, the U.S. Environmental Protection Agency established an MTBE limit of 20–35 parts per billion in water, resulting in a phase-out of MTBE in gasoline in the U.S. The MTBE experience makes it unlikely that other oxygenates will be used for gasoline blending, with the exception of ethanol.

Automobile manufacturers have taken initiatives to identify fuel quality needs and have proposed that fuel specifications be standardized around the world. Their proposal was first published as the World-Wide Fuel Charter in 1998; a revision was published in April 2000 (4). The members of the World-Wide Fuel Charter are the

European Automobile Manufacturers Association, the Alliance of Automobile Manufacturers, the Engine Manufacturers Association, and the Japan Automobile Manufacturers Association. Associate members are from Canada, China, Korea, Mexico, the Philippines, South Africa and Thailand.

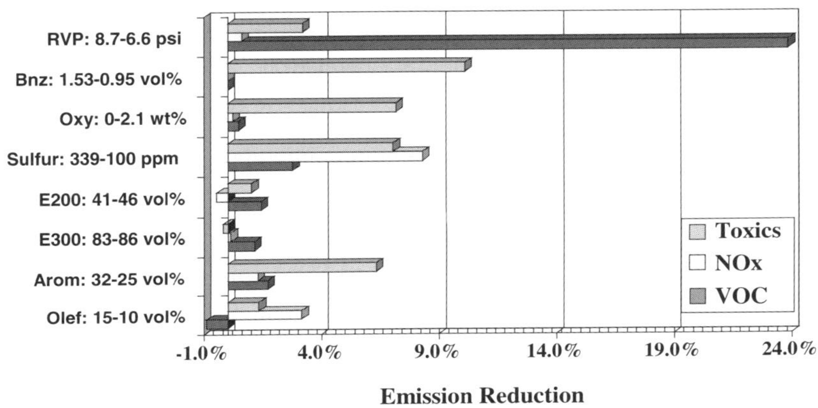


**Fig. 4.1** Sensitivity of emissions to gasoline parameter

The auto industry's World-Wide Fuel Charter proposes four different categories of fuel quality, based on the degree of emission control required. Category 1 in Table 4.2 is for gasoline markets with no requirements, or minimal requirements, for emission control. Category 2 is for markets with stringent requirements, representing the emission control technology required on vehicles in North America, Europe and Japan in 2000. Categories 3 and 4 are for markets with advanced requirements for emission control.

**Table 4.2** Gasoline specifications proposed in the world-wide fuel charter

| Fuel Parameter | Emission Control Stringency | | | |
|---|---|---|---|---|
| | **Category 1 Minimal** | **Category 2 Stringent** | **Category 3 Advanced** | **Category 4 Further Advanced** |
| Octane, (R+M) / 2 | | | | |
| Low | 86.5 | 86.8 | 86.8 | 86.8 |
| Mid | 90.0 | 90.0 | 90.0 | 90.0 |
| High | 93.0 | 93.0 | 93.0 | 93.0 |
| Oxygen, wt-% max | 2.7 | 2.7 | 2.7 | 2.7 |
| Benzene, vol-% | 5.0 | 2.5 | 1.0 | 1.0 |
| Aromatics, vol-% | 50.0 | 40.0 | 35.0 | 35.0 |
| Olefins, vol-% | NA | 20.0 | 10.0 | 10.0 |
| Sulfur, ppm | 1000 | 200 | 30 | 5-10 |

With the exception of low-octane Category 1, the proposed octane specifications for all three grades of gasoline are equivalent for all categories, reflecting the desire of both the auto industry and the consumer for good engine performance. However, octane must be maintained while decreasing benzene, aromatics and olefins, which are high-octane components of gasoline. In addition, in order to protect advanced catalytic converters and to permit NOx reduction, the auto companies are proposing that sulfur be reduced to 30 parts per million in Category 3 and to 'sulfur-free' in Category 4. Table 4.3 compares Category 4 proposals of the World-Wide Fuel Charter with specifications established by regulatory agencies in the U.S., Japan and Europe.

**Table 4.3** Gasoline specifications in selected regions

| Fuel Parameter | US RFG 2005 | California 2003 | Japan 2005 | Europe 2005 | WWFC Category 4 |
|---|---|---|---|---|---|
| RVP, kPa | <51.7 | 48.2 | 72 | 60 | 45-60 |
| Distillation | | | | | |
| E100, vol-% | | | | 46 | 50-65 |
| E150, vol-% | | | | 75 | |
| T50, °C | 77-110 | 99 | | | 77-100 |
| T90, °C | 185 max | 149 max | | | 130-175 |
| End Point, °C | 221 max | | | | 195 max |
| Oxygen, wt-% max | 2.7 | 2.2 | 0.5 | 2.7 | 2.7 |
| Benzene, vol-% | 1.9 | 0.8 | 1.0 | 1.0 | 1.0 |
| Aromatics, vol-% | 25-30 | 25 | | 35.0 | 35.0 |
| Olefins, vol-% | <20 | 6.0 | | 18 | 10.0 |
| Sulfur, ppm | 30 avg<br>80 max | 20 | 30 or 50? | 50 | Sulfur-Free* |

*5-10 ppm

The WWFC proposes RVP and distillation specifications for five classifications of ambient temperature ranges. The specifications in Table 4.3 are for ambient temperatures above 15°C, which is the WWFC's highest temperature classification. The state of California, an environmental leader in the U.S., established 2003 specifications more demanding than those of the U.S. government for 2005. The U.S. is on the low side of Category 4 vapor pressure specifications; Japan exceeds Category 4; and Europe is at the high end. Distillation specifications vary by region. Europe specifies E100 and E150, which are the percentages distilled over at 100°C and 150°C, respectively. The U.S. specifies T50, T90 and end point. T50 and T90 are the temperatures at which 50% and 90% of the gasoline has distilled over, respectively. California and Japan require lower oxygenate levels than Category 4. MTBE has been banned in California, where ethanol is the likely substitute of choice. California has more stringent 2003 specifications than Category 4 for benzene, olefins and aromatics. The sulfur specification is 20 ppm, which is lower than Category 3 but not sulfur-free. Europe's 2005 specifications are close to Category 4, with the exception of the higher 50 ppm sulfur requirement.

The $C_6$ to $C_8$ aromatic components of gasoline are shown in Table 4.4. Produced primarily by catalytic naphtha reforming, the aromatics have very high octane and relatively low RVP (5, 6).

**Table 4.4** Octane and RVP for aromatics

| Aromatic | (R+M)/2 | RVP, kPa |
|---|---|---|
| Benzene | 94 | 22.2 |
| Toluene | 118 | 7.1 |
| Ortho-xylene | 111 | 1.8 |
| Meta-xylene | 135 | 2.2 |
| Para-xylene | 136 | 2.4 |
| Ethylbenzene | 116 | 2.6 |

Olefinic compounds are major sources of octane for gasoline pools. The $C_5$ and $C_6$ olefins in Table 4.5 all have high octanes, ranging from 95 for 1-hexene to 158 for 2-methyl-2-butene (5). The $C_5$ olefins also have very high vapor pressure (6), 98.6–182.1 RVP, a property of significance for adjusting the RVP and distillation properties of gasoline pools. Meeting gasoline specifications for aromatics and olefins, while maintaining octane, is a complex issue for refineries.

**Table 4.5** Octanes for $C_5$ and $C_6$ olefins

<table>
<tr><th>Straight Chain</th><th>(R+M)/2</th><th>Mono-Branched</th><th>(R+M)/2</th><th>Di-Branched</th><th>(R+M)/2</th></tr>
<tr><td rowspan="3">1-Pentene<br>2-Pentene</td><td rowspan="3">113.5<br>144.0</td><td>3-Methyl-1-butene</td><td>127.0</td><td rowspan="3"></td><td rowspan="3"></td></tr>
<tr><td>2-Methyl-1-butene</td><td>139.0</td></tr>
<tr><td>2-Methyl-2-butene</td><td>158.0</td></tr>
<tr><td rowspan="5">1-Hexene<br>2-Hexene<br>3-Hexene</td><td rowspan="5">95.0<br>124.3<br>128.5</td><td>3-Methyl-2-pentene</td><td>121.5</td><td rowspan="5">3,3-Dimethyl-1-butene<br>2,3-Dimethyl-1-butene</td><td rowspan="5">129.0<br>138.0</td></tr>
<tr><td>4-Methyl-2-pentene</td><td>127.5</td></tr>
<tr><td>3-Methyl-1-pentene</td><td>141.5</td></tr>
<tr><td>4-Methyl-1-pentene</td><td>144.0</td></tr>
<tr><td>2-Methyl-2-pentene</td><td>153.5</td></tr>
</table>

In Figure 4.2, distillation specifications from Table 4.3 are plotted versus the percent of gasoline distilled over. The 2005 European E100 and E150 data fall in a reasonable pattern with the 2005 U.S. T50, T90 and end point data. The U.S. T50 is specified as a range from 77–110°C. California has a relatively low 2003 T90 specification, which decreases the slope of the distillation curve. Also in Figure 4.2 are two distillation envelopes (4) proposed by the World-Wide Fuel Charter for ambient temperatures >15°C and <15°C, respectively. The WWFC distillation specifications are identical for Categories 2, 3 and 4. The charter calls for lower T10s in summer than in winter, but the T90s and end points are the same for all seasons. California falls within the charter envelope, but Europe and the U.S. fall outside.

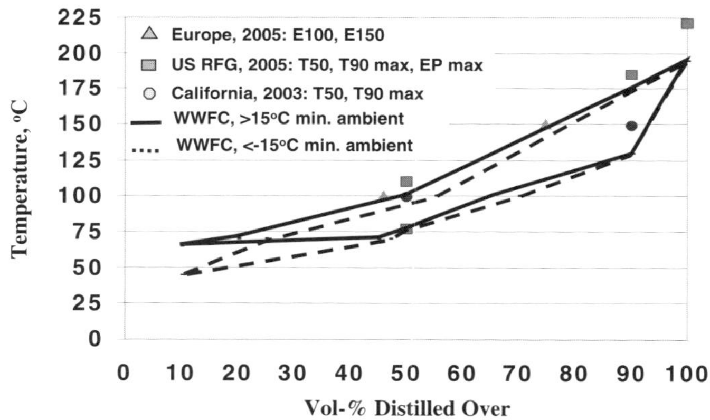


**Fig. 4.2** Gasoline distillation specifications

The World-Wide Fuel Charter also proposes a new standard set of specifications for diesel fuel, again depending on the stringency of required emission control technology. Table 4.6 compares WWFC Category 4 with specifications for the U.S., Japan and Europe. The U.S. allows relatively low cetane but requires very low sulfur in 2006. Japan and Europe demand higher cetane but allow higher sulfur than the U.S. The WWFC Category 4 proposal for high cetane and very low levels of aromatics and sulfur is a difficult goal to achieve.

**Table 4.6** Diesel specifications in selected regions

| Fuel Parameter | US 2006 | Japan 2005–2008 | Europe 2005 | WWFC Category 4 |
|---|---|---|---|---|
| Cetane No., min. | 40 | 45 | 51 | 55 |
| Cetane Index, min. | | | | 52 |
| Density @ 15 °C, kg/m$^3$ | | | | |
| Min. | | | | 820 |
| Max. | 880 | | 845 | 840 |
| Distillation | | | | |
| T90, °C | 338 | 330–350? | | 320 |
| T95, °C | | | 360 max | 340 |
| Final bpt, °C | 336 | | | 350 |
| Sulfur, ppm | 15 | 50 | 50 | Sulfur-Free* |
| Aromatics, wt-% | 35 | | | 15 |
| Polyaromatics, wt-% (di+tri) | | 2? | 11 max | 2.0 |

*5-10 ppm

Figure 4.3 is a graph of cetane number as a function of carbon number and type of hydrocarbon (7). Straight-chain paraffinic molecules have the highest cetane numbers, and multiple-ring aromatics have the lowest cetane numbers. FCC light cycle oil, which contains multiple-ring aromatics, is commonly used as a component of diesel fuel. As fuel specifications become stricter, it will become less desirable to blend polyaromatics into diesel fuel. An opportunity for catalytic researchers is the selective saturation and ring opening of polyaromatics, converting them to hydrocarbons with higher cetane numbers.

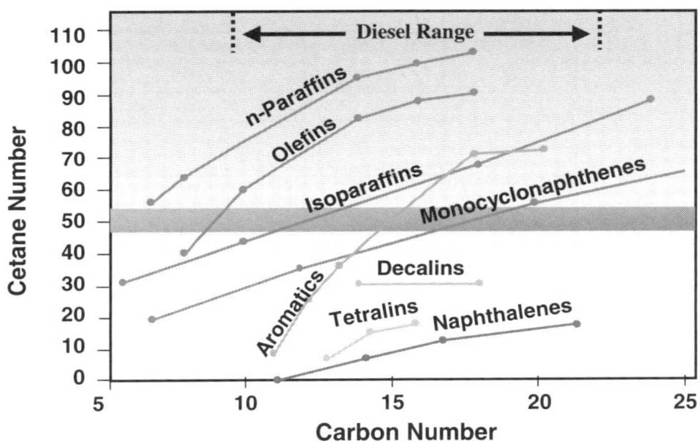


**Fig. 4.3** Cetane contributor

## 4.4 Fuels processes

Process technology must contribute to improved air quality in the 21st century by meeting the challenge of more severe gasoline and diesel specifications. Reformulated gasolines must remain high in octane with reduced levels of sulfur, olefins and aromatics. Diesel fuels also must have lower levels of sulfur and aromatics, with cetane being maintained or increased.

Fuels refineries contain a number of technologies for converting crude oil into blending stocks for transportation fuels. Table 4.7 is an example of gasoline blending stocks for a typical U.S. refinery. Typical blending octanes are listed, along with relative percentages of volume, sulfur, olefins and aromatics in the refinery's gasoline pool. The refinery contains fluid catalytic cracking, reforming, alkylation, isomerization, hydrocracking and coking processes.

FCC gasoline is an important gasoline blending stock world-wide. In Table 4.7, the FCC naphtha has a blending octane of 86 and comprises 38% of the gasoline pool. It contributes 98% of the pool sulfur, 98% of the pool olefins, and 35% of the pool aromatics. The heavy naphtha product from the hydrocracker, along with heavy straight-run hydrotreated naphtha, is feedstock to the reformer. Reformate contributes about the same volume as the FCC naphtha, but is much higher in octane due to its high aromatic content. The reformate contributes about 64% of the pool aromatics. The alkylate and isomerate streams are smaller in volume but high in octane. The remaining four blendstocks represent only 9% of the volume and 1–2% of the sulfur, olefins or aromatics. It is clear that the properties of the FCC naphtha must be modified in order to meet revised gasoline specifications.

**Table 4.7** Typical U.S. gasoline blending pool

| Gasoline Blendstock | Blending Octane | Per Cent of Pool | | | |
|---|---|---|---|---|---|
| | | Volume | Sulfur | Olefins | Aromatics |
| FCC Naphtha | 86 | 38 | 98 | 98 | 35 |
| Reformate | 95 | 36 | - | - | 64 |
| Alkylate | 92 | 13 | - | - | - |
| Isomerate | 95 | 5 | - | - | - |
| Light Straight-Run Naphtha | 75 | 3 | 1 | - | - |
| Light Hydrocracked Naphtha | 80 | 2 | - | - | - |
| MTBE | 110 | 2 | - | - | - |
| Coker Naphtha | 80 | 1 | 1 | 2 | 1 |

### *4.4.1 Fluid catalytic cracking*

FCC technology has undergone continuous improvements in efficiency since its introduction in the 1940's. Figure 4.4 uses feedstock utilization as a measurement of FCC efficiency. The vertical axis is octane barrels of naphtha per 100 barrels of feed, as calculated from the product of naphtha octane and vol-% naphtha yield. Amorphous catalysts were used in the early years of FCC and provided good octane but a relatively low yield of gasoline. The introduction of zeolite catalysts provided a step increase in octane barrels. The high activity of zeolite catalysts permitted lower residence times, leading to engineering improvements such as extended risers. Ultra-stable Y zeolites, ZSM-5 additives, and many engineering innovations have further increased octane barrels. Over a period of 50 years, the emphasis in FCC technology was on yield and octane. In the 1990s, new fuel specifications compelled refiners to increase their focus on other FCC gasoline quality issues, namely sulfur and olefin contents.

Figure 4.5 is a boiling point distribution of sulfur compounds and olefins in a typical FCC naphtha. The data were obtained by distilling the naphtha into 10 different cuts and then analyzing each cut for sulfur and olefin contents. The average sulfur content of the naphtha is 850 ppm but increases from 20 ppm in the lightest fraction to 6500 ppm in the heaviest fraction. The average olefin content is 37 vol-% but decreases from 58 vol-% in the lightest fraction to 7 vol-% in the heaviest fraction. A refiner could not use this naphtha for 38 vol-% of the gasoline pool and still satisfy California's 2003 gasoline specifications. One approach for meeting the 20 ppm sulfur limit is to discard 50% of the naphtha from the heavy fraction. Similarly, the 6.0 vol-% olefin limit could be achieved by rejecting 15% of the naphtha from the light fraction. These losses of FCC yield would not be economically viable, and consequently other approaches are necessary for reducing the sulfur and olefin contents of FCC gasoline.

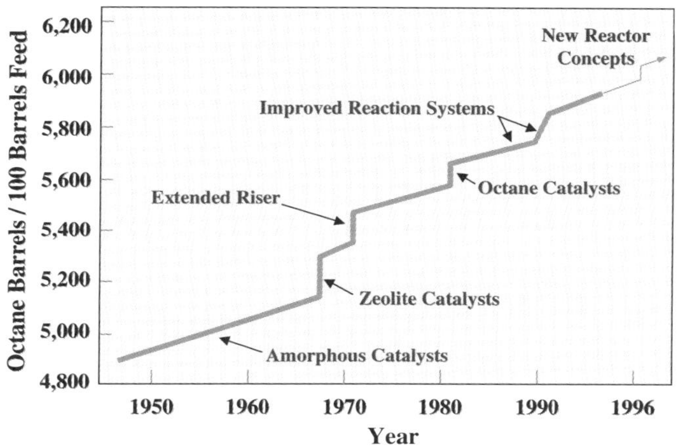


**Fig. 4.4** FCC: Efficiency improvements via catalysis and engineering

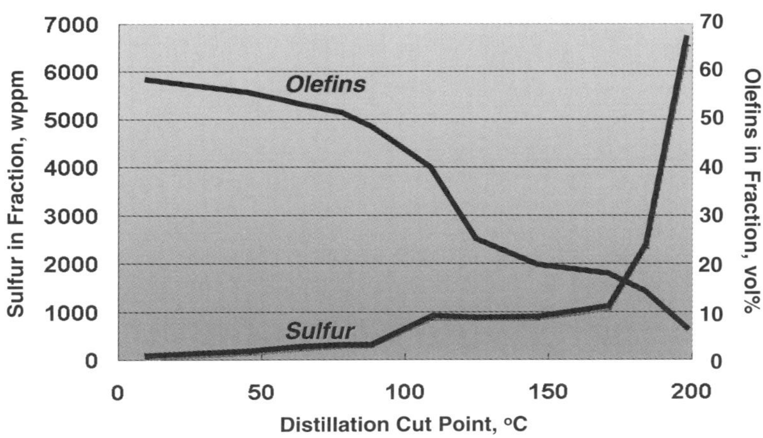


**Fig. 4.5** Sulfur and olefin distribution in typical FCC naphtha

Improvements in the quality of FCC gasoline can be attained by pretreating the vacuum gas oil feedstock, by catalyst and process modifications within the FCC unit, by treating the gasoline product, or by combinations of these strategies. Hydrodesulfurization of vacuum gas oil is proven technology, but it is costly (8). In some situations, however, it might be a preferred solution for sulfur management. Within the FCC unit itself, there are catalyst alternatives to consider, including zeolites with a high content of rare-earth metals to promote hydrogen transfer reactions for saturating olefins, catalyst additives to capture sulfur, and additives to selectively crack olefins to LPG. If product treatment is the desired approach, there are many choices available (10). BP, CDTech, IFP, ExxonMobil, Phillips, UOP/PDVSA-INTEVEP and others offer post-treatment processes for improving the quality of FCC gasoline while minimizing the loss of gasoline yield.

One option from UOP for olefin reduction is the revamp of an FCC unit to RxCat™ technology (10). In the RxCat process, Figure 4.6, a portion of coked catalyst is recycled to mix with regenerated catalyst at the bottom of the riser reactor. This feature allows the unit to run at a higher catalyst-to-oil ratio and a lower catalyst contact temperature. Moreover, ZSM-5 additive is more effective with RxCat because coked ZSM-5 retains more activity than coked Y zeolite.

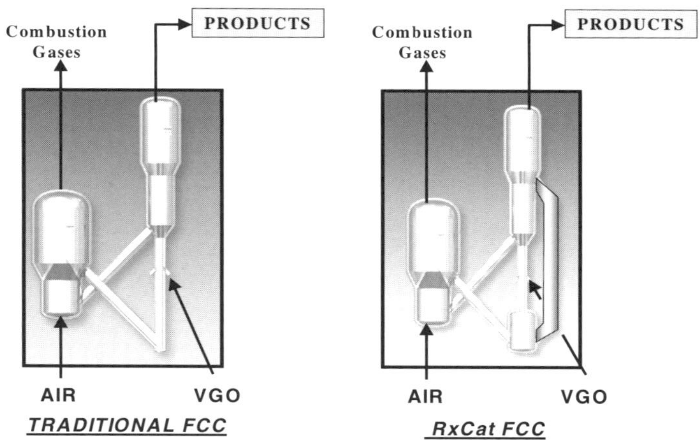


**Fig. 4.6** Traditional FCC vs. RxCat

Table 4.8 shows the shifts in product yields that can be realized with RxCat technology relative to traditional FCC, with an equal level of ZSM-5 additive in both reactors. The RxCat gasoline yield is 0–0.5 wt-% higher than the gasoline yield from a traditional FCC unit, and the olefin content of RxCat gasoline is 25% lower. The yields of LPG and light olefins increase, and the yields of heavy products decrease.

RxCat technology continues the 50-year historical extension of improvements in FCC technology, with gasoline olefin reduction as an additional measure of efficiency.

**Table 4.8** RxCat vs. traditional FCC

| RxCat Products | RxCat Yield Shifts, wt-% |
|---|---|
| Dry Gas ($C_1$ & $C_2$) | -0.5 |
| LPG ($C_3$ & $C_4$) | +0.5 |
| Propylene | +0.7 to +1.5 |
| $C_4$ Olefins | +0.5 |
| Gasoline* | 0 to +0.5 |
| Light Cycle Oil | -1.3 |
| Clarified Oil | -0.5 to –1.5 |

* 25% less olefins

*4.4.2 $C_5$ /$C_6$ isomerization*

Isomerization of $C_5$ and $C_6$ paraffins is a refinery process that generates high octane for the gasoline pool, while contributing no olefins or aromatics (11). Branched $C_5$ and $C_6$ paraffins have much higher octanes than normal pentane and normal hexane, as shown in Table 4.9 (5). Furthermore, $C_5$ and $C_6$ isomers can be separated by distillation, and flow schemes can be employed to recycle lower-octane isomers to the isomerization reactor.

**Table 4.9** Octanes and boiling points for $C_5$ and $C_6$ paraffins

| Paraffin | (R+M)/2 | Boiling Point, °C |
|---|---|---|
| Isopentane | 102.0 | 27.9 |
| Normal Pentane | 64.0 | 36.1 |
| 2,2-Dimethylbutane | 93.0 | 49.7 |
| 2,3-Dimethylbutane | 101.0 | 58.0 |
| 2-Methylpentane | 80.0 | 60.3 |
| 3-Methylpentane | 83.0 | 63.3 |
| Normal Hexane | 20.5 | 68.7 |

Flow Scheme A in Figure 4.7 is a once-through isomerization reactor in which $C_5$/$C_6$ feedstock is converted to an equilibrium mixture of isomers. The octane of the reactor effluent is typically in the range of 78 to 84, depending on the temperature of the reactor. In Flow Scheme B, a de-isohexanizer, DIH, provides recycle of methyl pentanes and normal hexane, resulting in a net product with higher octane. A de-isopentanizer, DIP, is added in Flow Scheme C, and the DIH is modified to send pentanes to the DIP. Fresh feed is also directed through the DIP, where isopentane is separated as an overhead product. Flow Scheme C yields the highest-octane isomers as the net product, i.e. isopentane plus the two dimethylbutanes. Flow Scheme D incorporates a zeolitic adsorptive separation process instead of distillation. In this case, the zeolitic molecular sieve selectively adsorbs normal paraffins from the reactor effluent. The normal paraffins are recycled to the isomerization reactor.

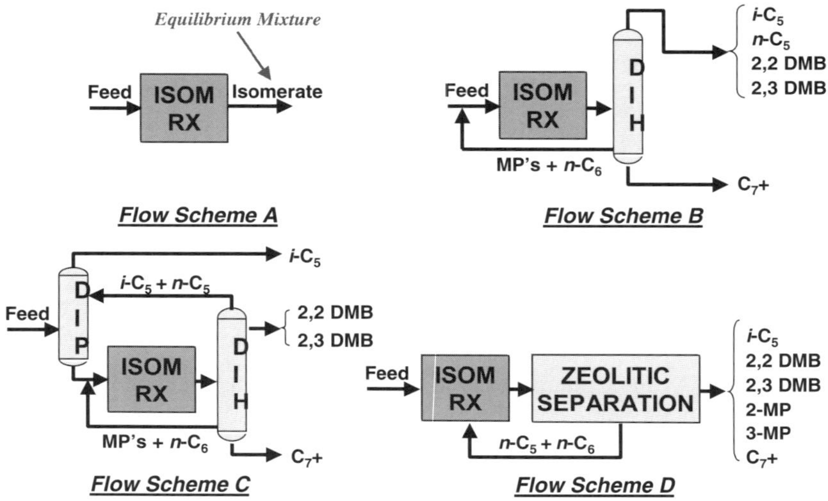


**Fig. 4.7** $C_5/C_6$ isomerization flow scheme

The temperature of isomerization controls equilibrium isomer composition, and thereby product octane. Figure 4.8 is a plot of isopentane in the $C_5$ product as a function of temperature. The data are from pilot plant runs with three types of commercial UOP isomerization catalysts. The feedstock was a 50/50 mixture of normal pentane and normal hexane, containing about 6% cyclics. The I-8™ and I-80™ catalysts are very active at a low temperature, where equilibrium isopentane content is highest. The acid functions in I-8 and I-80 are chlorided aluminas. The zeolitic catalyst, HS-10™, requires relatively high temperatures of operation. The LPI-100™ catalyst contains sulfated zirconia as the acid function and falls in the middle of the temperature range (12). Due to the equilibrium constraints, a lower temperature operation yields a higher octane product. The I-8 and I-80 catalysts yielded Research Octane Numbers of 82–84, as compared to 80–82 for LPI-100 catalyst and 78–80 for HS-10.

Product octanes by various combinations of flow schemes with UOP catalysts and processes are given in Figure 4.9. The TIP™ process, with HS-10 catalyst, incorporates an IsoSiv™ process to increase product RON from a range of 78–80 to 87–89, depending on feedstock composition. IsoSiv is an adsorptive separation process with zeolitic molecular sieves that operates at high temperature and in the vapor phase. Alternatively, the Penex$^{TM}$ process, with I-8 or I-80 catalyst, can be combined with a Molex$^{TM}$ process. Product RON in this case is 88–90. The Molex process is a member of UOP's family of Sorbex$^{TM}$ processes. Sorbex applies a simulated moving bed of adsorbent to separate desired products at high purity and recovery from the liquid phase (13). The Penex process with DIP/DIH fractionation instead of with the Molex process yields the highest RON, 90–92. DIH columns can be used with any of the catalytic processes to enhance product octane. The Par-Isom$^{TM}$ process, with LPI-100 catalyst, yields a product RON of 80–82. Combining a DIH with the Par-Isom process increases RON to 85–87. One advantage of a molecular sieve separation is the absence of naphthenes in the recycle stream. In a fractionation flow scheme, naphthenes are recycled to the reactor. Although the catalysts have ring-opening functionality, additional catalyst volume is required to efficiently recycle the naphthenes to extinction.

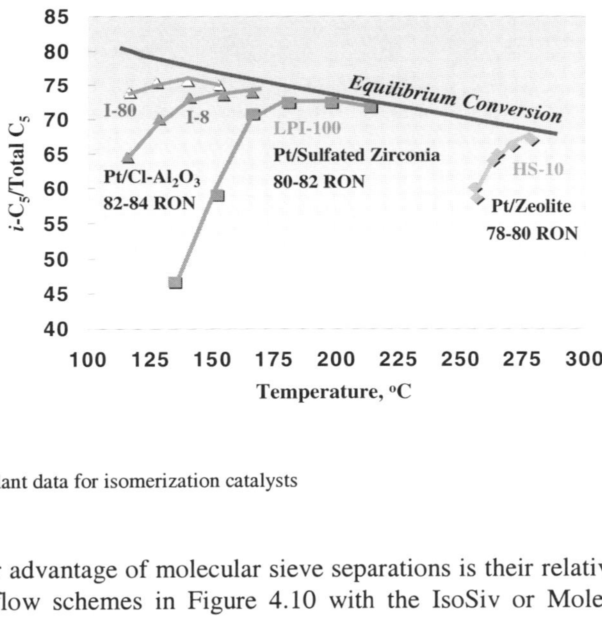


**Fig. 4.8** Pilot plant data for isomerization catalysts

Another advantage of molecular sieve separations is their relatively low energy costs. The flow schemes in Figure 4.10 with the IsoSiv or Molex process have substantially lower operating costs than flow schemes with fractionation. Although the Penex/DIP/DIH flow scheme has a RON advantage over the Penex/Molex flow scheme, the operating costs of the fractionation flow scheme are over twice as high.

With the need to reduce olefins and aromatics in gasoline, $C_5/C_6$ isomerization has become increasingly important to a fuels refinery. The goal is to maximize gasoline octane with recycle flow schemes and with catalysts that are active at low temperatures. Catalysts also must have high selectivity to isomerization in order to maximize gasoline yield. Zeolitic molecular sieves can be utilized to recycle normal paraffins at high energy efficiency compared to fractionation.

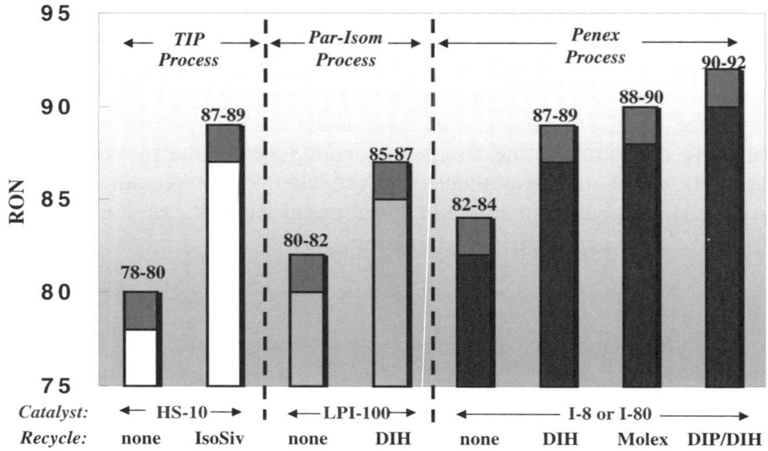


**Fig. 4.9** A comparison of $C_5/C_6$ isomerization catalysts and flow scheme

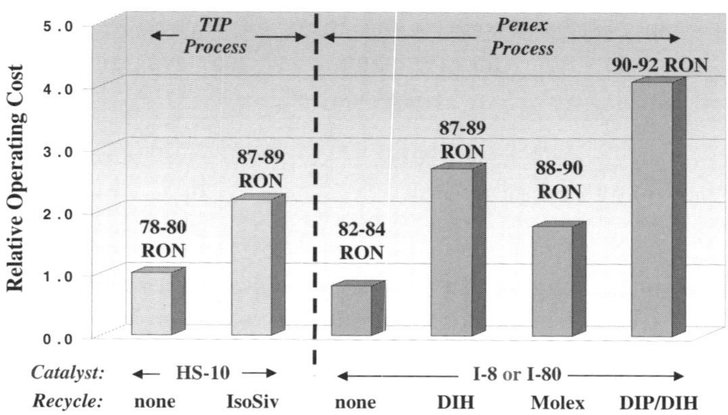


**Fig. 4.10** Relative operating costs of $C_5/C_6$ isomerization flow schemes

*4.4.3 Hydrocracking*

Hydrocracking is an extremely versatile petroleum refining process. The acid and metal functions of the catalyst can be designed to tailor the product slate to the needs of a particular refiner. Figure 4.11 depicts how catalysts can be designed to maximize diesel fuel, jet fuel or naphtha, and to provide flexibility if the desired product slate varies with time. The primary acid function of a hydrocracking catalyst has traditionally been amorphous silica-alumina if maximum heavy distillate is desired. As the desired product slate moves towards naphtha, zeolite becomes the primary acid function. The catalyst activity increases in accordance with increasing and/or modifying the zeolite content. Altering the zeolitic component also increases selectivity to naphtha and LPG.

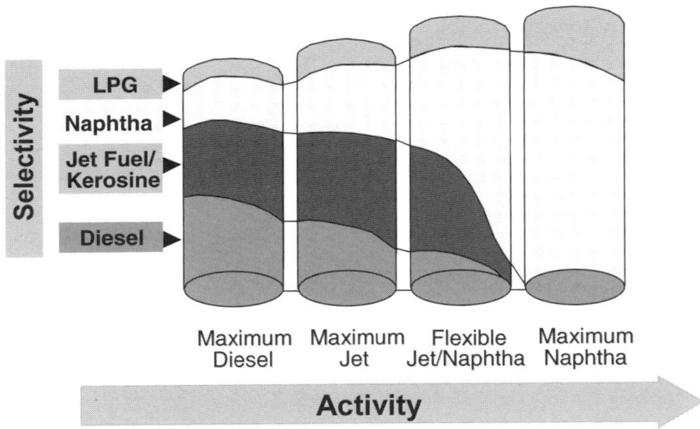


**Fig. 4.11** Hydrocracking catalysts designed to optimize product slate

As in the case of FCC catalysts, there have been many historical improvements in the efficiency of hydrocracking catalysts. Advances in zeolite acid chemistry and metal functionality have led to new generations of catalyst families. A more selective catalyst at a given activity leads to more efficient utilization of petroleum feedstocks. A higher activity catalyst at a given selectivity frequently leads to lower operating costs, thereby reducing energy consumption in a refinery (14).

The DHC-8™ and HC-24™ catalysts in Figure 4.12 are second-generation, state-of-the-art catalysts for maximum diesel and maximum naphtha, respectively. In 2000-2001, UOP introduced a third generation of hydrocracking catalysts that further improve the efficiency of the hydrocracking process (15). HC-110™ is a third-generation maximum-diesel catalyst that provides up to 3 vol-% higher distillate yield than DHC-8 at the same activity. Another distillate catalyst, HC-

115™, offers the same high yield as DHC-8, but with an 8°C advantage in activity. HC-170™ is a maximum-naphtha catalyst that provides up to 1.5 vol-% higher naphtha yield than HC-24, at similar activity. These improvements were made possible through the discovery of new zeolitic materials and through an improved balance of metal and acid functions within the catalyst. It is anticipated that even higher performance curves will be developed in the future as a result of further research into zeolite synthesis and modification.

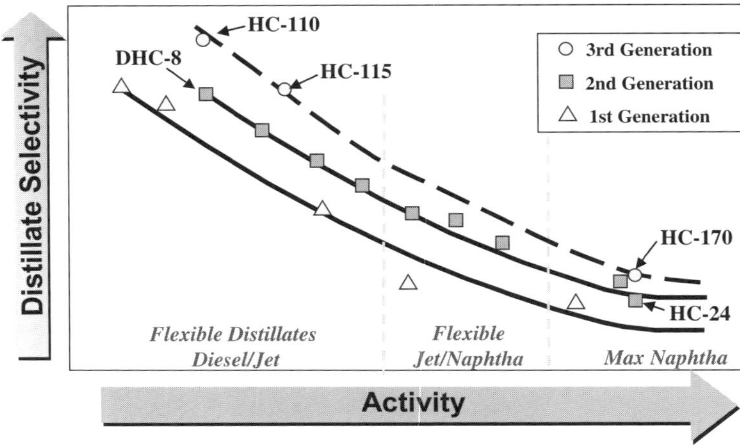


**Fig. 4.12** Improvements in hydrocracking catalyst efficiency

In addition to continuous developments in hydrocracking catalyst technology, there have been many enhancements in process technology. These combined innovations have further improved the efficiency of feedstock utilization and have reduced operating costs. An example is HyCycle™ Unicracking™ technology, a UOP process that represents a step improvement in hydrocracking. Figure 4.13 is a schematic flow diagram of the HyCycle process. Fresh feed to the process is directed first to a hydrotreating reactor, where it is combined with the effluent of the hydrocracking reactor. A unique feature of the flow scheme is a separator/finisher vessel, which operates at the same pressure as the reactor and provides separation of the reactor effluent into desired products and unconverted oil. There is a catalyst bed in the vessel, or alternatively external to the vessel, where the separated distillate and naphtha products are immediately subjected to an additional hydrogenation step in the vapor phase. This hydrogenation further upgrades the quality of the distillate products for a given operating pressure.

The liquid phase of the separator/finisher, which contains hydrotreated feed plus unconverted effluent from the hydrocracking reactor, is recycled to the hydrocracker. A small amount of this recycle oil is routed to a fractionator to purge heavy materials from the system, in order to avoid excess accumulation of heavies in the recycle stream. The novel design of the fractionator allows lighter recycle oil to be stripped from the heavies. The net yield of unconverted oil heavies is 0.5 vol-% or less, compared to 3+ vol-% in conventional high-conversion hydrocrackers, providing a further increase in product yields and a positive impact on the process economics.

The combined feed to the hydrocracking reactor has been stripped of ammonia and hydrogen sulfide, and so provides a relatively clean environment and a high partial pressure of hydrogen for the hydrocracking catalyst. From a process economy viewpoint, the recycle stream can be relatively large compared to fresh feed because most of it essentially remains at high pressure in the process. The lower per-pass conversion of feed increases the selectivity of the process to diesel production, compared to a more traditional hydrocracking flow scheme.

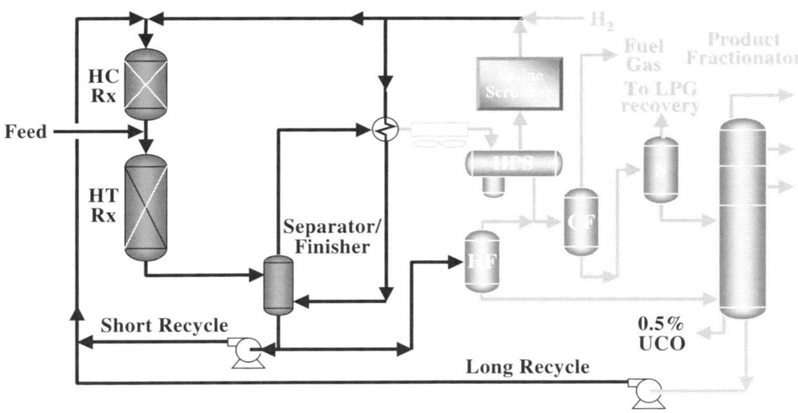


**Fig. 4.13** HyCycle unicracking flow scheme

Data from UOP hydrocracking pilot plants are summarized in Figure 4.14, a graph of weight percent of a given distillation fraction versus the average cut point temperature. Most of the vacuum gas oil feed is outside the diesel boiling range. The reference product distribution is from a standard hydrocracking flow scheme operating in maximum diesel mode at a pressure of 13,900 kPa. For comparison, the same catalyst was evaluated in a HyCycle flow scheme at a lower pressure, 12,200 kPa. The lower operating pressure translates to 10% lower operating costs. HyCycle yielded 10–15 wt-% more diesel product and maintained the sulfur content of the diesel at less than 10 ppm. HyCycle utilizes lower per-pass conversion to minimize undesired reactions that form gaseous products and precursors to catalyst coking. The lower operating pressure shifts equilibrium to a more favorable regime for selective ring opening reactions and higher quality diesel fuels. In a maximum-naphtha operation, HyCycle produces more naphtha and improved selectivity to $C_7^+$ naphtha, compared to standard hydrocracking.

Hydrocracking is indeed a very versatile refinery process. In 21st century fuel scenarios, hydrocracking is a process of choice for meeting specifications for high quality diesel fuel. Naphtha from hydrocrackers continues to be good feedstock for the catalytic reformers that provide octane to gasoline pools. Zeolites are the acid function of choice for rugged service in hydrocracking catalysts. Ongoing innovations in zeolite chemistry and acid/metal functionality will lead to further improvements in hydrocracking technology, for better feedstock utilization and enhanced energy efficiency.

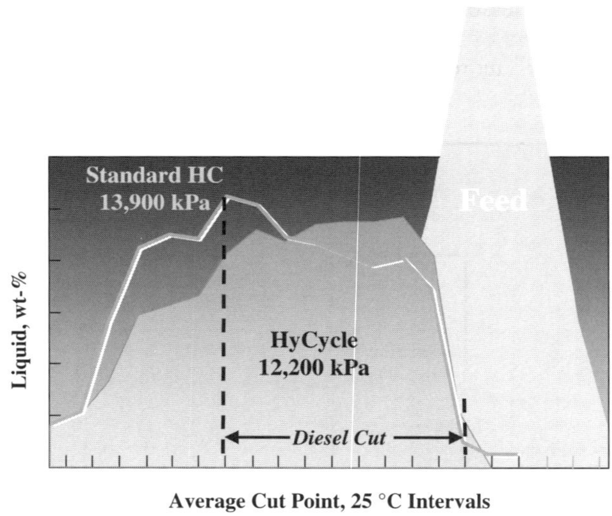


**Fig. 4.14** Yields from HyCycle vs. standard hydrocracking

## 4.5 Petrochemical processes

Advances in efficiency for the manufacture of fuels and petrochemicals improve air quality by reducing emissions of carbon dioxide into the atmosphere. When the fuels are consumed, the impact on air quality can be measured in a relatively straightforward manner because all fuels are burned. Cleaner fuels produce fewer harmful emissions during combustion. The positive impact of petrochemicals usage on air quality is not as obvious, but nevertheless can be very significant.

Petrochemical products are intermediates for the manufacture of plastics. It is of course desirable to minimize fuel consumption during the manufacture of petrochemicals so as to minimize carbon dioxide and other emissions. However, a more important factor is the conservation of energy that occurs by substituting plastics for other materials in consumer goods.

One example of the impact of plastic substitution is the use of plastics as high-performance components of automotive parts. In the U.S., passenger vehicles discharge about 20% of the nation's carbon dioxide emissions (16). Fuel economy is directly proportional to vehicle weight, and a 10% reduction in vehicle weight. improves fuel economy by about 7% (17). Because of plastics and other lightweight materials, a 2001 sports utility vehicle weighs less than the automobiles of the 1970s. On average, a kilogram of plastic can replace two to three kilograms of other materials in automobiles (18). Vehicles manufactured in North America in 2000 contained on average about 116 kilograms of plastic. This is expected to increase by 20%, to about 139 kilograms per vehicle, by 2010 (17).

Another important application for plastics is thermal insulation. Without plastic insulation, major appliances would consume 30–50% more energy. From 1972 to 1995, the energy consumption of major appliances decreased by up to 40%, and energy efficiency increased 50–100% (19).

The recycling of plastics also contributes to energy efficiency. Soft drink bottles made from polyethylene terephthalate reached a recycling rate of 41% in the U.S. in 1995 (19). Recycling is becoming widespread, and many studies are in progress to develop new and improved recycling practices.

Petrochemicals are the raw materials for plastics, making it possible for plastics substitution to have a positive impact on air quality. Zeolitic catalysts and adsorbents are essential to efficiently producing petrochemicals via environmentally sound process technologies.

### *4.5.1 Benzene-derived petrochemicals*

The benzene-derived petrochemicals in Figure 4.15 are intermediate feedstocks for styrenic and phenolic plastics. In the styrenics chain, ethylbenzene is dehydrogenated to styrene, to be used as polystyrene monomer or as a copolymer with acrylonitrile and butadiene. In the phenolics chain, cumene is an intermediate for making phenol. Bisphenol A is the condensation product of two moles of phenol and acetone. Phenol and Bisphenol A are used to manufacture resins and polycarbonates. Phenol and cyclohexane are the starting materials for the manufacture of nylon 6.

Friedel-Crafts alkylation of benzene was first commercialized for ethylbenzene and cumene in the 1940s. Aluminum chloride is the Friedel-Crafts catalyst, and the process is operated in the liquid phase. Several alternatives to aluminum chloride technology were developed later, but zeolitic catalysts are a rather recent introduction. UOP began using zeolitic catalysts in the 1990s.

UOP introduced the Alkar$^{TM}$ process for ethylbenzene production in 1966. The Alkar process operates in the vapor-phase with boron trifluoride on an alumina support as the catalyst. By the 1980s, about 15% of the world's ethylbenzene was being produced by the Alkar process. In the 1990s, UOP collaborated with ABB Lummus to develop the EBOne™ process. EBOne operates in the liquid phase with zeolitic catalysts. Three generations of zeolitic catalysts have been commercialized for EBOne, the most recent being EBZ-500™, which was introduced in 1996 (20).

In the case of cumene, UOP introduced a liquid-phase process in the 1940s to compete with aluminum chloride technology. The catalyst is SPA, a solid phosphoric acid catalyst in which the phosphoric acid is supported on silica. Many improvements were made to the SPA catalyst and process over the years, leading to 70% of the world's cumene being produced with SPA by the 1990s. In 1996, UOP introduced the Q-Max™ process, featuring a zeolitic catalyst and operating in the liquid phase (21). A new Q-Max catalyst, QZ-2001™, was introduced in 2001.

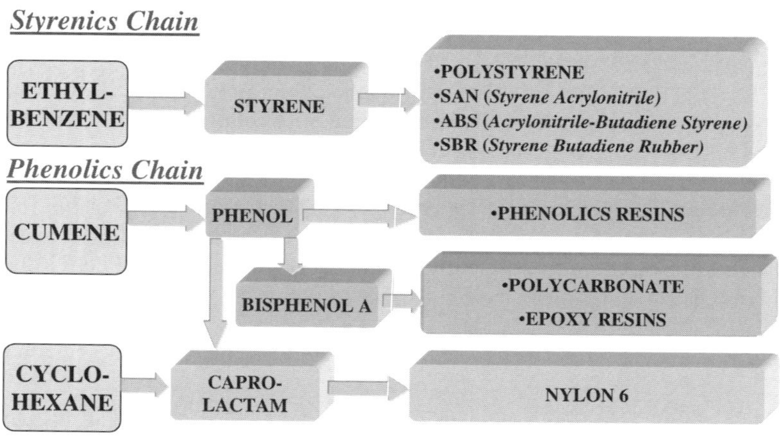


**Fig. 4.15** Benzene-derived petrochemicals for styrenics and phenolics

In the manufacture of ethylbenzene and cumene, the cost of benzene feedstock is a major factor in the overall economics. Thus, it is critical to have efficient technology for the alkylation of benzene. Zeolitic catalysts have the advantages of achieving higher purity and higher yield of product relative to aluminum chloride and SPA catalysts. Table 4.10 compares purities and yields and also shows the breakdown of impurities for both ethylbenzene and cumene. In both cases, extremely high purities can be achieved, 99.96 and 99.97%, respectively. The product yields are also extremely high, 99.6% and 99.7%.

**Table 4.10** Benzene alkylation catalyst comparisons

| Ethylbenzene Product | Technology | | Cumene Product | Technology | |
|---|---|---|---|---|---|
| | $AlCl_3$ | EBOne | | SPA | Q-Max |
| Purity, wt-% | 99.85 | 99.96+ | Purity, wt-% | 99.92 | 99.97 |
| Yield, wt-% | 98.5 | 99.6 | Yield, wt-% | 95.0 | 99.7 |
| Typical Impurities, ppm | | | Typical Impurities, ppm | | |
| Xylene | 10 | <10 | Cymene | <1 | 5 |
| *n*-Propylbenzene | 40 | 20 | Butylbenzene | 200 | 10 |
| Cumene | 100 | 50 | Diisopropylbenzene | <1 | 10 |
| Toluene | 30 | 50 | Nonaromatics | 300 | 10 |
| Nonaromatics | 50 | 120 | Ethylbenzene | 50 | 20 |
| Other Alkylaromatics | 1270 | <150 | *n*-Propylbenzene | 200 | 250 |

The higher yields represent improvements in alkylation efficiency and better utilization of the benzene feedstock. The higher purities improve the efficiencies of downstream technologies, so that ultimately less energy is consumed in the styrenic and phenolic chains for producing plastic products. Another important factor is that the zeolitic catalysts are more environmentally friendly materials than aluminum chloride and SPA.

### *4.5.2 Xylene-derived petrochemicals*

Xylenes are the petrochemical building blocks for the polyesters in Figure 4.16. Para-xylene is oxidized to PTA, purified terephthalic acid, which is esterified with ethylene glycol to PET, polyethylene terephthalate. There is a similar processing route for oxidizing meta-xylene to PIA, purified isophthalate acid. PIA is used in unsaturated polyester resins and as an additive to PET for improving product properties. Ortho-xylene is used primarily to manufacture phthalic anhydride, but markets are emerging for 2,6 DMN. Various manufacturing technologies create polyester films, fibers, plasticizers and resins

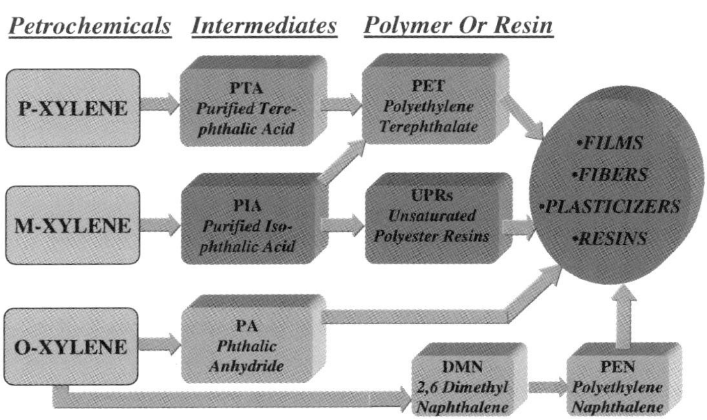


**Fig. 4.16** Xylenes to polyesters

High-purity benzene and xylenes are products of aromatics complexes having several interconnected processes and unit operations (22). In 1998, the market demand for benzene, on a world-wide basis, was 27.4 million metric tons per year, mostly for styrene. By comparison, the demand for para-xylene was 16.1 million metric tons. Ortho-xylene demand was lower, at 3 million metric tons. The market for meta-xylene was even lower, at about 300,000 metric tons. Because of these relative market requirements, most aromatics complexes are designed for benzene and para-xylene. Depending on local situations, they may also produce ortho-xylene, which can be separated by fractionation, and/or meta-xylene. Process units that can be integrated into UOP aromatics complexes are described in Figure 4.17.

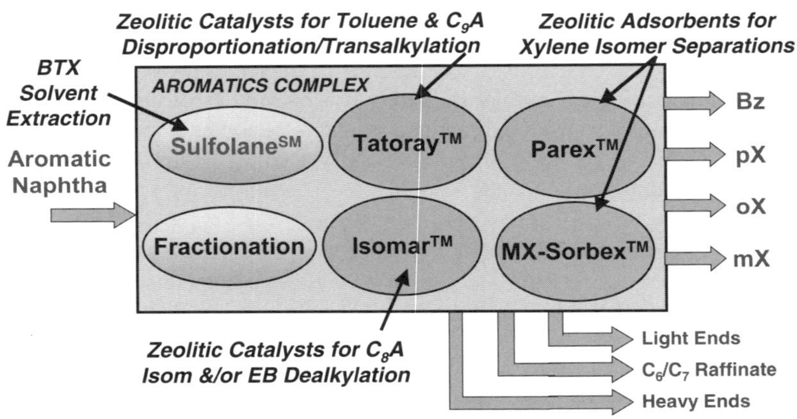


**Fig. 4.17** Process options for a UOP aromatics complex

The feed to an aromatics complex is normally a $C_6^+$ aromatic naphtha from a catalytic reformer. The feed is split into $C_8^+$ for xylene recovery and $C_7^-$ for solvent extraction. The extraction unit recovers pure benzene as a product and $C_7^+$ aromatics for recycling. A by-product of extraction is a non-aromatic $C_6^+$ raffinate stream. The complex contains a catalytic process for disproportionation and transalkylation of toluene and $C_9^+$ aromatics, and a catalytic process for isomerization of $C_8$ aromatics. Zeolitic catalysts are used in these processes, and catalyst selectivity is a major performance factor for minimizing ring loss and formation of light and heavy ends. The choice of isomerization catalyst is dependent on whether it is desired to isomerize ethylbenzene plus xylenes to equilibrium or to dealkylate ethylbenzene to benzene while isomerizing the xylenes. Para-selectivity may also be a desired feature, depending on the required product mix. Zeolitic adsorbents are used to separate para-xylene and meta-xylene at high purity and recovery.

The Parex™ and MX-Sorbex™ processes are both members of UOP's family of Sorbex™ processes. The MX-Sorbex process for the separation of high-purity meta-xylene was introduced in 1998. Five MX-Sorbex units were commercialized between 1998 and 2001. The market for meta-xylene is expected to grow to 800,000 metric tons per year by 2009 (23). The Parex process was introduced in 1971 for the separation of para-xylene, and 71 units were commercialized by 2001.

In the years following the initial commercialization of the Parex process, UOP developed substantial improvements in the technology with respect to energy efficiency and feedstock utilization. Figure 4.18 characterizes these improvements in terms of relative heat duties and product purities. A 'heavy-desorbent' Parex process was first commercialized in 1971 with ADS-3™ adsorbent and diethylbenzene desorbent. In 1972, a 'light-desorbent' Parex process was commercialized with ADS-4™ adsorbent and toluene desorbent. The heat duties in Figure 4.18 are over 50% lower for the 2001 versions of these processes. Also, para-xylene purity was 99.9% in 2001, compared to 99.3% in the 1970s. In addition, the recovery of para-xylene increased from 90% at 99.3% para-xylene purity in the 1970s to 99% at 99.9% para-xylene purity in 2001. These improvements in efficiency are a result of combining innovations in process design with more effective combinations of zeolitic adsorbents and desorbents, ADS-27™ with D-1000™ desorbent, and ADS-40™ with toluene desorbent.

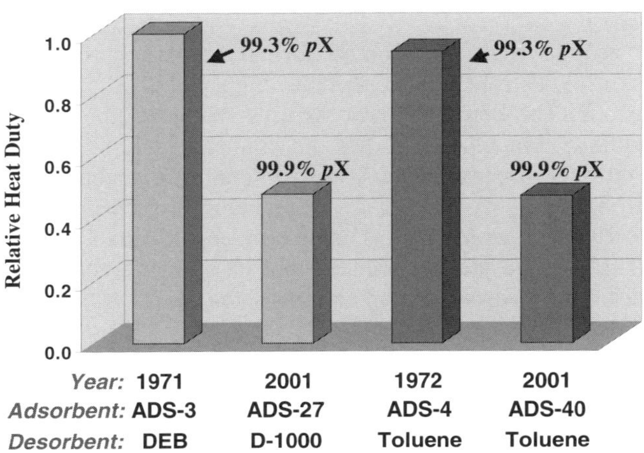


**Fig. 4.18** Advances in parex technology

### *4.5.3 Alternate feedstocks for aromatics complexes*

The primary source of feedstock for aromatic complexes is naphtha from a catalytic reformer. Naphtha reforming catalysts contain metal and acid functions, and the acid functions characteristically are not zeolitic. For paraffin feedstocks, however, zeolitic catalysts are employed to produce aromatics in two UOP processes, the RZ Platforming™ process and the Cyclar™ process. In the RZ Platforming process, $C_6$ and $C_7$ paraffins are processed over a non-acidic zeolite containing a metal function (24). The feed source can be hydrotreated light straight-run paraffins and/or the raffinate by-product from a BTX solvent extraction unit. Table 4.11 contains typical yields. Aromatic yields are greater than 60 wt-%, and hydrogen yields are 4–5 wt-%.

The Cyclar process converts $C_3$ and $C_4$ paraffins to aromatics via a bifunctional zeolitic catalyst (25). Typical aromatic yields in Table 3.11 are 70 wt-%, with 6 wt-% hydrogen yields. The light hydrocarbon products can be used as fuel for the plant. Cyclar can be considered for remote locations as an alternative to flaring LPG, or when refrigerating LPG for shipment is not economical.

**Table 4.11** Typical products from RZ platforming and cyclar

| Products, wt-% | RZ Platforming | Cyclar |
|---|---|---|
| Hydrogen | 4–5 | 6 |
| Light Ends | 9–11 | 24 |
| $C_6^+$ Paraffins + Naphthenes | <25 | – |
| Aromatics | 60+ | 70 |

Another UOP zeolitic route to naphtha for an aromatics complex is the PetroFCC™ process (26). PetroFCC also provides high yields of light olefins for petrochemical feedstock. The PetroFCC process uses RxCat technology and high levels of ZSM-5 additive. Figure 4.19 shows the dramatic yield shift that can be obtained with the PetroFCC process, relative to traditional FCC technology. The $C_3$ yield increases from 6.5 to 24 wt-%, with 22 wt-% being propylene. The $C_4$ yield increases from 11 to 19 wt-%, with 14 wt-% being butylenes. There is also a 6 wt-% yield of ethylene. The naphtha yield is reduced, but its aromatic content is 80–90 wt-%, making it a desirable feedstock for an aromatics complex.

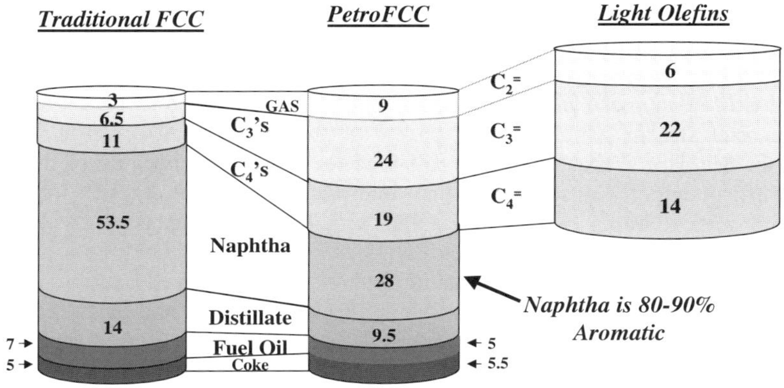


**Fig. 4.19** Yields from PetroFCC vs. traditional FCC, Wt-%

The PetroFCC process changes an FCC unit from a gasoline-producing device to a source of petrochemical feedstocks. It can provide both aromatic naphtha to an aromatics complex and light olefins to a polyolefins complex. Feedstock utilization can be further improved by recycling the raffinate by-product from BTX solvent extraction back to the PetroFCC reactor for conversion to light olefins.

Another UOP zeolitic process that produces petrochemical feedstocks is the MaxEne™ process (27). The MaxEne process, another member of the Sorbex family of processes, separates $C_5$ to $C_{11}$ full-range naphtha into an extract stream containing more than 90 wt-% normal paraffins and a raffinate stream containing over 99 wt-% non-normals, namely isoparaffins plus naphthenic and aromatic hydrocarbons. The high normal-paraffin content of the extract makes it a preferred feedstock for a naphtha steam cracker, and the absence of normal paraffins in the raffinate makes it a preferred feedstock for catalytic reforming.

Figure 4.20 displays material balances for a MaxEne flow scheme and for a conventional flow scheme with fractionation separating the full-range naphtha. The conventional flow scheme includes the fractionator, a steam cracker and a catalytic reformer. The fractionator has a capacity of 100 weight units of naphtha. The cracker is designed for a feed rate of 34.1 weight units, and the reformer is designed for a capacity of 56.8 weight units. In the conventional flow scheme, the major products are olefins to be used for petrochemical production, 9.1 weight units of straight-run gasoline with a research octane number of 72, and reformate containing 96.4 wt-% aromatics to be used as feed to an aromatics complex.

The addition of a MaxEne process to the flow scheme provides several benefits, beginning with an increase in throughput of 18%. With this higher feed rate, the MaxEne extract stream matches the capacity of the cracker, 34.1 weight units. However, the MaxEne extract is a preferable feed for the cracker because of its high normal paraffin content. Consequently, the yield pattern from the cracker, in Table 4.12, is improved. Olefins increase by 15%, with ethylene increasing by almost 50%. In addition, LPG increases by 32%.

The MaxEne raffinate is fractionated into a light gasoline fraction and a reformer feed fraction. The gasoline fraction is 28.1 weight units and has a research octane number of 82. Compared to the original flow scheme, there is a three-fold increase in the gasoline product, and it has much higher octane because the normal paraffins have been removed. The reformer feed is 0.7 weight units lower in the MaxEne flow scheme, but the aromatic product is 0.5 weight units higher due to the higher-quality feed to the reformer. The MaxEne case yields a reformate containing 97.6 wt-% aromatics. These improvements in feedstock utilization and process efficiency are the result of the unique advantages of the zeolitic separation.

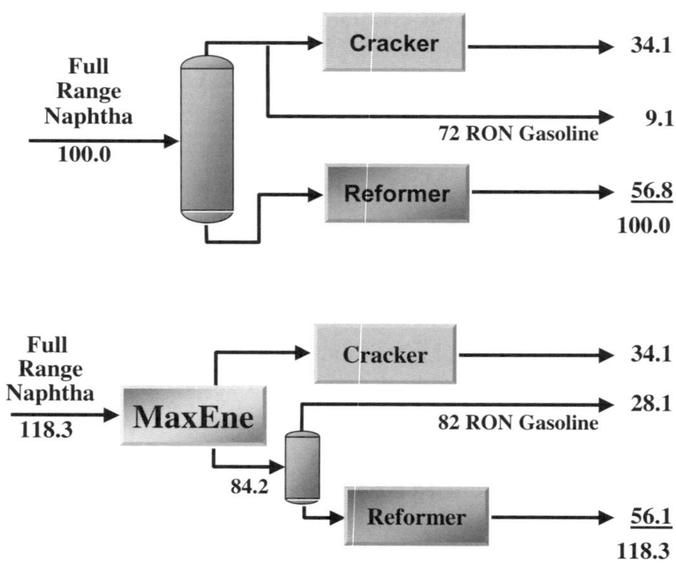


**Fig. 4.20** A comparison of flow schemes and material balances

**Table 4.12** Product distributions from Fig. 4.20 flow schemes

| Product Stream | Without MaxEne | With MaxEne |
|---|---|---|
| ***Steam Cracker Product*** | | |
| Light Ends | 5.8 | 5.8 |
| Ethylene | 9.8 | 13.0 |
| Propylene | 5.1 | 5.1 |
| Butenes | 2.9 | 2.3 |
| LPG | 1.9 | 2.5 |
| Pyrolysis Gasoline | 8.6 | 5.4 |
| TOTAL | 34.1 | 34.1 |
| ***Reformer Product*** | | |
| Light Ends | 6.5 | 5.9 |
| Aromatics | 48.5 | 49.0 |
| Non-Aromatics | 1.8 | 1.2 |
| TOTAL | 56.8 | 56.1 |

## 4.6 Summary

Regulatory agencies are addressing air quality through stringent specifications on the composition of fuels. Concerns about emissions of volatile organic compounds, toxics, and nitrous oxides have led to restrictions on many fuel properties, including benzene, aromatics, sulfur, oxygenates, vapor pressure and distillation. Refiners must comply with these new regulations while maintaining the required octane and cetane of their gasoline and diesel products.

The use of zeolites in fluid catalytic cracking, $C_5$ /$C_6$ paraffin isomerization, and hydrocracking processes, has facilitated significant efficiencies in the production of fuels. FCC naphtha is 30 to 40% of the gasoline pool in a typical fuels refinery. However, new gasoline specifications are causing refiners to seek options for desulfurizing FCC naphtha, while minimizing yield loss. Branched $C_5$ and $C_6$ paraffins from isomerization are high-octane gasoline blending components. Isomerization, coupled with separation and recycling of lower-octane isomers, is a proven process strategy for maximizing the octane of isomerate. Hydrocracking catalysts and processing conditions can be designed for product slates ranging from maximum diesel/jet to maximum naphtha. New diesel specifications will require essentially complete removal of sulfur and reductions in aromatics, while preserving high cetane.

Zeolites are integral components of petrochemical refineries that produce benzene, xylene isomers, ethylbenzene and cumene. These aromatics must be high in purity for downstream conversion to polyesters and styrenic or phenolic based plastics. Catalytic processes for producing aromatics employ zeolites for isomerization, disproportionation, transalkylation, alkylation, and dealkylation.

Zeolites are used in separation processes for extracting p-xylene and m-xylene at high purity and recovery. There are zeolitic processes for converting $C_3$-$C_7$ paraffins into aromatics. In addition, zeolitic processes can co-produce aromatics and chemical-grade light olefins, or co-produce superior feedstocks for catalytic reformers and naphtha crackers.

Zeolites hold great promise for continuing to improve the energy efficiency of refinery processes, thus reducing greenhouse gases. Zeolites make possible the manufacture of higher-quality fuels, resulting in reduced air pollution. Zeolites facilitate the manufacture of plastics. The energy efficiencies gained through increased usage of plastics in applications such as lighter-weight vehicles and thermal insulation will further enhance air quality in the 21$^{st}$ century.

## References


1. Flanigen E.M., *Studies in Surface Science and Catalysis* **137** (2001) 11–35.
2. Pujado P.R., Rabo J.A., Antos, G.A. and Gembicki S.A., *Catalysis Today* **13** (1992) 113–141.
3. Sherman J.D., *Proc. Natl. Acad. Sci. USA* **96** (1999) 3471–3478.
4. Blum C., Cooper J.S., Keller G. and Yoshioka S., *World-Wide Fuel Charter* (European Automobile Manufacturers Association, Brussels, 2000) 1–48.
5. *ASTM Special Technical Publication No. 225* (American Society for Testing Materials, Philadelphia, 1958) 8–21.
6. *ASTM Special Technical Publication No. 109A* (American Society for Testing Materials, Philadelphia, 1963) 2–26.
7. *Diesel Fuel Specifications and Demand for the 21$^{st}$ Century* (UOP LLC, Des Plaines, 1998) 37.
8. Romanow-Garcia S., *Hydrocarbon Processing*, **79**, 9 (2000) 17–19.
9. *Hydrocarbon Processing*, **79**, 11 (2000) 113–123.
10. Upson L.L. and Nelson E.C., *RxCat Technology for more FCC Gasoline* (AKZO Nobel Catalysts Symposium, Noordwijk aan Zee, 1998).
11. Rice L.H., Kuchar P.J. and Gosling C.D., *Tutorial: Upgrading Light Naphtha and Refinery Light Ends* (AIChE Spring National Meeting, Houston, 1999).
12. Bullen P.J. et al, *Light Paraffin Isomerization Using Sulfated Metal Oxide Catalysts* (AIChE Spring National Meeting, Houston, 1999).
13. Meyers R.A., *Handbook of Petroleum Refining Processes* (McGraw-Hill, New York, 1997) 10.45–10.51.
14. Rabo J.A., *Catalysis Today* **22** (1994) 201–213.
15. Kalnes T.N. et al, *Unicracking Innovations Deliver Profit* (NPRA 2001 Annual Meeting, New Orleans, 2001).

16. *Emissions of Greenhouse Gases in the United States 1998* (DOE/EIA-0573(98), Washington DC, 1999).
17. Best J.R., *2000 Annual Automotive Plastics Report* (Market Search Inc., Toledo, 2000).
18. Woodward, K., *Metal Center News* (February, 1999) 6.
19. *Plastics and the Environment-Quick Facts* (The Society of the Plastics Industry, 2001) 2.
20. Pohl S.L., *EBOne Liquid-Phase EB Technology: Unmatched Experience and Commercial Improvement* (Styrene Conference 2000, Lexington, 2000).
21. Jeanneret J.J. et al, *The UOP Q-Max Process: Setting the Pace for Cumene Production* (The 22nd Annual DeWitt Petrochemical Review, Houston, 1997).
22. Meyers R.A., *Handbook of Petroleum Refining Processes* (McGraw-Hill, New York, 1997) 2.30–2.62.
23. Kirkpatrick Awards, *Chemical Engineering*, **106**, 12 (1999) 96.
24. Solis J.J., Moser M.D. and Ibanez F.J., *RZ Platforming Process Improves Profitability* (European Oil Refining Conference and Exhibition, Lisbon, 1997).
25. Gosling C,D., Gray G.L. and Jeanneret J.J., *Produce BTX from LPG with Cyclar* (CMAI World Petrochemical Conference, Houston , 1995).
26. Houdek J.M., Hemler C.L., Upson L.L and Pittman R.M.,*PetroFCC - New Paradigm?* (European Refining Technology Conference, Rome, 2000).
27. Foley T.D., Greer D.W. and Pujado P.R., *MaxEne Process: Increased Ethylene Yield in Naphtha Crackers* (AIChE Spring National Meeting, Houston, 2001).

CHAPTER 5

# ADVANCES IN FLUID CATALYTIC CRACKING

E.T. HABIB, JR., X. ZHAO, G. YALURIS, W.C. CHENG, L.T. BOOCK
*Grace Davison, 7500 Grace Drive, 21044, Columbia, Md., USA*

J.-P. GILSON
*Laboratoire de Catalyse et Spectrochimie, ISMRA-CNRS,*
*6, Bd du Maréchal Juin, 14050 Caen Cedex, France*

The Fluid Catalytic Cracking (FCC) process remains the primary molecular weight reduction method practiced in modern petroleum refineries. While originally designed for cracking the overhead stream from vacuum distillation units, known as vacuum gas oil, most FCC units currently operate with some higher boiling vacuum distillation bottoms (resid) in the feed. Designing catalysts to tolerate the high level of metal contaminants in the resid, while still maintaining high conversion and selectivity, is a key issue of FCC catalyst design.
While FCC feedstocks are becoming heavier and more metal contaminated, new demands are also being placed on the products produced. Demand for propylene is increasing more rapidly than steam cracking, the traditional source of propylene for chemical applications, can supply. Consequently, increasing quantities of propylene are being produced in the FCC process for use in chemicals and plastics. Highly propylene selective catalysts have been developed to meet this challenge. These catalysts, referred to as 'FCC olefin additives', are generally used in admixture with more traditional FCC catalysts.
More stringent clean air requirements are also being imposed upon the FCC process. $NO_x$ emissions, produced by combustion of the coke during FCC catalyst regeneration, are being more strictly limited. FCC $NO_x$ reduction additives have been developed to minimize or reduce the amount of this pollutant produced by the FCC process.
The impact of more stringent clean air requirements continues beyond the refinery gates, as lower gasoline sulfur levels are being mandated to reduce automobile emissions. Gasoline produced by the FCC process is the primary source of sulfur in the refinery gasoline pool. Fortunately, FCC catalyst technology is again providing an answer in the form of low sulfur gasoline FCC catalysts and catalyst additives.

## 5.1 Introduction

The Fluid Catalytic Cracking (FCC) process lies at the heart of the modern refinery (1) as outlined in Chapter 4 of this book. The position of the FCC unit in the refinery is schematically sketched in Figure 5.1.

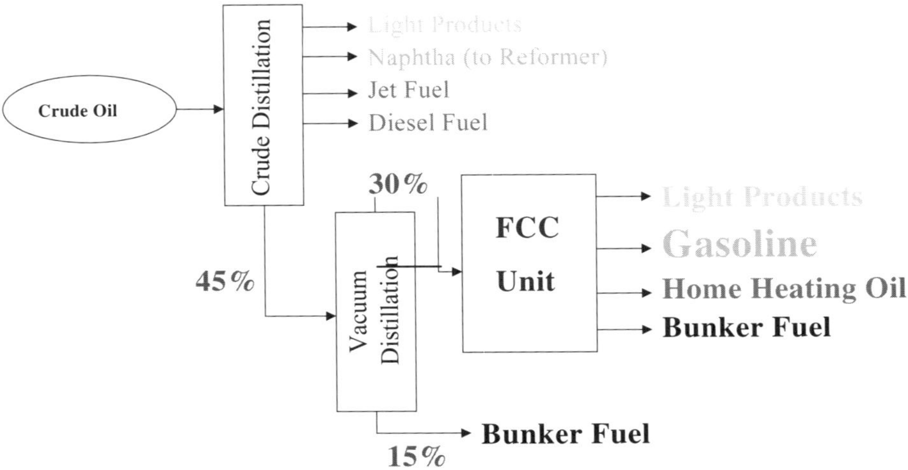


**Fig. 5.1** Simplified schematic of an oil refinery

More than 350 units of various sizes and configurations are dispersed in refineries throughout the world and consume catalysts at an average daily rate of 1400 tons/day. A detailed account of the FCC history, science and technology is found in exhaustive books and publications (2, 3). The objective of this chapter is to stress the role FCC continues to play regarding clean technologies, processes and products.

The Fluid Catalytic Cracking (FCC) process remains the primary molecular weight reduction method practiced in modern petroleum refineries, but its role is strongly evolving. Originally designed for the production of aviation fuel during World War II, the FCC process has been primarily operated to produce large volumes of gasoline (4) for the booming post-World War II economies, but now has to cope with the changing needs and wishes of modern societies. Three issues will be discussed in this chapter in the context of environmental technologies:

1. The challenges posed by the processing of heavy feedstocks (Resids)
2. Meeting the ever increasing demand for propylene feedstock for petrochemicals
3. The control of NOx emissions of the FCC unit and the reduction of sulfur in the FCC gasoline

Other issues in modern FCC are treated in detail elsewhere (5); they are for instance the particulate, CO and SOx emissions from the FCC unit. The modern FCC unit, Figure 2, is made up of a reactor zone where oil and regenerated catalyst (in the form of ca. 70 μm fluidizable spheres) are contacted, sometimes with added steam. The cracking reaction (T ≅ 810 °K, P ≅ 2 bar, Residence Time ≅ 3-10 sec) increases the number of moles and the volume of products, lifting catalyst and oil upwards in the reactor (called the riser) for a few seconds. Because the endothermic cracking reaction takes place in an adiabatic reactor, the catalyst temperature decreases from the bottom to the top of the riser. The coke-deactivated catalyst is then steam stripped of the adsorbed hydrocarbons and separated in cyclones from the fluid phase. The catalyst is transferred to the regenerator where it is reactivated (T ≅ 1000 °K, P ≅ 2 bar, Residence Time ≅ 15 min) by the burning of the coke deposited on the catalyst. The exothermic nature of this reaction releases heat that is transferred by the catalyst to the reaction section. The unit is therefore heat balanced. While the riser operates in a reducing atmosphere, the regenerator atmosphere is oxidizing and the FCC unit offers an opportunity to operate reactions along reducing-oxidizing cycles. The technique of SOx reduction in the FCC is a clever example where this opportunity is fully exploited (5). The catalyst inventory is mainly located in the regenerator and the FCC unit allows a fast changeover of catalyst. Catalyst loss by attrition alone requires a daily fresh catalyst addition of about 0.5% of the inventory, and daily addition rates of 1% are common. In addition to the FCC catalyst proper (see Chapter 3 of this book), other fluidizable particles (Additives) can be added to impart new functionality. Use of such Additives further increases the flexibility of the FCC unit by affording fast entry (and exit) to the unit. These features are summarized in Figure 2.

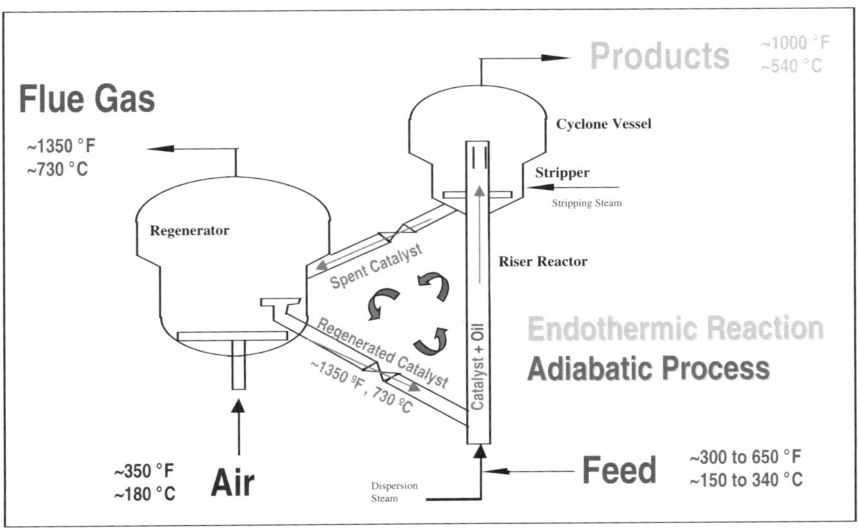


**Fig. 5.2** Typical commercial FCC unit

The topics discussed in this chapter cover various aspects of the FCC unit and affect different parts of the FCC process:

- The changes in **feedstock** properties (heavier and more metal loaded, the so-called 'Resid') due to the whitening of the barrel (lower demand for fuel and bunker oils and increased demand for cleaner products)
- The ever increasing demand for light olefins (propylene) **products**
- The need to reduce the **sulfur** level of the **gasoline** produced in the FCC unit
- The need to reduce the **emissions of the FCC** unit itself (NOx emitted from the regenerator)

## 5.2 Resid cracking challenges

### *5.2.1 The problem*

While originally designed for cracking the overhead stream from vacuum distillation units, known as vacuum gas oil (4), most FCC units currently operate with some higher boiling vacuum distillation bottoms (Resid) in the feed. Table 5.1 illustrates the difficult challenges faced by refiners, process licensors and FCC catalysts producers: the resid feeds are heavier (lower API gravity), contain many more metals like Ni and V as well as more polyaromatic hydrocarbons prone to form coke on the catalysts (Conradson Carbon Residue, or CCR).

**Table 5.1** Properties of FCC feedstocks – Gasoil and resid

| | Gas Oil A | Gas Oil B | Resid | Total FCC Feed |
|---|---|---|---|---|
| **API Gravity** | 22.7 | 22.5 | 19 | 25 |
| **K Factor** | 12.01 | 11.57 | 12 | 11.9 |
| **Sulfur, wt%** | 0.62 | 2.59 | 0.8 | 0.6 |
| **Ni, ppm** | 0.2 | 0.8 | **7.5** | **1.5** |
| **V, ppm** | 0.3 | 0.6 | **4** | **1.3** |
| **CCR, wt%** | 0.89 | 0.25 | **5.7** | **1.1** |

Providing catalysts that tolerate the high level of metal contaminants in the resid, while still maintaining high conversion and selectivity, is a key issue of FCC catalyst design.

*5.2.2 Handling the heavy metals: Ni, V and Fe*

Ni and V are deposited on the FCC catalyst from high boiling metal organic compounds (Asphaltenes) present in the feed. Fe comes from two sources: the feed, (the source of Ni and V), but also from particulate Fe released by the corrosion of equipment. The organics are strongly adsorbed on the catalyst particles, burned off in the regenerator while the metal is oxidized. Oxidized V, for instance, is highly mobile within and between catalyst particles. The FCC unit operating conditions will strongly affect the (adverse) activity of the metals deposited on the catalyst; the Na level on the FCC particles will further exacerbate the effects of V and Fe.

5.2.2.1 Ni tolerance

Ni is a strong dehydrogenation catalyst and it promotes the formation of $H_2$ in the products and coke on the FCC catalysts. Ni is, however, relatively immobile and does not transfer from particle to particle; intraparticle diffusion occurs, however, and is a function of the catalyst addition rate (average residence time in the FCC unit). Figure 5.3 illustrates two extreme cases: a high catalyst turnover unit and a low catalyst turnover unit. It is clear that in the high turnover unit, Ni tends to accumulate on the outer rim of the catalyst, while in the other unit, Ni has had the opportunity to diffuse and is distributed more uniformly through the particle because the average catalyst age is longer.

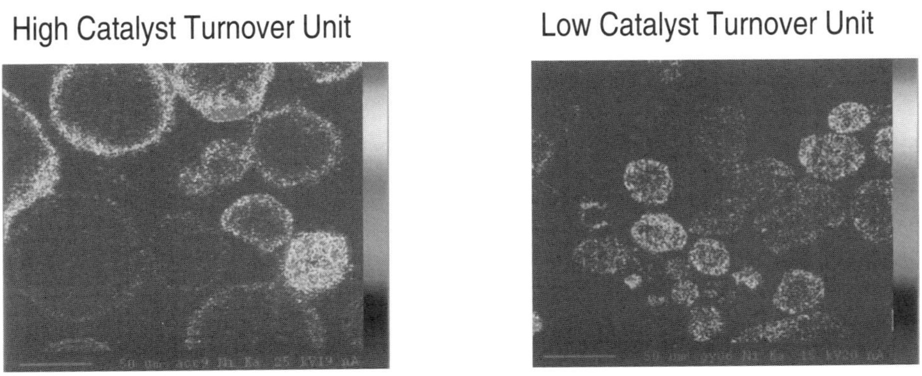


**Fig. 5.3** Ni Distribution on equilibrated FCC catalysts: High and low catalyst turnover units

It is well known (6) that Sb compounds are effective at passivating Ni through alloy formation; these Ni-Sb alloys are relatively inactive for catalyzing dehydrogenation. Ni can also be passivated by reaction with $Al_2O_3$ to form Ni aluminates having much lower dehydrogenation activity. The problem can therefore also be tackled at the level of the binder, where specialized $Al_2O_3$ based matrices have been designed to function according to:

$$\mathbf{NiO + xAl_2O_3 \rightarrow NiAl_2O_4.(x\text{-}1)Al_2O_3}$$

The effect of such proprietary $Al_2O_3$ matrices is highlighted in Figure 5.4 with the example of the LCM/SAM-200 matrices marketed by Grace Davison: $H_2$ production (as well as coke, not shown) is lowered compared to a more classical matrix.

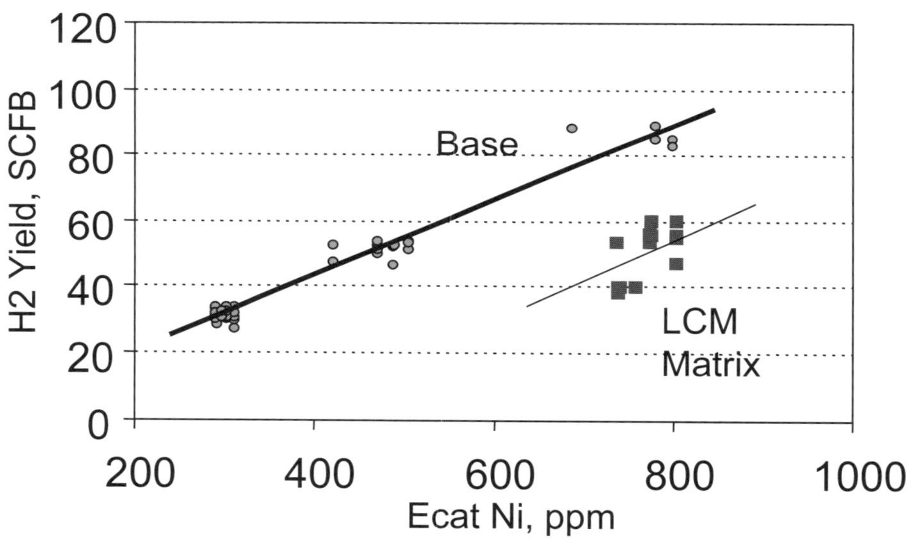


**Fig. 5.4** Reduction of $H_2$ production through the use of specialized $Al_2O_3$ matrices: LCM matrix from Grace Davison

5.2.2.2 V Tolerance

V is oxidized and hydrolyzed in the FCC regenerator as follows:

$$\mathbf{2\ VC_xH_y + (2.5 + 2x + y/2)\ O_2 \rightarrow V_2O_5 + 2x\ CO_2 + y\ H_2O}$$

$$\mathbf{V_2O_5 + 3\ H_2O \leftrightarrow 2\ H_3VO_4}$$

Oxidized V (Oxidation state: +5) has a greater mobility due to its higher vapor pressure, and the vanadic acid produced is very destructive towards the zeolite. This is a fundamental difference between Ni and V deactivation: while Ni promotes undesirable side reactions ($H_2$ production and coke), V is also lethal for the zeolite. In the case of the FAU zeolite, the following irreversible reaction takes place and destroys the zeolite:

$$\mathbf{H_3VO_4 + H_xAl_xSi_{(192-x)}O_{384}}\ (\textit{FAU}) \rightarrow \mathbf{AlVO_4 + (x\text{-}1)/2\ Al_2O_3.(192\text{-}x)\ SiO_2}\ (\textit{Amorphous}) + \mathbf{(3+x)/2\ H_2O}$$

The presence of Na on the catalyst further enhances the destructive ability of V. Figure 5.5 shows that both Na and V are distributed uniformly on equilibrated FCC catalyst particles and therefore damage the entire FCC particle.

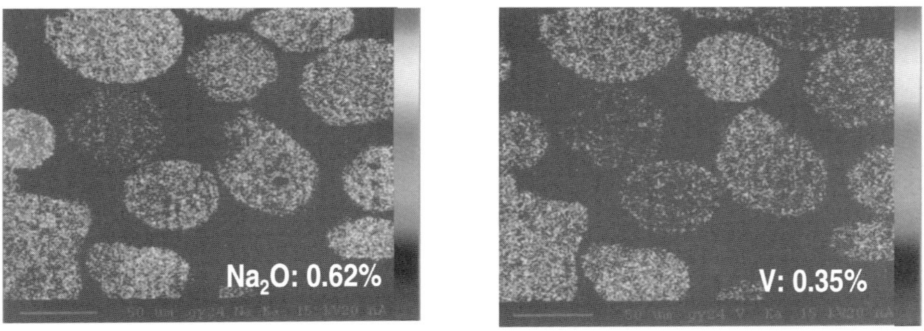


**Fig. 5.5** Na and V distribution on FCC particles

In addition, oxidized V also catalyzes coke and $H_2$ formation through dehydrogenation as described in the case of Ni. V is therefore a significant contributor to catalyst deactivation, Figure 5.6. In addition, the more oxidized the V, the more destructive the effect; if, therefore, the regenerator is operated in the more efficient full burn mode (lower levels of CO emissions), the destruction of the zeolite is increased, Figure 5.7.

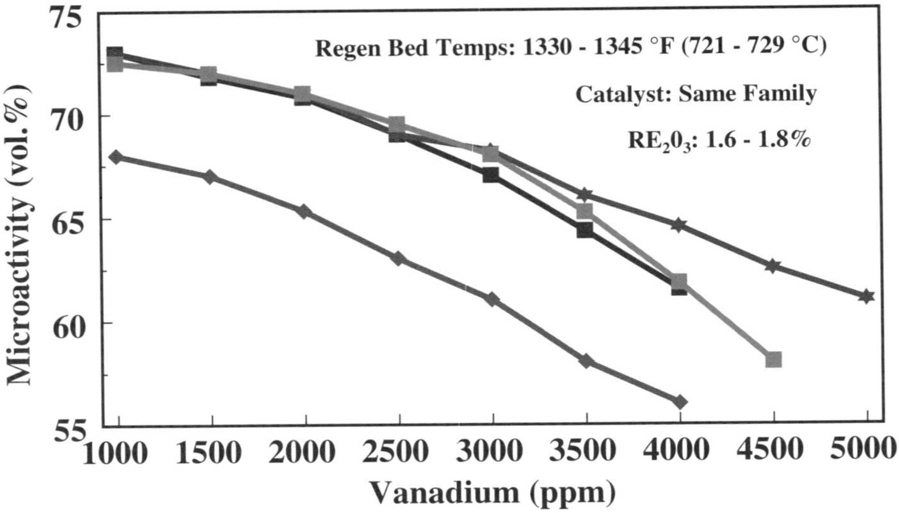


**Fig. 5.6** MAT activity evolution vs. V level

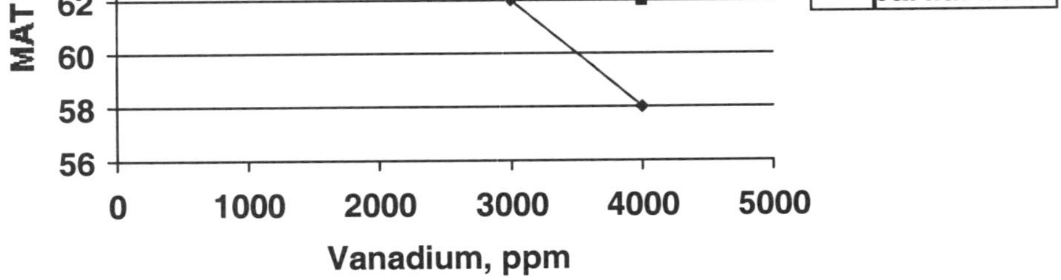


**Fig. 5.7** V effect and burning mode in the regenerator

The solution to the problem is two-fold:

- Use more stable zeolites to better resist the $H_3VO_4$ attack (lower the Al and Na contents, for instance)
- Use Vanadium traps: These are solid bases such as MgO, CaO, SrO, $BaTiO_3$, or $La_2O_3$ that react with $H_3VO_4$ to form stable vanadates

The $La_2O_3$ based traps are the most effective: they are commercially proven, not poisoned by sulfate and have the advantage of being incorporated either as a separate particle (additive technology) or as an integral catalyst component (within the zeolite-containing FCC particle). They interact with vanadium according to the following schemes:

$$\mathbf{La_2O_3 + 2\ H_3VO_4 \rightarrow 2\ LaVO_4 + 3\ H_2O}$$

$$\mathbf{LaOCl + H_3VO_4 \rightarrow LaVO_4 + HCl + H_2O}$$

The efficiency of a Vanadium trap additive is illustrated in Figure 5.8 (Si reflects the presence of the zeolite based catalyst while La and V are present on the same additive particle, the V-trap) and the catalytic effect demonstrated in Figure 5.9. The effectiveness of V-traps is particularly difficult to test in the laboratory, because the level of vanadium mobility in commercial units is difficult to simulate, and the competitive reaction to form sulfate is not taken into account by most laboratory testing.

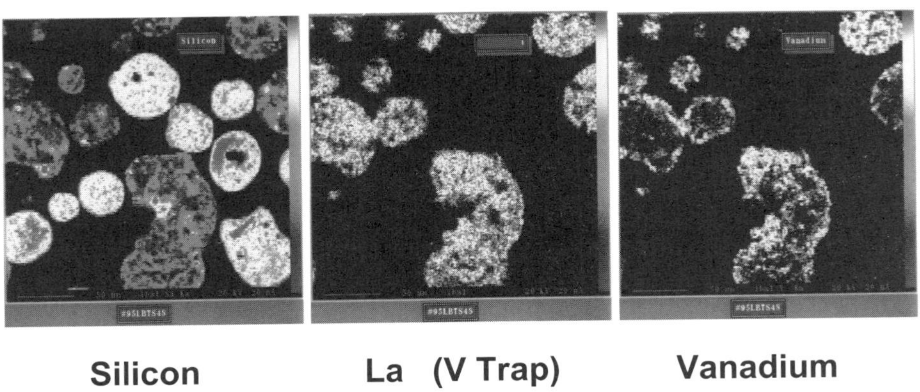


**Fig. 5.8** Electron microprobe imaging of a commercial FCC catalyst with a V-trap additive

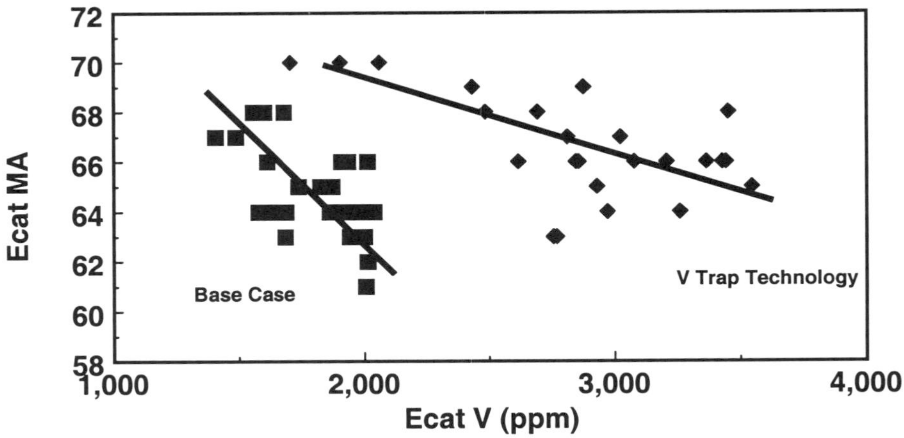


**Fig. 5.9** Effect of a V-trap on a commercial FCC catalyst (Ecat = Equilibrium Catalyst, MA = Microactivity test (7))

### 5.2.2.3 Fe deactivation

Understanding of the effect of Fe contaminants on the FCC catalyst is relatively new and comes from direct observations in the field: dramatic losses in activity sometimes coupled with a loss in bottoms cracking selectivity that appears to correlate with decreased ABD (Average Bulk Density measured in g/cc) due to catalyst particles sticking to each other (8). The catalyst sticking appeared in turn to be correlated with increasing Fe content on the equilibrium catalyst. Although these observations are not universal, two field results are presented in Figure 5.10.

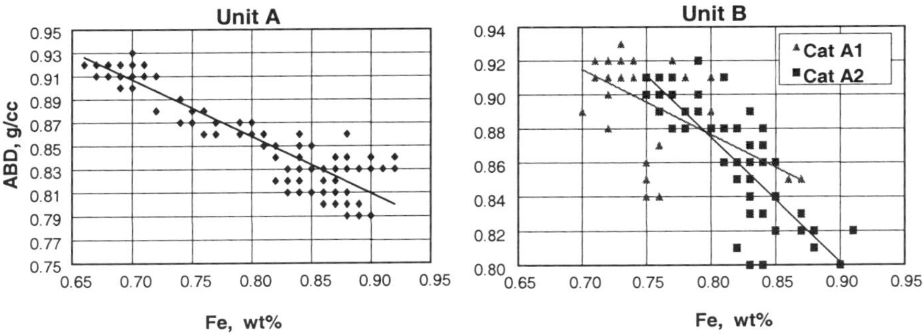


**Fig. 5.10** Effect of Fe content on FCC catalysts ABD – Field observations

These observations are reminiscent of the severe start-up problems experienced by the Hycon Process (Royal Dutch/Shell). In the HDM (HydroDeMetallization) section of this process, operating in a bunker flow mode (9), Fe originating from the feed (Vacuum Resid) or the upstream equipment reacts with the catalyst and leads to an agglomeration of the $SiO_2$ based HDM catalyst. The fused particles prevent a smooth operation in a bunker mode. In the case of FCC catalysts, EDX (Figure 5.10), SEM (Figure 5.11), Optical Microscopy (Figure 5.12) and XPS (Table 5.2) show that Fe is deposited on the catalyst particles' outer surface in the form of nodules, increasing the surface rugosity.

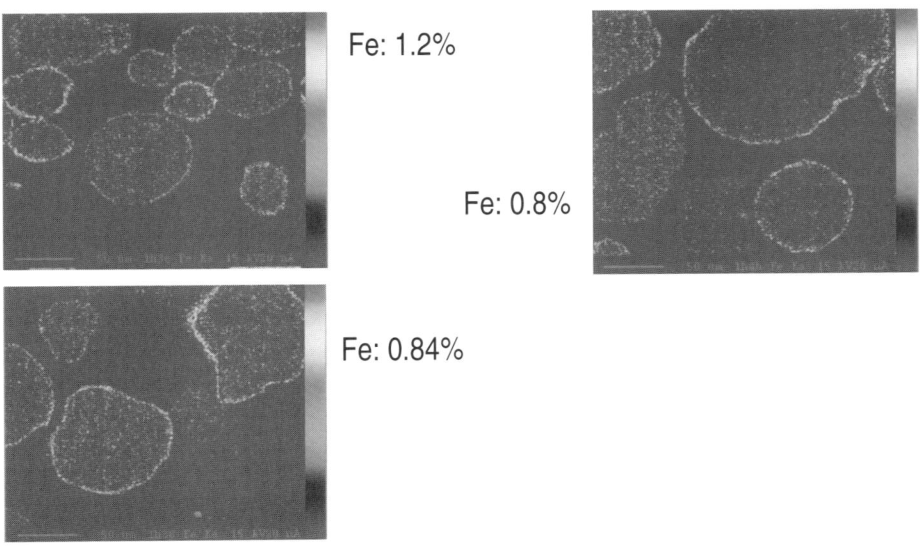


**Fig. 5.11** EDX map of Fe deposited on 3 different FCC catalysts

**Fig. 5.12** SEM picture of Fe contaminated FCC ($SiO_2$ matrix) particles

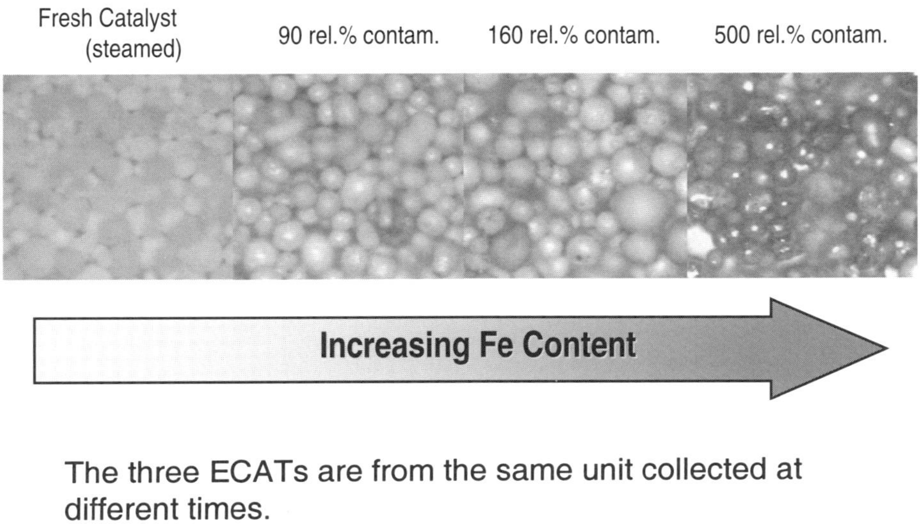


**Fig. 5.13** Optical microscopy view of increasingly contaminated FCC particles

**Table 5.2** XPS measurement shows an excess of Fe on the surface of FCC particles

**High Fe Unit ECAT**

| | **Suface Atomic Concentration** (%) | **Suface Weight Concentration** (%) | **Bulk Weight Concentration** (%) | **Surface Enrichment Ratio** |
|---|---|---|---|---|
| **O** | 68.24 | 49.23 | | |
| **Na** | 0.49 | **0.50** | **0.31** | **1.6** |
| **Al** | 7.53 | **9.16** | 23.71 | **0.4** |
| **Si** | 14.69 | **18.61** | 23.47 | **0.8** |
| **Ca** | 0.36 | **0.66** | **0.11** | **5.8** |
| **Ti** | 0.19 | 0.42 | 0.66 | **0.6** |
| **V** | 0.07 | 0.17 | 0.06 | **2.8** |
| **Ni** | 0.29 | 0.77 | 0.16 | **4.8** |
| **Fe** | 8.14 | **20.49** | **1.57** | **13.1** |

An examination of the phase diagrams of $SiO_2$-FeO (Figure 5.14a) and $Al_2O_3$-FeO (Figure 5.14b) shows that Fe oxides form a eutectic with both $SiO_2$ and $Al_2O_3$ and therefore lower their melting points. From this data, it is already clear that $Al_2O_3$ binders should be preferred over the $SiO_2$ ones because of the resulting higher eutectic temperature.

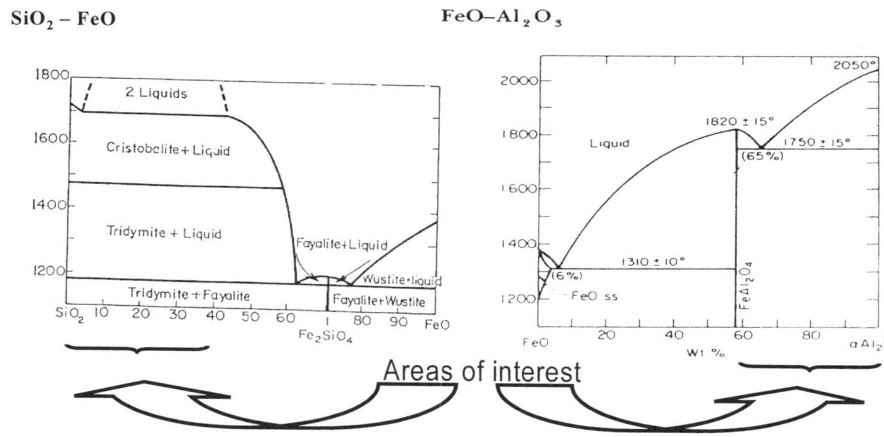


**Fig. 5.14** Presence of low temperature eutectics on the $SiO_2$-FeO (a) and $Al_2O_3$-FeO *(10)* (b) Phase diagrams (11)

A closer examination of the ternary diagrams ($Na_2O$-FeO-$SiO_2$ and $Al_2O_3$) indicates that:

- *Pure $SiO_2$ binders*: The temperature at which the first liquid phase appears (eutectic) can be reduced to ***less than 500°C***, i.e. below the reaction temperature of most risers. The ternary diagram also shows that even small amounts of Fe and Na result in the formation of low temperature liquid phases (Figure 5.15)
- *Addition of $Al_2O_3$ to $SiO_2$ binders*: Figure 5.16 illustrates that adding $Al_2O_3$ to a $SiO_2$-FeO-$Na_2O$ system results in an increase of the temperature corresponding to the appearance of the first liquid phase. The eutectic appears at 771°C in the $SiO_2$ rich phase while the temperature is increased up to 998°C on the $Al_2O_3$ rich phase in the ternary phase diagram

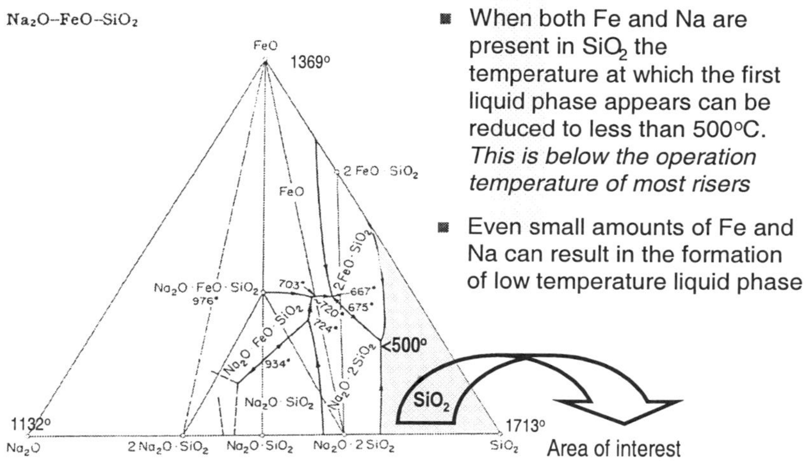


**Fig. 5.15** Ternary phase diagram of $Na_2O$-FeO-$SiO_2$ (12)

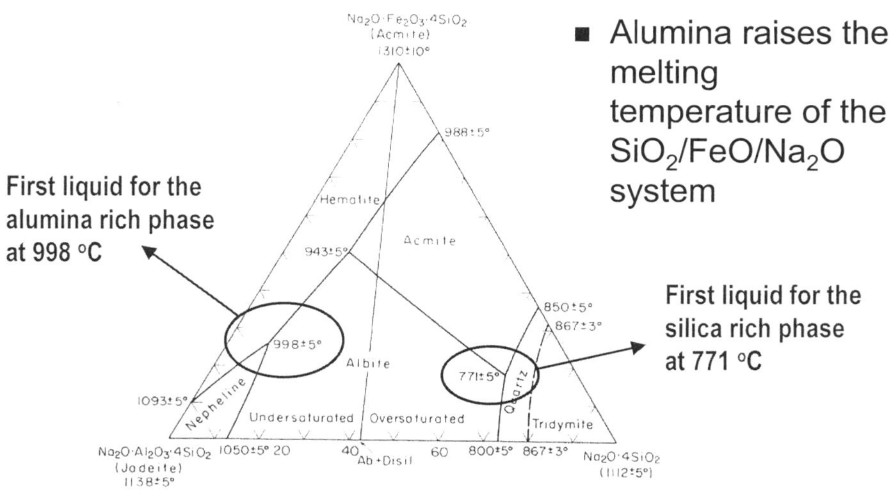


**Fig. 5.16** Effect of adding $Al_2O_3$ to the $Na_2O$-FeO-$SiO_2$ system (13)

A pictorial representation of the mechanisms leading to catalyst deactivation by Fe for $SiO_2$ bound and $Al_2O_3$ bound catalysts is sketched in Figures 5.17 and 5.18 respectively.

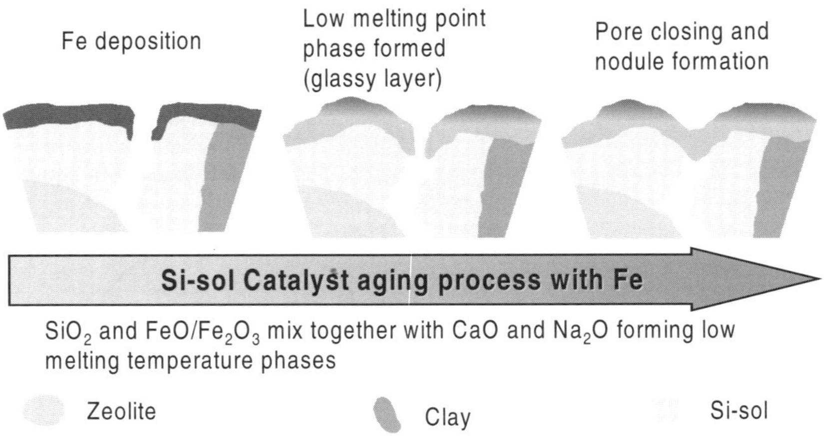


**Fig. 5.17** Mechanism of catalyst deactivation by Fe for $SiO_2$ bound FCC catalysts

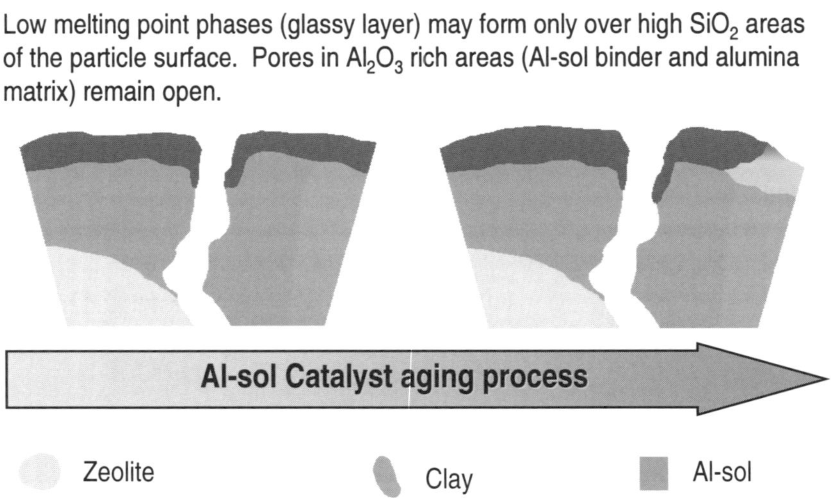


**Fig. 5.18** Mechanism of catalyst deactivation by Fe for $Al_2O_3$ bound FCC catalysts

Such an analysis of the problem leads to recommendations to address this nagging issue:

- *Feedstocks*: use low Fe, Na and Ca feeds and reduce its acid content to lower equipment corrosion
- *FCC unit*: minimize the regeneration temperature
- *FCC catalyst*: use $Al_2O_3$ bound catalysts instead of $SiO_2$ bound ones

### *5.2.3 Bottoms cracking and the effect of pore structure*

The matrix plays a critical role in the selective cracking of the bottoms fraction when resid-containing feedstock is processed (9). The generic roles of the matrix are to pre-crack large molecules and adsorb Ni and V preferentially to protect the zeolite component of the FCC particle. In an ideal situation, macropores should lead to mesopores and these to the (zeolitic) micropores. The features and catalytic effects of the various levels of porosity are as follows:

- *Small Pores (<100 Å)*: they develop a high surface area (~ 600 $m^2/cm^3$ of pore volume) and possess a high bottoms cracking activity, but have selectivity towards $H_2$ and coke
- *Medium Pores (100 Å – 1000 Å)*: they have a reasonably high surface area (~130 $m^2/cm^3$ of pore volume), are effective for capture of unvaporized feedstock liquids and display good bottoms cracking with low coke and $H_2$ generation
- *Large Pores (>1000 Å)*: they have a low surface area (~ 15 $m^2/cm^3$ of pore volume) with a corresponding low activity

### *5.2.4 Putting it together*

The conclusions to be drawn from this analysis of the challenges for processing resid, are that dedicated resid cracking catalysts should contain:

1. A highly stable zeolite
2. An optimized matrix pore structure in order to crack large molecules to useful products
3. High levels of specialized $Al_2O_3$ in the matrix to be Ni and Fe tolerant
4. $La_2O_3$-based V-traps, either as additive or an integral part of the FCC particle

In addition, the complex problem of unit design (feed nozzles, regenerators, strippers, etc.) and operation has to be addressed properly (15) in order to further extend the limits (higher Ni, V, Fe contaminants as well as more refractory feedstocks) of feeds processable by FCC (16).

## 5.3 On purpose propylene

### *5.3.1 The problem*

While FCC feedstocks are becoming heavier and more metal contaminated, new demands are also being placed on the molecules produced. Worldwide demand for propylene continues to increase at a rate of 5-6% per year, mostly to meet the requirements for polypropylene, Figure 5.19. FCC currently provides about 30% of the world propylene demand.

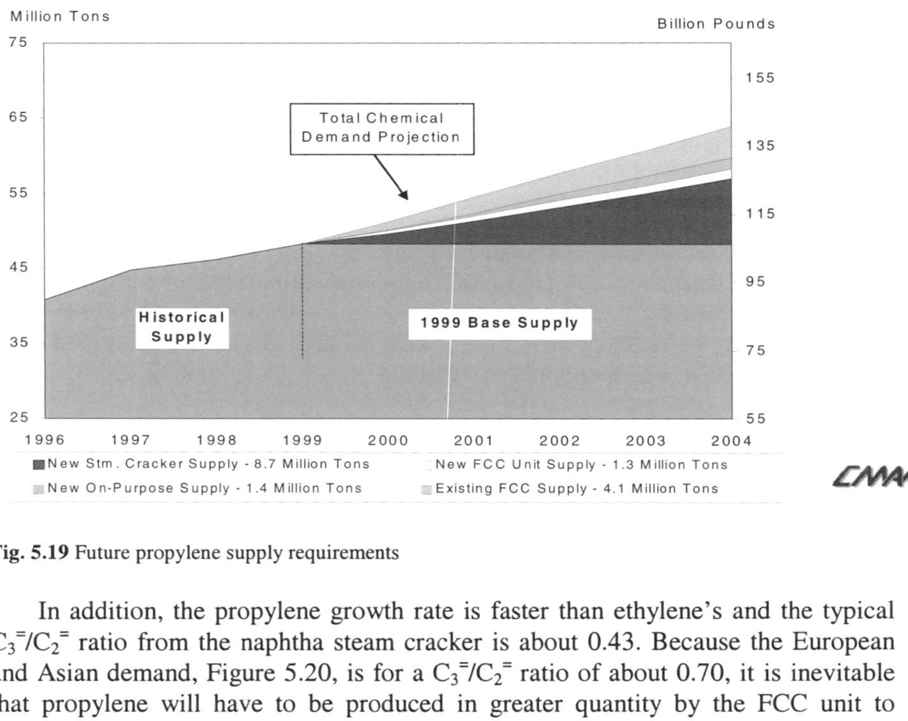


**Fig. 5.19** Future propylene supply requirements

In addition, the propylene growth rate is faster than ethylene's and the typical $C_3^=/C_2^=$ ratio from the naphtha steam cracker is about 0.43. Because the European and Asian demand, Figure 5.20, is for a $C_3^=/C_2^=$ ratio of about 0.70, it is inevitable that propylene will have to be produced in greater quantity by the FCC unit to supply the chemical and plastic industries.

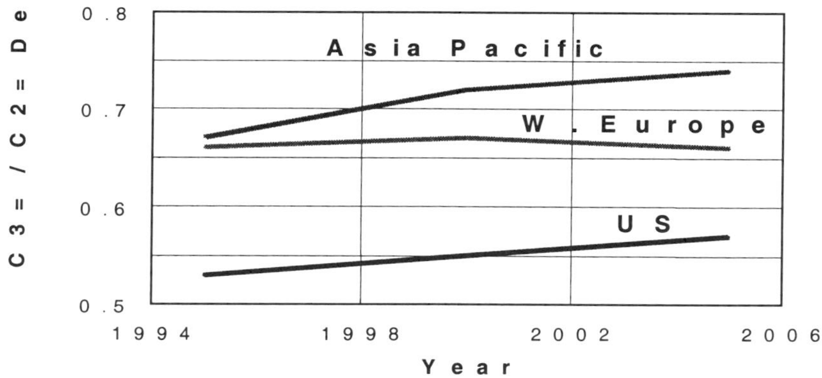


**Fig. 5.20** Worldwide propylene and ethylene growth rates

### *5.3.2 FCC olefin additives*

Highly propylene selective catalysts have been developed to meet this challenge. These catalysts, referred to as 'FCC olefin additives', are generally used in admixture with more traditional FCC catalysts. These additives are based on the MFI (H-ZSM-5) zeolite and the effect of the additive level, which, combined with higher temperatures increases $C_3$ (Figure 5.21) and $C_4$ (Figure 5.22) olefinicity. For instance, at a temperature of 566 °C and 32% Olefin additive, the propylene yield can reach 15% of the feed.

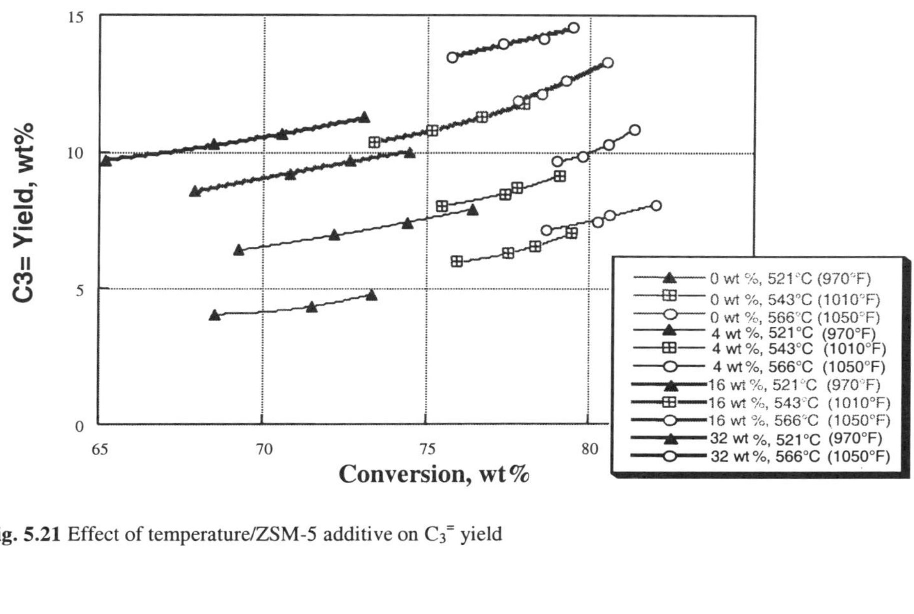


**Fig. 5.21** Effect of temperature/ZSM-5 additive on $C_3^=$ yield

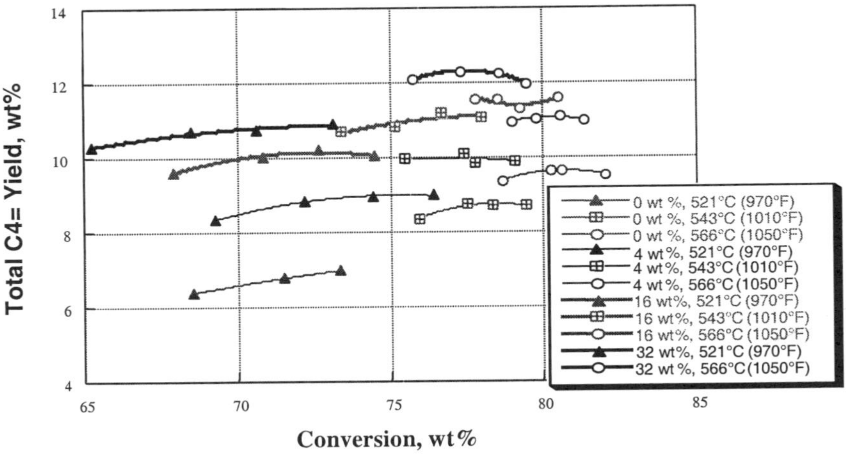


**Fig. 5.22** Effect of temperature/ZSM-5 additive on $C_4^=$ yield

The extra $C_{3\text{-}4}$ olefins production comes at the expense of the $C_6^+$ olefins as illustrated in Figure 5.23. This figure also reveals that the $C_5$ olefin level is barely affected by changing the level of additive in the FCC catalyst inventory. Finally, the hydrogen transfer properties (17) of the base catalyst (US-FAU = low Hydrogen Transfer, RE-FAU = high hydrogen transfer) also affect light olefin yields.

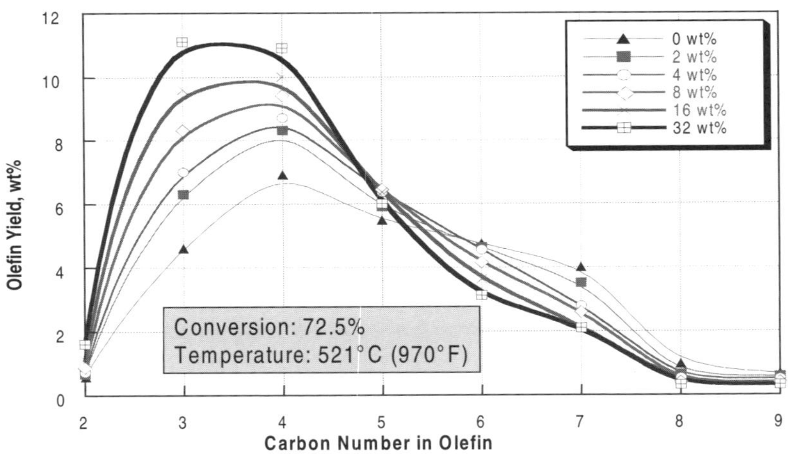


**Fig. 5.23** Effect of ZSM-5 additive on olefin distribution

## 5.4 Future challenges: NOx emission control and gasoline sulfur reduction

### *5.4.1 $NO_x$ emission control*

A nitrogen balance around an FCC unit, Figure 5.24, shows that about 50% of the feed nitrogen ends up in the coke deposited on the catalyst. The burning of the nitrogen containing coke during the regeneration, generates primarily $N_2$, but also NO and smaller amounts of $NO_2$ and $N_2O$. The total $NO_x$ levels are in the range of 50-200 ppm and are increasingly being controlled by various regulations (5). In contrast to the case of the gasoline and diesel engines, operating at higher temperatures, the source of $NO_x$ is not $N_2$ from the air blown into the regenerator (thermal $NO_x$), but originates solely from the feed. Therefore, a solution to the problem could be the (expensive) hydrodenitrogenation, HDN, of the feed.

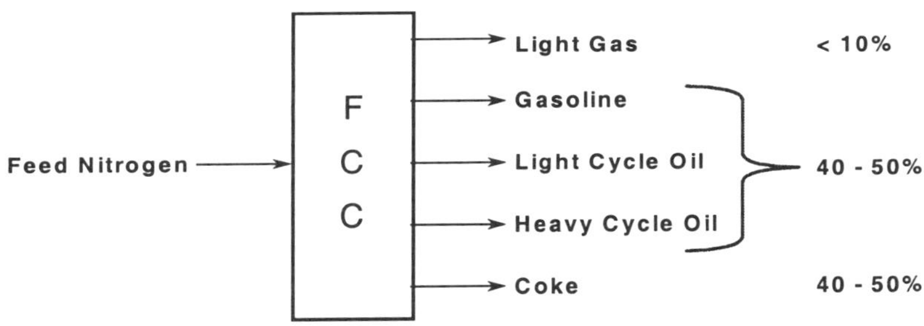


**Fig. 5.24** Nitrogen balance around an FCC unit

The mechanism of $NO_x$ formation has been studied in detail and can be summarized as follows (Figure 5.25):

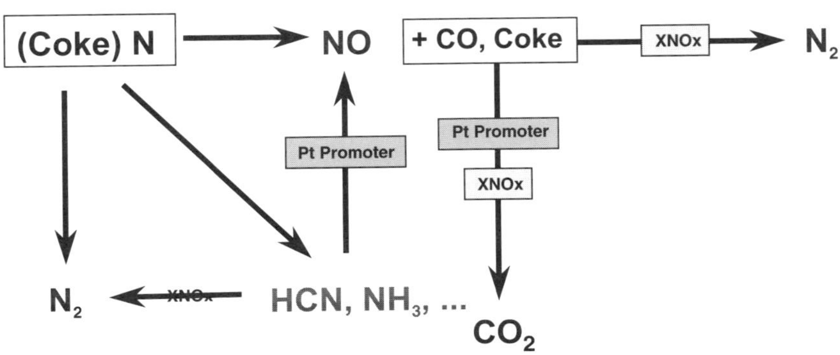


**Fig. 5.25** Mechanism of $NO_x$ formation (18)

According to this scheme, NO can be reduced to $N_2$ with the help of reductants such as CO. The combustion promoters ($Pt/Al_2O_3$) used to minimize CO production catalyze its highly exothermic transformation to $CO_2$, but deplete the regenerator of a valuable reducing agent for the $NO_x$. In 1996 however, a modified combustion promoter, $XNO_x^{TM}$, was introduced and shown to be able to reduce NO emissions by more than 50% by selectively catalyzing the reaction steps leading to molecular nitrogen. This NO reduction was achieved while maintaining an excellent CO combustion promoting activity. The results from a commercial use of the $XNO_x^{TM}$ are shown in Figure 5.26.

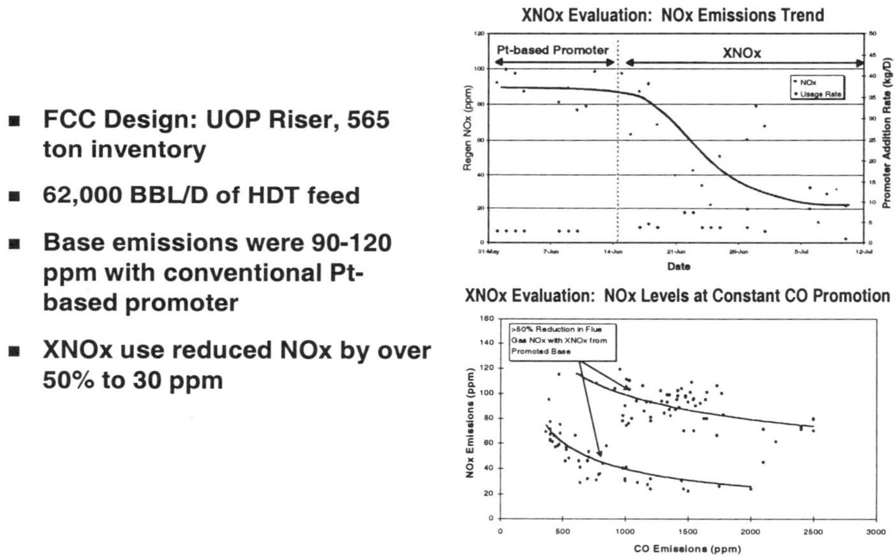


**Fig. 5.26** $XNO_x^{TM}$ commercial example (20)

Another material, DENOX$^{TM}$, was introduced commercially in 1997 and is, contrary to the $XNO_x^{TM}$, a single function $NO_x$ reduction additive (no CO oxidation activity) that decreases $NO_x$ by more than 50%. The field results are summarized in Figure 5.27.

- **FCC Design: UOP Modified Riser w/ 80 ton inventory**
- **14,500 BBL/D of VGO and**
- **Feed contains 650 ppm N**
- **Base emissions were 165 ppm with convention Pt-based promoter**
- **DENOX use reduced NOx by 58% to 70 ppm**

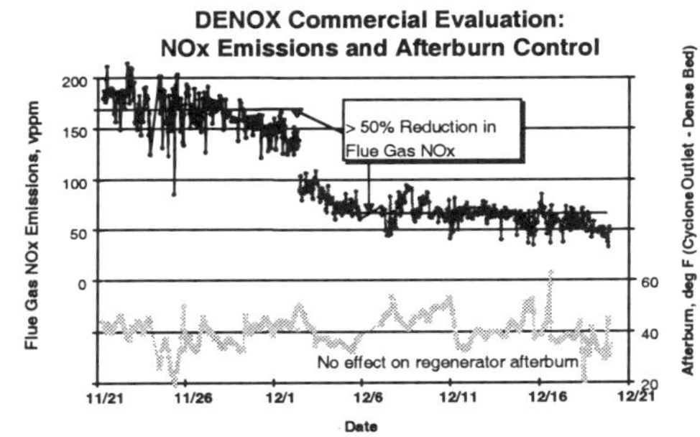


**Fig. 5.27** DENOX™ commercial example (20)

The economic advantages of such technologies are obvious over feed pretreatment and post-treatments (no new capital required, minimum maintenance costs).

## *5.4.2 FCC gasoline sulfur issues*

### 5.4.2.1 The problem

Gasoline sulfur regulations are becoming more stringent and it is expected that a level of 30 ppm sulfur in gasoline will become the limit in the near future in the EU and the US. FCC gasoline accounts for about a third of a typical gasoline pool in the US; however it contributes to over 90% of the total gasoline sulfur, the remaining 10% originating from straight run naphtha. The sulfur in the gasoline is concentrated (up to 50%) in the higher boiling region (193$^+$ °C). Moreover, only about 2-5% of the sulfur in the FCC feedstock ends up in the gasoline as summarized in Figure 5.28.

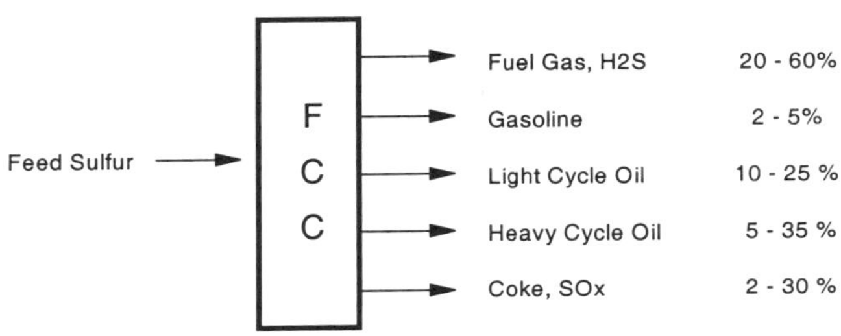


**Fig. 5.28** Sulfur distribution in the FCC products

The sulfur compounds in the FCC gasoline typically comprise mercaptans, thiophene, $C_{1-4}$ substituted thiophenes, thiophenol, $C_{1-2}$ substituted thiophenols, tetrahydrothiophene and benzothiophene, and are best analyzed by a Gas Chromatograph (GC) equipped with an Atomic Emission Detector (AED), Figure 5.29.

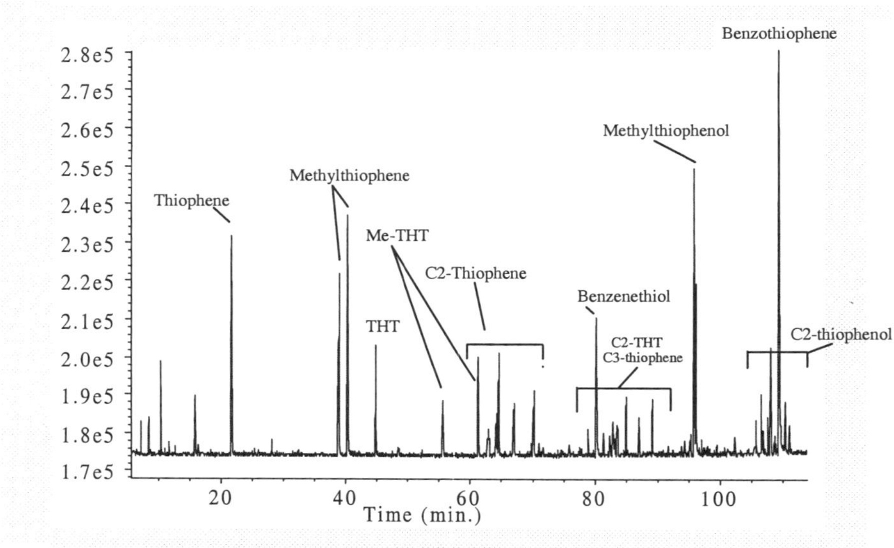


**Fig. 5.29** GC-AED analysis of a typical FCC gasoline

Various options are offered to the refiner to reduce the S content of the FCC gasoline, namely:

- *Hydrotreat the FCC feed.* This solution is technically feasible and allows a very smooth operation of the FCC unit, but is costly (new capital and operating costs are added)
- *Hydrofinish the FCC gasoline.* Processes exist to tackle this problem (See Chapter 4 of this volume) but always result in octane barrel losses and also require the erection of new units
- *Reduce the FCC gasoline cut point.* This solution is very effective at removing a large portion of sulfur in the gasoline, but the overall gasoline yields are obviously decreased
- *Lower the FCC unit riser temperature.* This solution indeed lowers the S level (up to 7%) in the gasoline range, but many units would require large temperature drops (30 °C) leading to octane losses and operating difficulties (19)

- *Use an FCC catalyst with a High Hydrogen Transfer potential.* Replacing a US-FAU (UltraStable FAU, highly dealuminated) by a REUS-FAU (Rare Earth exchanged US-FAU stabilized at a higher Al content, i.e. a higher site density) leads to lower sulfur levels (6% lowering), but often gives unacceptable changes in product quality such as decreased RON (19)
- *Using a Gasoline Sulfur Reduction Catalyst (GSR) in the FCC unit.* Even if only partly successful, such a strategy would reduce S at a relatively low capital cost (no new unit required) and operating costs would be minimized

5.4.2.2 The gasoline sulfur reduction additive

The GSR technology developed by Grace Davison (5) uses a solid Lewis acid (ZnO based) as an additive at the level of about 10% of the total FCC catalyst. It has been shown in laboratory tests to be capable of reducing sulfur in the gasoline by up to 25% without affecting other yields. It operates by cutting mostly the light ($C_2^-$) substituted thiophenes. The sulfur removed by the GSR technology ends up as $H_2S$, an easily processed gas in a refinery environment (Claus unit). GSR technology has also been tested in commercial units.

That this emerging technology deserves more scrutiny, as was detailed in the recently published report by Corma (20) outlining the chemistry of sulfur removal during FCC operations. A better understanding of the process chemistry and the role of specific catalysts should allow the FCC unit itself to contribute to a lowering of the sulfur in the gasoline pool.

## 5.5 Conclusions and perspectives

The FCC process remains the primary commercial means for lowering the molecular weight of petroleum products. The gasoline it produces suffers from many drawbacks (sulfur level, aromaticity, olefinicity, volatility, octane sensitivity, etc.) but the FCC unit and the associated catalysts have proven that they are up to the level of these numerous challenges (14). The FCC field is far from being mature and it will have to meet increasing demands for light products, low sulfur fuels and the processing of ever-dirtier feedstocks. All these challenges must be met while simultaneously reducing the environmental impact.

Catalyst technologies have been and will continue to be a key part of achieving these objectives. These new technologies offer exciting areas of research to both academic and industrial researchers and the impact of their research will contribute to a better environment through cleaner processes and less damaging products.

## References


1. a) Martino G., Courty P. and Marcilly C., in *Handbook of Heterogeneous Catalysis* Ed. Ertl G. et al., (VCH Wiley, Weinheim, 1997) 1801;
   b) Marcilly C. *Stud.Surf. Sci. Catal.* **135** (2001) 37.
2. Magee J.S. and Mitchell M.M., *Stud.Surf. Sci. Catal.* **76** (1993).
3. Von Ballmoos R., Harris D.H., Magee J.S., in *Handbook of Heterogeneous Catalysis,* Ed. Ertl G. et al., (VCH Wiley, Weinheim, 1997) 1955.
4. Venuto P.B. and Habib E.T. Jr, *Fluid Catalytic Cracking with Zeolite Catalysts*, (Marcel Dekker, New York, 1979).
5. Cheng W.C. et al., *Catal. Rev.-Sci.Eng.,* **401** (1998) 39.
6. Nielsen R.H. and Doolin P.H., *Stud.Surf. Sci. Catal.* **76** (1993) 339.
7. Young G.W., *Stud.Surf. Sci. Catal.* **76** (1993) 257.
8. Yaluris G. et al., NPRA Annual Meeting (2001) AM-01-59.
9. Scheffer B., van Koten M.A., Röbschläger K.H. and de Bok, F.C., *Catal. Today,* **43** (1998) 217.
10. N. L. Bowen and J. F. Schairer, Am. J. Sci., 5th Ser., 24, 200 (1932).
11. Novokhatskii I.A., Belov B.F., Gorokh A.V., and Savinskaya A.A., *Russ. J. Phys. Chem.*, 39 [11] 1498 (1965).
12. Carter P.T., and Ibrahim M., *J. Soc. Glass Technol.,* 36 (1952) 156.
13. Bailey D.K., and Schairer J.F., *J. Petrol.,* 7, Pt. 1, (1966) 125.
14. Sie S.T., *Stud. Surf. Sci. Catal.,* **85** (1994) 587.
15. Upson L.L., Hemler C.L. and Lomas D.A., *Stud.Surf. Sci. Catal.* **76** (1993), 385.
16. Naber J.E., Akbar M., *Hydrocarbon Technology International* (1989/1990) 37.
17. Scherzer J., *Stud.Surf. Sci. Catal.* **76** (1993) 145.
18. Peters A.W. et al., NPRA Annual Meeting (1998), AM-98-43.
19. Wormsbecher R.F., Weatherbee G.D., Kim G. and Dougan T.J., NPRA Annual Meeting (1993), AM-93-55.
20. Corma A., Martinez C., Ketley G. and Blair G., *Appl. Catal. A*, **208** (2001) 135.

CHAPTER 6

# HYDROCRACKING

J.A.R. VAN VEEN
*Shell Research & Technology Centre, Amsterdam*
*P.O. Box 38000, 1030 BN, Amsterdam, The Netherlands*

## 6.1 Introduction

Hydrocracking is one of the major conversion processes in the oil refinery. It usually converts a rather heavy, low quality feedstock into lighter, highly valuable transportation fuels, under an appreciable hydrogen pressure. Single or multiple catalysts systems are normally employed in fixed-bed reactor systems. These catalysts are dual-functional in nature: they contain both acidic cracking, and hydrogenation functions. Hydrocracking is predominantly suited to producing middle distillates with excellent product properties. Jet and Diesel fractions can be obtained with very low sulphur contents (*i.e.* below 20 ppm) and very good combustion properties (kerosine smoke points above 25 mm, and cetane numbers above 55). The obvious reason for this is the relatively high hydrogen pressure used, typically above 100 bar, which results in high removal rates of hetero-atoms (sulphur, nitrogen) contained in the feedstock, and deep saturation of aromatic compounds. Other typical characteristics of hydrocracking are its flexibility in varying product slate, depending to a large extent on the type of catalysts used, and its potential to produce very good quality feedstock for lube base oil manufacturing, ethylene crackers and fluid catalytic crackers.

Nowadays, the world hydrocracking (HC) capacity is about 200 million tons per annum (~120 HC units) – the main driving forces for the further expansion of this process are (1): (i) a steady, but continuing growth in middle distillates consumption in all parts of the world, (ii) introduction of ever more stringent automotive fuel specifications (sulphur, aromatics, 95%vol recovery temperature) as prescribed in the USA and Europe, (iii) processing synergy in combining catalytic cracking and HC, and (iv) increasing development of mild hydrocracking technology (2, 3).

The origin of HC might be traced back to the late 1920s when IG Farben developed a process for converting lignite coal into gasoline. However, it was not until 1958 that the first modern oil-feedstock-based hydrocrcacker was commercialised by Chevron (4). Initially, hydrocrackers were primarily built in the USA and were designed to convert relatively low value, mostly aromatic feedstocks (e.g. catalytic cracked cycle oils) into gasoline components and still lighter products. Starting in the 1970s and continuing, the demand for transportation fuels shifted to a broader range of products, particularly middle distillates such as kerosine, jet fuel and diesel fuel, with stronger emphasis on the latter products in HC Units outside the USA. All this of course entailed continuous developments in catalyst and process technology. In this chapter we will concentrate on the catalytic aspects, but the main process configurations will be described first. We lean heavily on the chapter on HC and dewaxing in the Handbook of Heterogeneous Catalysis (5); many further details can be found in an excellent book on HC by Scherzer & Gruia (6).

## 6.2 Process configurations

Several basic line-ups for the HC process have been developed, see Fig. 6.1. In all of them the liquid oil feedstock and gaseous hydrogen are passed in down flow over the fixed catalyst beds (so-called trickle flow (7)). Typical operation pressures are in the 70-200 bar range (at the lower pressures one has partial conversion, or mild HC), and applied temperatures are in the 300-450 °C range. Some typical hydroprocessing reactions are collected in Table 6.1. Usually, the removal of S and N goes virtually to completion – the hydrogenation of aromatics takes place to a considerable extent as well, but this will obviously depend strongly on the catalyst and conditions applied.

**Table 6.1** Typical Hydroprocessing Reactions (6)

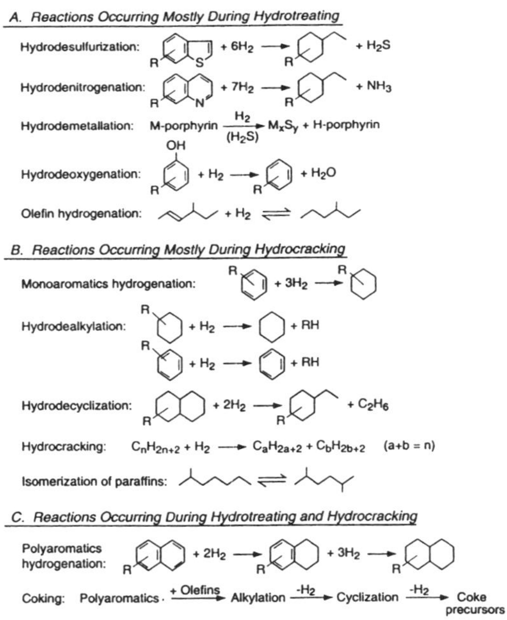

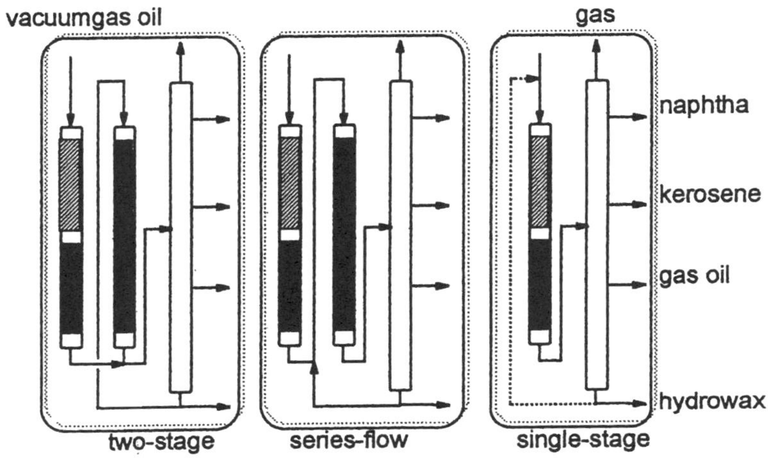


**Fig. 6.1** Hydrocracking modes of operation (8)

The second-stage and series-flow configurations are characterised by the presence of two separate stages (reactors). The first stage is then primarily for feedstock pretreatment (i.e. removal of hetero-atoms, mainly), while the (hydro-)cracking proper is carried out in the second stage – although there is a tendency to shift some of the cracking into the first stage. In two-stage operation the effluent of the first and second stages are combined and sent to a fractionator for product recovery (this also removes the $NH_3$ and $H_2S$ formed in the first stage). Since, o.a. for selectivity reasons, the conversion in the second stage is less than 100%, a heavy liquid fraction remains, and this is recycled to the second stage. The application of a liquid recycle feed is unique to hydrocracking (8). In series-flow operation the full first stage effluent (i.e. including the formed $NH_3$ and $H_2S$) is routed to the second stage. The heavy liquid fraction can again be recycled to the second stage. Because of the presence of $NH_3$, which poisons the acidic cracking function of a HC catalyst, in the second stage, this configuration only became possible with the advent of more active, more ammonia-resistant, in a word zeolite-based, catalysts.

The most simple HC process scheme consists of a single reactor containing one or more (stacked) catalysts. All the required reactions, both the hetero-atom removal and the cracking conversion, have to proceed in this one stage. Originally no recycle of unconverted oil was applied in the configuration because of insufficient catalyst activity. Currently, with the availability of more active catalyst systems, a recycle operation is possible, but the once-through process is still quite widely applied – the reason being that the heavy liquid ('hydrowax') can be favourably used as feedstock for lube oil production, etc. Here again, the ammonia produced in the first part of the reactor is passed over the dedicated cracking catalyst, thus poisoning it to a certain extent. As will be pointed out below, the eventual presence of ammonia strongly influences the hydrogenation/cracking balance in the bifunctional HC catalyst, with important consequences for the product selectivity.

As to feedstocks, the most common one used outside North America is straight-run (*i.e.* directly from the crude distiller) vacuum gas oil with a 350-600 °C boiling point range. In the USA, lighter feedstocks are generally processed, including cracked materials (i.e. obtained from a preceding conversion step, e.g. catalytic and thermal crackers). As in that part of the world, the demand for naphtha (e.g. C5-180 °C) is also more important relative to that for kerosene (~180-250 °C) and Diesel (~250-370 °C) than elsewhere, the catalysts applied in the USA tend to be somewhat different from those used in Europe, the Middle and the Far East. In general, the lighter the feed, and the lower the nitrogen and aromatic contents, the easier it is to process. Feeds with endpoints surpassing 620 °C cannot as yet be processed in the basic line-ups discussed above (see further section 4 'Outlook', below).

## 6.3 Catalyst systems

Catalysts used in the first stage for feedstock pretreatment are generally nickel-molybdenum and nickel-tungsten based systems, supported on [semi-]amorphous inorganic oxides such as [gamma-] alumina and amorphous silica-alumina. Up till now, alumina has been the preferred carrier for these systems; addition of phosphorus promotes the HDN activity (9). As a result, the pretreat catalysts usually consist of NiMoP/alumina, and for reaction temperatures below some 380 °C, in the presence of sulfur, no systems with higher HDN activity have been found (10). As mentioned above, the proper HC catalysts are dual-functional and therefore choices need to be made as to which (de)hydrogenation and (acidic) cracking function to apply. The common dual-functional alternatives are shown in Table 6.2.

**Table 6.2** Bifunctional hydrocracking catalysts

| | Hydrogenation function | Acidic function (support) | |
|---|---|---|---|
| ↓ Increasing Hydrogenation Power | Ni/Mo | $Al_2O_3$ | Increasing Acidity |
| | Ni/W | $Al_2O_3$/halogen | |
| | Pt/ Pd | $SiO_2/Al_2O_3$ | |
| | | Zeolites | |
| | (low-S conditions ) | | |

As is evident from the discussion in section 6.2, HC catalysts are exposed to very different environments, depending on process configuration, and therefore vary markedly in their composition. For example, single-stage bottom-bed, and first-stage bottom-bed catalysts operate at relatively high temperatures in relatively high partial pressures of $NH_3$ and $H_2S$ and quite high concentrations of organic nitrogen. Under these severe conditions such catalysts are required to exhibit a high HDN performance and, particularly in modern hydrocrackers, also provide significant cracking activity. Series-flow catalysts are exposed to less organic nitrogen but are required to achieve quite high conversion performance levels in the presence of $NH_3$ and $H_2S$. By contrast, two-stage catalysts have the least severe requirements, normally operating in the absence of $NH_3$ and with moderate to little $H_2S$.

Despite these significant differences in catalyst requirements, the basic design rules for the various HC catalysts would appear to be quite similar. Of the hydrogenation functions mentioned in Table 6.2 one can say that, in general, Ni/Mo and Ni/W mixed sulfides are applied – the former usually in situations were some extra HDN activity is required, and the latter where hydrogenation performance is of paramount importance. In some cases where the concentrations of sulphur are low (i.e. two-stage conditions), a (noble) metal like Pd may be used, a very powerful hydrogenation catalyst. Such Pd-promoted catalysts are generally more active than the mixed-sulfide containing ones, and the products are also very well hydrogenated. On the debit side are the high sensitivity to sulphur, and the cost of the metal (and of keeping the hydrogen-sulphide pressure low).

As to the acidic functions mentioned in Table 6.2, alumina, even after fluoridation (11), is obviously the least active in cracking. But since alumina is such an excellent support for the hydrotreating function, these catalysts do have merits in hydrocracking of e.g. nitrogen-rich feeds, as in mild hydrocracking.

More acidity is provided by amorphous silica-aluminas (ASA), a well-known base material since the early days of fluid catalytic cracking. They can be prepared in a variety of ways, and over a range of alumina contents – they usually have a rather high surface area (say, 400-500 $m^2/g$). At first the acidity increases with alumina content, but beyond 25%w the cracking activity decreases again, as shown by Ward (12). Since with varying alumina content, its dispersing power for the hydrogen function (in particular the base-metal ones) also changes, it is not easy to define the optimum ASA from first principles. Improved metal-emplacement routes can still lead to improved catalysts (1). Often ASAs are considered to have a rather small number of acid sites of high strength (we come back to this below, Fig. 6.9) – this makes them less acidic than zeolites, and also more prone to poisoning, by e.g. nitrogen compounds.

Zeolites provide in general by far the highest cracking activity. Compared with ASAs they also show a more stable performance. On the other hand, they may suffer from mass-transport limitations, leading to lower selectivities to desired products and/or increased gas make. These aspects will be further discussed below.

### *6.3.1 HC catalyst design rules*

A great deal of study has been devoted to unravelling the chemistry of the HC process (13). A simplified mechanistic scheme, adequate for basic design purposes, is presented in Fig. 6.2. This scheme is primarily derived from model studies related to the hydroconversion of normal paraffins (4-19). The first step in this mechanism is the dehydrogenation of (normal) paraffins to produce olefin intermediates that subsequently react on the (Brönsted) acid sites to form adsorbed carbenium ions (or, strictly speaking, alkoxides). These carbenium ions can then undergo the usual acid-catalysed reactions, viz. first isomerisation and subsequently beta-scission cracking. The isomerised and/or cracked species are then desorbed from the acid site and hydrogenated to the corresponding paraffin. In addition, olefins can react in bi-molecular H-transfer reactions with naphthenes in a process that may ultimately lead to coke (20, 21). A lot of the details of these paraffin cracking reactions, including the fundamental kinetics, have been unravelled, not only for single molecules, but also for mixtures, and for naphthenic compounds (cyclic paraffins) (18, 22, 23).

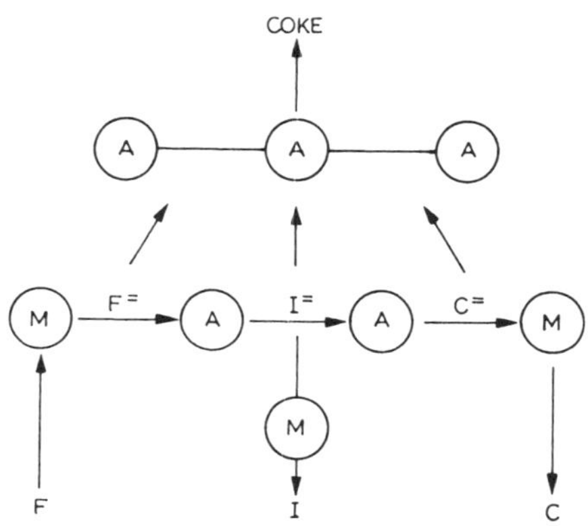


**Fig. 6.2** Simplified reaction scheme for hydrocracking.
F= feed molecule, I= isomerised molecule,
C= cracked molecule, M= metal site, A= acid site (19)

From the above scheme, Fig. 6.2, an obvious design parameter is the relative importance of the (de)hydrogenation and the cracking functions. When the former dominates over the latter, so-called 'ideal HC' conditions exist whereby consecutive reactions (after the isomerisation and the primary cracking steps) play a minor role. A strong (de)hydrogenation activity leads to a relatively high steady-state concentration of n-alkenes, which in turn enhances the rate of desorption of products from the acid sites through the mechanism of competitive adsorption, thereby maintaining low alkyl-carbenium ion catalyst residence times (24). For the conditions of ideal HC to prevail, the hydrogenation and cracking functions should also be in relatively close proximity, so that the equilibrium composition of olefins is maintained at the acid sites without distortion from diffusion limitations (25). Cornet et al. (26) suggest that in the case of NiMo/Y the two functions need to be in closer proximity than in the case of Pt/zeolite catalysts.

Non-ideal conditions occur, for example, when the rate of (de)hydrogenation is less than that of cracking, so that the rate of olefin desorption will be low relative to β-scission, leading to multiple cracking events, and lighter products. In this case a higher iso/normal ratio will be observed in the products than in the ideal case, since secondary isomerisation (i.e. of primary cracked products) is now possible. The results of a classical study of ideal vs. non-ideal HC are shown in Fig. 6.3 (cf. also ref. 27).

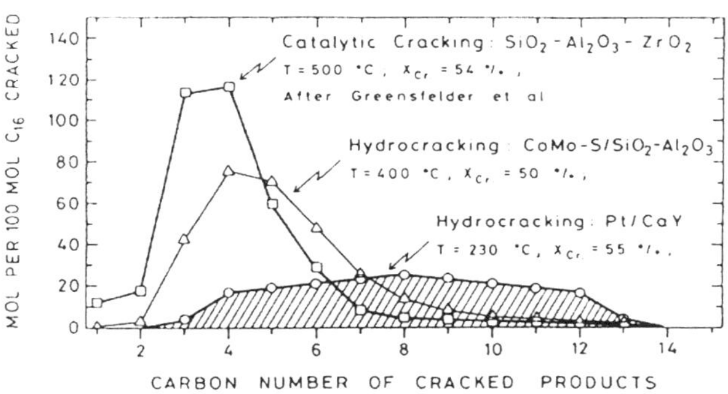


**Fig. 6.3** Molar carbon number distributions in catalytic cracking and hydrocracking of n-hexadecane at 50% conversion (24)

Qualitatively, the same principles can be applied in commercial practice (28). For example, a high naphtha selectivity requires a strongly acidic catalyst with a relatively weak (de)hydrogenation function – the resultant relatively high iso/normal ratio being beneficial for the naphtha quality too (higher octane numbers). By contrast, for middle-distillate selective catalysts the reverse is true and a relatively high ratio of (de)hydrogenation/acidity function is desired for both yield and product-quality requirements.

The effects of changes in operating conditions can be rationalised in a similar manner. For example, when Pd is used as the (de)hydrogenation function, the presence of $H_2S$ will significantly decrease the metal performance, resulting in a shift towards lighter products. Conversely, the presence of ammonia will suppress the acidity function leading to a shift towards heavier products, cf. Table 6.3. The base-metal hydrogenation function, which, by the way, is sulphided before use to form the active mixed sulphide phase, is relatively insensitive to $H_2S$, so that for base-metal catalysts only the ammonia partial pressure plays a determining role (the deactivation of the hydrogenation function by $NH_3$ is relatively minor).

**Table 6.3** Product from Pd on $SiO_2/Al_2O_3$ with California gas oil – effects of sulfur and nitrogen (from Sullivan and Scott (28)

| | Isohexanes / n-hexane | $C_{5+}$ yield, wt. -% | 138-288°C yield liq. vol. - % | $C_5$ - 83°C octane, F-1 Clear |
|---|---|---|---|---|
| Before addition of sulfur or nitrogen | 5 | 95 | 65 | 80 |
| 8 ppm nitrogen added to feed (catalyst equilibrated) | 3 | 96 | 71 | 78 |
| 100 ppm sulfur added to feed (after 170 volumes of feed containing sulfur) | 13 | 93 | 58 | 84 |

To a first approximation, it would not appear to matter whether one varies the cracking ability of a catalyst by varying the amount of cracking function in it, or by varying the ammonia partial pressure – also, it is the cracking function that in the first instance determines the activity of the catalyst – therefore, the simple mechanistic scheme predicts that there should be a general relationship between activity and selectivity, and this has in fact been observed in practice, Fig. 6.4 (29) – note that the hydrogenation function not only increases relatively upon decreasing the amount of cracking function or increasing the ammonia partial pressure, but also absolutely, as a result of the higher temperature requirement for a given conversion. It should be borne in mind, of course, that this selectivity-activity relation is only a trend, not a unique relation, and a well-designed catalyst can have a higher selectivity at the same activity, but nevertheless still.

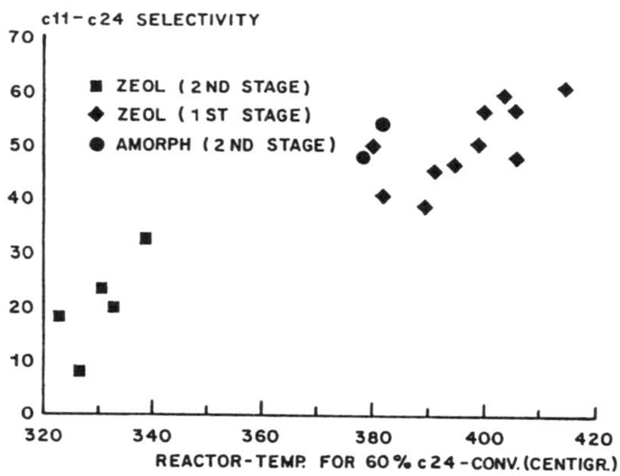


**Fig. 6.4** Middle-distillate selectivity versus activity (29)

On the basis of the mechanistic scheme, Fig. 6.2, one would also expect the HC activity to increase with the rate of formation of olefins. Further, since increasing the hydrogen pressure suppresses the dehydrogenation reaction, a negative order in hydrogen for the rate of hydrocracking, at least of n-alkanes, would be predicted. This has indeed been observed (2). In industrial practice, however, the hydrogen pressure needs, generally speaking, to be high, and the predicted negative order in $H_2$ pp is only very rarely sighted (we know of one possible case in a 160 → 180 bar experiment). This contrast is discussed in the next section.

### *6.3.2. Coke deactivation*

The reason why under representative HC process conditions catalyst performance increases with hydrogen partial pressure is due to the relatively high levels of coke precursors (polyaromatic compounds) that are prevalent in industrial feedstocks. Under these more practical operating conditions the primary role of the (de)hydrogenation function is to hydrogenate coke precursors and thereby minimise the steady-state coke levels on the catalyst which would otherwise severely suppress its activity. When focusing on virtually metals-free feedstocks, coke formation is the main cause of catalyst deactivation in hydrocracking, although it has been reported that some forms of coke could also be beneficial (30-32).

The strongly inhibiting effects of even simple aromatics on the conversion of n-alkanes has also been observed in some model HC experiments. For example, both 1-methylnaphthalene (33) and phenanthrene (34) have been shown to strongly inhibit the hydroisomerisation of n-heptane.

A key question related to the design of HC catalysts is the importance of the proximity, or as it is often termed, the intimacy of the hydrogenation function to the acidic sites. Studies based on model HC catalyst systems (25) led to the development of intimacy criteria that need to be satisfied for sufficient rates of diffusion of olefinic intermediates between acidic and hydrogenation functions to achieve equilibria. Similarly, intimacy rules will apply to aromatic intermediates in order to minimise coke formation. Literature data on this aspect are rather scarce, however.

There is, nevertheless, some evidence (35, 36), based in NiNaY and NiMo/alumina/Y model catalyst systems, that the amount of coke formed is reduced with increasing intimacy of mixing of the two functions at the submicron level. This concept is further supported by the reported relatively high performance of NiW/ASA (amorphous silica-alumina) cogel HC catalysts which, it is claimed, exhibit an excellent distribution of the NiW hydrogenation function throughout the catalyst particles (37).

It is, of course, one thing to want to emplace the hydrogenation function in a controllable manner, and another to be able to do it in practice. Industrial zeolite-Y based catalysts contain a binder, usually alumina or an ASA/alumina mixture. In this case one can direct Pd to the alumina phase by employing $H_2PdCl_4$ as the metal precursor in the pore-volume impregnation step (38), while the tetrammine complex, at near neutral pH, will preferentially exchange into the zeolite. On the other hand, NiW(Mo) will almost always prefer the binder, and the problem of having a substantial part of the NiW mixed sulphide inside the zeolite pores has not yet been solved, although some of it will in practice be found in the mesopores of zeolite Y, when present (see below). It is to be noted that (Ni)Mo prefers alumina even stronger than (Ni)W (39). It has been reported that NiW and Mo sulphides supported on ASA have higher hydrogenation activities than their alumina counterparts (40). To what extent the zeolite can have a similar (indirect?) effect remains a matter for consideration.

Coke is formed through aromatic condensation reactions (21) which on the molecular scale are spatially quite demanding. Zeolites, with molecular-sized pore systems, tend to be significantly less prone to coke formation than ASA-based catalysts, which do not have such spatially restricted pore structures. Correspondingly, in commercial operation zeolitic catalysts tend to be significantly more stable than ASAs (see Fig. 6.5).

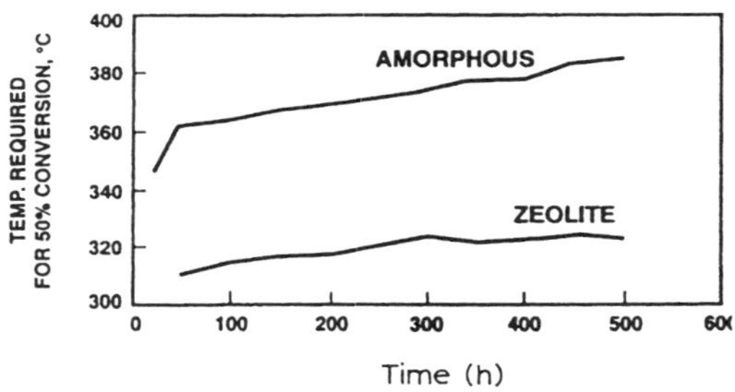


**Fig. 6.5** Deactivation rates of amorphous and zeolite catalysts (48)

### *6.3.3 Zeolite-based catalysts*

As indicated in Table 6.2, zeolites generally exhibit relatively high acidity and are therefore a preferred catalyst ingredient, particularly where high activity levels are required. To date, almost all commercial hydrocracking catalysts have been based on zeolite Y, although may others have been tried, e.g. L, mordenite, omega in the late 1960s, and later X, beta, ZSM-5 types (41). This is simply due to the fact that, of all the sufficiently acidic zeolites, zeolite Y has the widest pores – given the fact that the feed to be converted has a boiling range of ~370-600 °C, zeolites with smaller pores would simply not be able to adsorb the molecules to be converted. The latter also tend to suffer from severe diffusional limitations under HC conditions, which is manifested by an increase in the relative importance of secondary HC reactions.

This effect has been demonstrated, for example, by studying the effect of zeolite pore size on the hydroconversion of n-heptane (19, 42). As shown in Table 6.4, there is a dramatic reduction in the isomerisation/cracking ratio with decreasing pore size which can be attributed to the decreased mobility of the n-heptene intermediate. Under commercial hydrocracking conditions, this effect manifests itself as a marked increase in gas formation, which represents a severe economic penalty: under conditions in which a dealuminated-Y based catalysts yielded only 5%wop gas ($C_1$-$C_4$), catalysts based on dealuminated beta and ZSM-12 produced 25 and 35%wop gas, respectively. This not to say that zeolite Y has no diffusion problems at all: even Y has problems of access to molecules in the higher boiling range, and it is for this reason that sometimes an ASA is added: it is much less acidic, but it does not induce any diffusion limitations, thus serving to convert the species that the zeolite cannot cope with (see Fig. 6.6 & cf. section 6.3.4).

**Table 6.4** Effect of zeolite pore system (n-heptane conversion over Pt/zeolite) (from Giannetto et al. [30]).

| Zeolite | Isomerisation / Cracking |
|---|---|
| FAU (USY) | 20 |
| BETA | 2.5 |
| OFF | 1.1 |
| MOR | 0.6 |
| MFI (ZSM-5) | 0.3 |

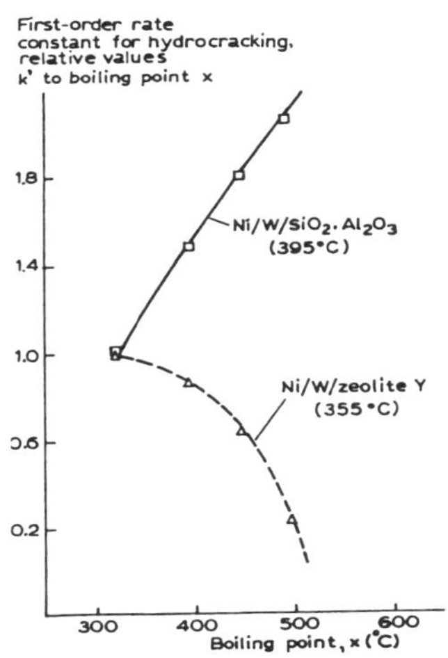


**Fig. 6.6** First-order HC rate constant as a function of boiling-point fraction for NiW/zeolite Y and NiW/ASA catalysts with a Middle-East flashed distillate feedstock (72)

Although the development of zeolitic HC catalysts has so far been largely limited to the Y-type structure, this crystalline material can be modified and tailored to an astonishing degree for particular applications. These modification techniques usually involve some combination of steaming and/or calcination, ion-exchange and leaching processes (43). These techniques provide a means of controlling the amount of aluminium located in the tetrahedral framework sites – the primary Brönsted acid sites – but also the amount of extra-framework aluminium and the mesopores that are created during the dealumination process. The experimental space set up by these techniques (see Fig. 6.7 – parameters are synthesis conditions, [crystallite size, SAR, crystallinity], depth of i.e., steaming/calcination T, steam pp, residence time, ammonium exchange acidified or not, how much acid, how to combine the various process steps) is so enormous that to this day new, more effective modifications of Y continue to be found. In essence this comes down to a fine tuning, c.q. tailoring, of the number and distribution of acidic sites, their intrinsic acidity (distribution), and their accessibility, while preserving a high crystallinity (6, 44). Remember that the acid site consists of a hydroxyl bridging a framework Al & Si; but while, then, the amount of aluminium located in the tetrahedral framework sites primarily determines the concentration & strength of the acidic sites, it does not do so exclusively (45, 46).

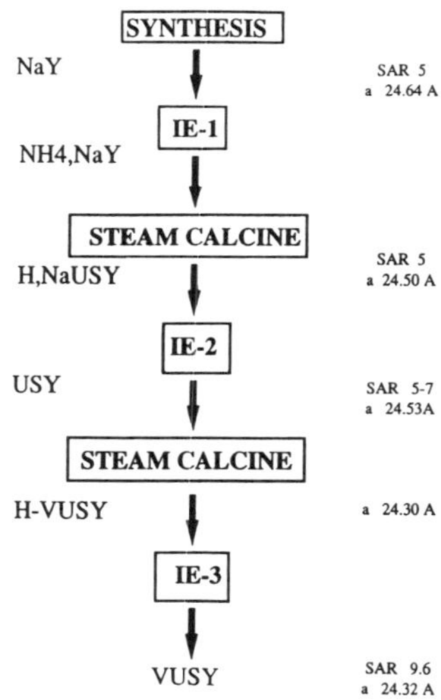


**Fig. 6.7** Schematic representation of the zeolite Y dealumination process steps

In general, the zeolite-Y crystallographic unit cell contracts with the removal of framework aluminium and this parameter is therefore a rough measure of acid-site density (47). It has been shown, for example (48), that the middle distillate selectivity (kerosene) of HC catalysts can be related to the unit cell size (see Fig. 6.8). This increase in kero selectivity with decreasing number of acidic sites is entirely consistent with the mechanistic scheme. Reducing the acid-site density while maintaining a constant hydrogenation power should reduce secondary cracking, as is observed. Perhaps surprisingly, the overall catalyst activity remains relatively high despite the reduction in acid-site density. This can be understood, at least in part, as being due to a compensating effect of reduced coke formation, as the hydrogenation function now has to keep less sites clean. Conversely, if naphtha is the desired product, a high concentration of acid sites is desired, and USY the cracking function of choice.

Although useful, the unit cell parameter is not always the most accurate measure of acid-site density. For example, studies by Bezman (47) showed that HC catalysts based on zeolite-Y materials with similar unit-cell dimensions had significantly

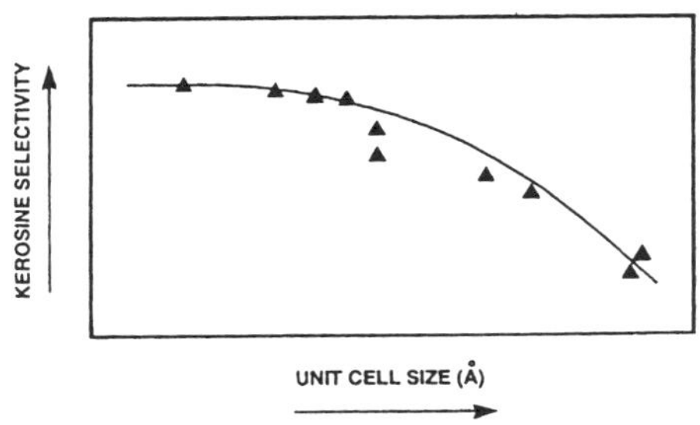


**Fig. 6.8** Influence of unit-cell size on product selectivity (48)

different performances. The application of a more comprehensive range of techniques, including IR, solid-state NMR and sodium back-exchange, is often necessary to provide a more complete characterisation of acid strength and site density. Even so, e.g. the role of extra-framework aluminium in modifying zeolite acidity continues to be discussed. In our opinion, there does appear to be an extra-framework Al species that enhances the intrinsic acidity of zeolite Y Brönsted sites (49): as indicated in Fig. 6.9, EFAl-free zeolite Y contains sites with a lower intrinsic ability to hydroisomerise n-heptane, a Brönsted-acid catalysed reaction (see Ch. 6.1), than does regular (ultrastabilised) Y. This has, of course, consequences for HC catalyst design: when applying materials with enhanced acidity one needs to provide a stronger hydrogenation function to balance this acidity. Fig. 6.9 also includes some data points pertaining to ASA materials: they tend to indicate that the intrinsic acidity of ASA sites is as strong as that of the 'enhanced' Y materials, but their number is just much lower, thus explaining the rather large difference in cracking activity between the two materials.

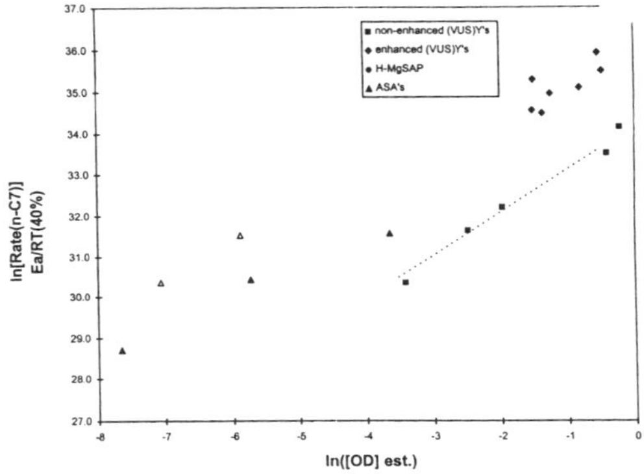


**Fig. 6.9** Log-log plot of n-heptane hydroisomerisation activity versus number of strong acid sites (as measured by H/D exchange method) (49)

To explore the effect of still weaker acid strengths than that available in EFAL-free zeolite Y, but retaining the FAU structure, one could either evaluate a material like SAPO-37 or try and substitute Fe(3+) or Ga(3+) in (US)Y (cf. ref. [50]). The problem with SAPO-37 (apart from its water-sensitivity) is, however, that it has very large crystallites leading to mass-transfer limitations (see below). And although we were able to synthesise a nice RE-Ga-USY material, it turned out to be unstable under HC testing conditions (51). Indeed, a little hydrogen is already sufficient to dislodge Ga(3+) from the framework (Fig. 6.10).

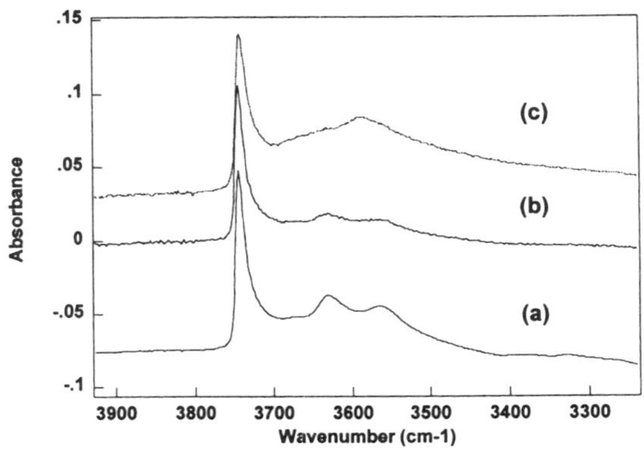


**Fig. 6.10** FTIR spectra of US-GaY (a) as such; (b) after contact with pyridine (15 micro) at 673 K; (c) after subsequent contact with hydrogen (10 kPa) at 673 K for 90 mins

Another important property of stabilised zeolite-Y materials is their mesoporosity (52). As previously discussed, aluminium is removed from the framework during steaming and the resulting lattice defects are (partially) healed by silica which migrates from other parts of the structure. This mass transfer of silica within the zeolite creates a secondary, larger pore structure: the so-called mesopores as illustrated in Fig. 6.11 (52). The mesoporosity has been shown to reduce mass-transport limitations during hydrocracking and thereby suppress secondary cracking. For heavy feeds the activity is also improved due to enhanced accessibility, as has been recently confirmed in a comparison between tetralin and atmospheric residue HC (53). A recent PQ patent describes stabilised Y materials with much enhanced mesoporosity (54), and it turns out that indeed such materials can lead to increased middle-distillate selectivity. It has even been claimed that the selectivity pattern of mesoporous zeolite-Y approaches that of an ASA (52), although this has not been well substantiated (37). As yet, the influence of zeolite-Y mesopore structure has not been considered in model studies. A comparison with SAPO-37 (FAU structure, weaker acidity than Y [55] is also instructive: its large (a few microns), mesopore-free crystallites lead to mass-transport hindrances, so that despite the relatively low

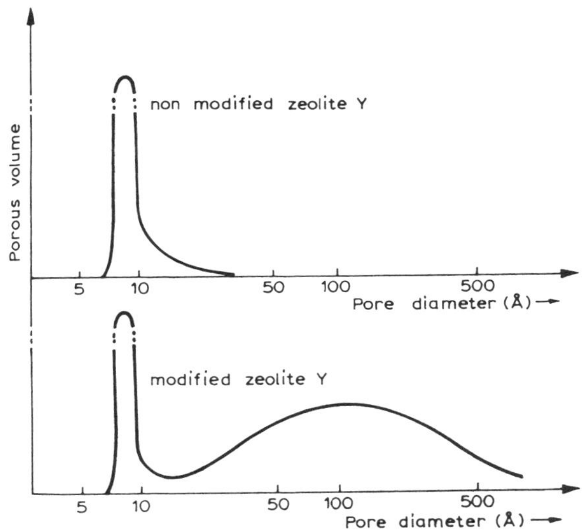


**Fig. 6.11** Creation of a secondary pore structure as a result of dealumination (52)

activity, dictated by the low intrinsic acidity, secondary cracking occurs to a considerable extent (high gas make).

In view of the above, it does not surprise that crystallite size is often considered to be an important issue. On the basis of the Thiele concept, smaller crystallites would be expected to generate less mass-transport limitations. Indeed, claims to that effect can be found in the literature (56). But, although zeolite-Y crystallites are large enough to hamper mass transport through its pores (say, 0.5 +/- 0.2 micron), the existence of a mesopore system in ultrastabilised versions may imply that the effective crystallite size is much smaller (57), and that the quest for small-particle Y is an unfruitful one.

### *6.3.4 Effect of recycle operation*

In many hydrocracker designs, complete conversion of the feed into lighter products is desired. This is normally achieved by incorporating a recycle to extinction (Fig. 6.1). The consequence of such a configuration is that the more refractory molecules that are not readily converted over the HC catalyst in a single pass will tend to build up in the recycle stream. Given the restricted access of the larger feed molecules to the zeolite-Y pore system (Fig. 6.6), one may expect bulky polynaphthenes to remain in the unconverted fraction of the recycle stream, when using a pure zeolitic catalyst. This is in fact what is observed, using Field Ionisation Mass Spectrometry as the analytical tool (Fig. 6.12 (a)). Thus, in operations with a zeolite-based HC catalyst, the heavier fraction of the feed will tend to build up in the recycle stream with a concomitant increase in the temperature required for conversion to extinction.

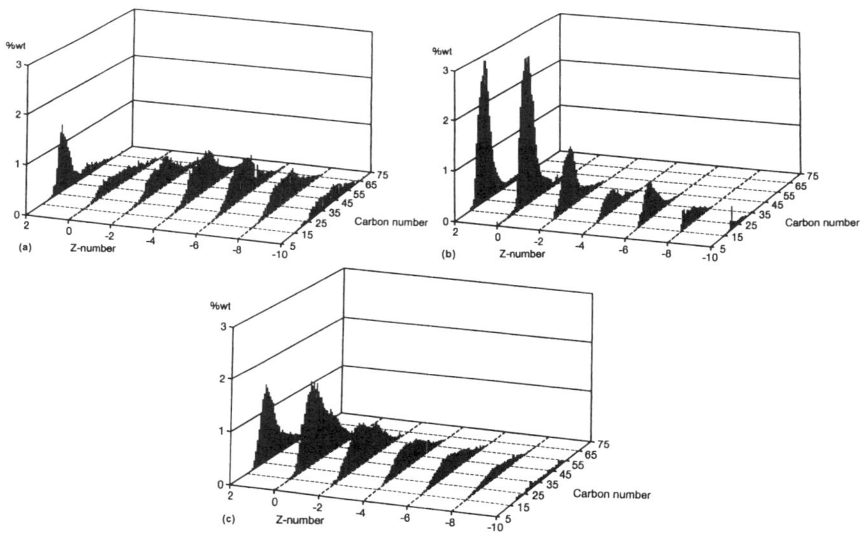


**Fig. 6.12** FIMS spectra of the unconverted material after hydrocracking treated VGO over (a) all-zeolite, (b) all- amorphous, and (c) composite catalysts (two-stage operation). Z indicates the hydrocarbon stoichiometry $C_nH_{2n+z}$

This, in turn, can lead to a deterioration of the product yield pattern (e.g. increased gas make) (58). Large-pore amorphous catalysts, on the other hand, follow the intrinsic cracking rate constants, and convert the heavy polynaphthenes preferentially, while leaving the relatively low-molecular-weight paraffins and mono-ring naphthenes more or less alone (Fig. 6.12 (b)). As mentioned above, however, such amorphous silica-alumina catalysts are significantly less active than zeolitic ones, more prone to N-poisoning, and have a relatively high propensity towards coke formation, resulting in a lower stability.

Modern two-stage catalysts are so-called composite ones, containing both a zeolite-Y and an ASA component, where they should be balanced so as to achieve both a high activity and a stable yield structure in recycle. The FIMS analysis of the unconverted material obtained over such a composite catalyst demonstrates that a more balanced conversion of the different feed components can be achieved (Fig. 6.12 (c)). And indeed, such catalysts perform quite satisfactorily in practice (59).

Bulky polyaromatic molecules tend to concentrate in recycle streams and are known to be prime coke precursors, as demonstrated in model studies by Latif (60). However, by means of a relatively small bleed stream, the concentration of these polyaromatic species can be contained at a low level, thereby minimizing their deleterious effect on catalyst stability. Alternatively, they can be substantially removed by adsorption on an active-carbon column.

## 6.4 Outlook

As mentioned before, most present-day commercial hydrocracking catalysts are composed of the components listed in Table 6.1, with zeolites limited to various modified forms of the Y structure. Recent developments have been mainly incremental improvements in the ASA and Y base materials and the control of catalyst architecture, sometimes augmented by improved metal emplacement routes. That better catalysts can still be found in this way can be exemplified in plots like Fig. 6.13. Improvement of the traditional mixed-sulphide hydrogenation functions has also been sought through promotion (by, e.g. Nb) or even through replacement (by, e.g. $RuS_x$-based systems), without apparent breakthroughs so far. Another line that has recently been taken is to include two zeolites (e.g. USY/VUSY) in the catalyst particles, with some benefits according to the patent literature, but it is unclear to what extent, if at all, such catalysts have reached commercial applications (61).

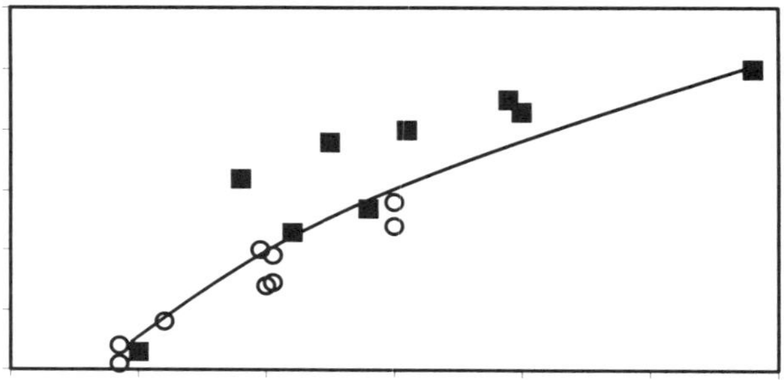


**Fig. 6.13.** Middle-distillate selectivity versus activity in series-flow operation: the next generation of catalysts. On the far right is an all ASA catalyst and going to the left the zeolite content increases

It is natural to expect, given a particular feed and the HC mechanism outlined above, that there will be a limit on the selectivity with which a desired product can be made. In the case of MD selectivity, Figs. 6.8 and 6.13 can be read as indicating that we are already in sight of this limit. However, to be sure, we would need to be able to determine the molecular composition of the feed, and to model the HC process in sufficient detail. In both these areas, good progress is being made (62, 63), but we are not yet there. However this may come out, the challenge to develop still more active and selective catalysts will remain with us for some time to come.

Another recent approach has been to tailor catalysts to fit a particular purpose or process configuration, mainly to shift more HC duty to the first stage. This has become possible due to the advent of high-nitrogen resistant zeolite-Y containing catalysts, which can be applied in the bottom bed of the first stage. For such stacked-bed configurations it is sometimes advantageous to have NiMo rather than NiW as the metal function to improve the denitrogenation activity. However, this is normally achieved at the cost of some loss in aromatics hydrogenation performance (64). Variations in one of the basic designs are also sometimes claimed to lead to important benefits, as in UOP's recently announced HyCycle Unicracking process, a modification of the two-stage configuration to improve the Diesel yield.

There is a major economic incentive to extend the current HC processes to enable heavier feedstocks to be converted to lighter, higher-value transportation fuels. Studies by Idemitsu indicate that iron-modified zeolite catalysts significantly enhance conversion when heavy oils such as long residue are hydroprocessed (65). Nevertheless, major technical barriers exist which make high conversions and product selectivities difficult to achieve with truly heavy feeds (end boiling points beyond 620 °C) – these include:

- the relatively high concentrations of polyaromatics lead to severe catalyst deactivation, exacerbated by the fact that the higher operating temperatures required with heavier feeds tend to shift the hydrogenation equilibria towards the aromatic side (this is obviously not good for the product properties either) – although higher hydrogen pressures could be applied, this generally imposes some severe economic penalties (e.g. beyond 170 bar)

- although zeolite-based catalysts offer the best prospects for heavy feeds conversion due to their relatively low coking propensity, their activity is suppressed due to the (very) restricted access of the heavier molecules to the zeolite pores – which may also entail a loss of product selectivity, as the zeolites now tend to be more active for cracking the initial products than the feedstock

One answer has been to switch to a rather different process design, viz. performing the HC in an ebullated bed, in which continuous withdrawal of deactivated catalyst from the system is practiced (LC-fining, H-Oil, T-Star, etc.). Because of the shorter average catalyst age it is easier to cope with catalyst deactivation. As the catalysts applied to date do not contain zeolites – indeed, these processes generally also profit from a relatively high thermal contribution to the cracking – further discussion of such designs is outside the scope of this chapter (66).

Another answer might be the use of zeolitic materials with wider pores. The development of such materials is in fact a thriving field. So far, however, attempts at hydrocracking with, e.g., VPI-5 and UTD-1, have not produced exciting results. Another class of compounds, ordered mesoporous materials like MCM-41/-48 and SBA-*n* (67), silica-aluminas with a regular pore structure with pore diameters in the range of 2-10 nm has recently come to the fore, but, as far as HC is concerned, they only appear to be very expensive ASAs.

A very recent development, however, is to have the walls of the regular pores consisit of small zeolitic nuclei (beta, Y), and these materials certainly merit close scrutiny (68, 69). Although the combination of 12-ring zeolites, however small their crystallites, with larger mesopores, does not bode too well for their stability, even if they were to achieve higher selectivities. Going to smaller mesopores [say, 1-1.5 nm] might also be interesting (70). Pillared clays have also come in for their share of attention (see, e.g. (71)), but did not emerge as commercial catalysts, one problem being that they coke up rather easily.

**Acknowledgement**

Over the years, the present author has greatly benefited from discussing the present subject with Drs. J.K. Minderhoud, J.W. Gosselink, and, especially, W.H.J. Stork.

**References**


1. Minderhoud J.K., Van Veen J.A.R. and Hagan A.P., *Stud. Surf. Sci. Catal.* **127** (1999) 3.
2. Dufresne P., Bigeard P.H. and Billon A., *Catal. Today* **1** (1987) 367.
3. Chen Q., Van den Oosterkamp P. and Barendregt S., *Petrol. Technol. Quart.* **4** (1999) 47.
4. Gary J.H. and Handwerk G.E., *Petroleum Refining: Technology & Economics* (Marcel Dekker, NY, 1994 [3rd edition]).
5. Maxwell I.E., Minderhoud J.K., Stork W.H.J. and Van Veen J.A.R., in *Handbook of Heterogeneous Catalysis*, Ed. Ertl G., Knözinger H. and Weitkamp J. (Wiley-VCH, 1997), Vol. 4, Ch. 3.13 (pp. 2017-2038).
6. Scherzer J. and Gruia A.J., *Hydrocracking Science and Technology* (Marcel Dekker, NY, 1996).
7. Sie S.T., *Revue de l'IFP* **46** (1991) 501.
8. Gosselink J.W. and Van Veen J.A.R., *Stud. Surf. Sci. Catal.* **126** (1999) 1.
9. Prins R., *Adv. Catal.* [forthcoming]
10. Blauwhoff P.M.M. et al., in *Catalysis and Zeolites – Fundamentals & Applications*, Ed. Weitkamp J. and Poppe L. (Springer-Verlag, Heidelberg, 1999), Ch. 7.
11. Douwes C.T. and 't Hart M., *Erdöl Kohle Erdgas Petrochem.* **21** (1968) 202.
12. Ward J.W., in *Preparation of Catalysts III*, Ed. Poncelet G., Grange P. and Jacobs P.A. (Elsevier, A'dam, 1983), 587.
13. Gates B.C., Katzer J.R. and Schuit G.C.A., *Chemistry of Catalytic Pocesses* (McGraw-Hill, NY, 1979).
14. Schulz H. and Weitkamp J., *Ind. Eng. Prod. Res. Dev.* **11** (1972) 46.
15. Weitkamp J., *Erdöl Kohle Erdgas Petrochem.* **31** (1978) 13.
16. Weitkamp J. and Dauns H., *ibid.* **40** (1987) 111.
17. Martens J.A., Jacobs P.A. and Weitkamp J., *Appl. Catal.* **20** (1986) 239, 283.
18. Jacobs P.A. and Martens J.A., *Stud. Surf, Sci. Catal.* **137** (2001) 633.
19. Guisnet M., Alvarez F., Giannetto G. and Perot G., *Catal. Today* **1** (1987) 415.
20. Sullivan R.F., Boduszinski M.M. and Fetzer J.C., *Energy & Fuels* **3** (1981) 603.
21. Guisnet M. and Magnoux P., *Appl. Catal. A* **212** (2001) 83.
22. Denayer J.F.M. *et al*, *Ind. Eng. Chem. Res.* **36** (1997) 3242; Martens G.G. et al., *J. Catal.* **195** (2000) 253.
23. Martens G.G., Thybaut J.W. and Marin G.B., *Ind. Eng. Chem. Res.* **40** (2001) 1832; cf. Berger R.J. et al., *CatTech* **5** (2001) 30, *esp.* Intermezzo 6.
24. Weitkamp J. and Ernst S., in *Guidelines for Mastering the Properties of Molecular Sieves*, Ed. Barthomeuf D. et al. (Plenum Press, NY, 1990), 343.
25. Weisz P.B., *Adv. Catal.* **13** (1962) 137.

26. Cornet D., El Qotbi M. and Leglise J., *Stud. Surf. Sci. Catal.* **106** (1997) 147.
27. Sie S.T., *Ind. Eng. Chem. Res.* **32** (1993) 403.
28. Sullivan R.F. and Scott J.W., *ACS Symp. Series* **222** (1983) 293.
29. Nat P.J., *Erdöl Kohle Erdgas Petrochem.* **42** (1989) 447.
30. Luo J.G. et al., *Chin. Chem. Lett.* **2** (1991) 963.
31. Reyniers M.-F. *et al, Appl. Catal. A* **202** (2000) 49, 65.
32. Amano H., Sato S., Takahashi R. and Sodesawa T., *Phys. Chem. Chem. Phys.* **3** (2001) 873.
33. Law D.V., Paper presented at the 7th Refinery Technology Meeting, Bombay, 6-8 Dec. 1993.
34. Mignard S. and Béroudiaux O., *React. Kinet. Catal. Lett.* **65** (1998) 185.
35. Welters W.J.J., PhD Thesis, Eindhoven, 1994.
36. Lemberton J.L., Touzeyido M. and Guisnet M., *Appl. Catal.* **54** (1989) 101; ibid. A **79** (1989) 115.
37. Habib M.M., Bezman R.D., Dahlberg A.J. and Krishna A.S., Paper presented at the XIth Refinery Technology Meeting, Hyderabad, 9-11 Feb. 2000.
38. Geus J.W. and Van Veen J.A.R., *Stud. Surf. Sci. Catal.* **123** (1999) 459.
39. Vissenberg M.J. et al., *J Phys Chem B* **104** (2000) 8456.
40. Robinson W.R.A.M., Van Veen J.A.R., De Beer V.H.J. and Van Santen R.A., *Fuel Proc. Technol.* **61** (1999) 103; Hensen E.J.M. et al., *J. Catal.* **199** (2001) 224.
41. Ward J.W., in *Applied Industrial Catalysis*, Ed. Leach B.E., Vol. 3 (Academic Press, 1984), Chapter 9.
42. Giannetto G. *et al, Ref. [24]*, 355.
43. Scherzer J., *ACS Symp. Series* **284** (1984) 157.
44. Szostak R., *Stud. Surf. Sci. Catal.* **137** (2001) 261.
45. Beaumont R. and Barthomeuf D., *J. Catal.* **26** (1972) 218, **27** (1972) 45; **30** (1973) 288.
46. Wachter W.A., in *Theoretical Aspects of Heterogeneous Catalysis*, Ed. Moffat J.B. (Van Nostrand Reinhold, 1990), 10.
47. Bezman R., *Catal. Today* **13** (1992) 143.
48. Hoek A. et al., *Oil & Gas J.* **1991**, April 22, 77.
49. Rigutto M.S. et al., Manuscript in preparation
50. Chu C.T.W. and Chang C.D., *J Phys Chem* **89** (1985) 1569; Post M.F.M. et al., *Stud. Surf. Sci. Catal.* **46** (1989) 365.
51. Maesen, Th.L.M. et al., *J. Phys. Chem. B* **104** (2000) 716.
52. Hennico A., Billon A. and Bigeard P.-H., *Hydrocarbon Technol. Int.* **1992**, 19; Maier L., Bigeard P., Billon A. and Dufresne P., NPRA Paper AM-88-76 (1988).
53. Sato K., Nishimura Y. and Shimada H., *Catal. Lett.* **60** (1999) 83; Sato K. et al., *J. Catal.* **200** (2001) 288.

54. Cooper D.A., Hastings T.W. and Hertzenberg E.P., US Patent **5601798** (1997).
55. Mostad H.B., Stöcker M., Karlsson A. and Rørvik T., *Appl. Catal. A* **144** (1996) 305; Corma A. and Martínez-Triguero, ibid. **118** (1994) 153.
56. Krishna R., *Erdöl Kohle Erdgas Petrochem.* **42** (1989) 194.
57. Janssen A.H., Koster A.J. and De jong K.P., *Angew. Chem. Int. Ed.* **40** (2001) 1102.
58. Yan T., *Ind. Eng. Chem. Proc. Des. Dev.* **22** (1983) 154.
59. Huizinga T., Theunissen J.M.H., Minderhoud J.K. and Van Veen J.A.R., *Oil & Gas J.* **1995**, June 26, 40.
60. Abdul Latif N., *Stud. Surf. Sci. Catal.* **53** (1990) 349.
61. Ward J.W., *Fuel Proc. Technol.* **35** (1993) 55.
62. Barman B.N., Cebolla V.L., Mehrotra A.K. and Mansfield C.T., *Anal. Chem.* **73** (2001) 2791.
63. Martens G.G. and Marin G.B., *AI Ch E J.* **47** (2001) 1607.
64. Minderhoud J.K. and Van Veen J.A.R., *Fuel Proc. Technol.* **35** (1993) 87.
65. Maxwell I.E. and Stork W.H.J., *Stud. Surf. Sci. Catal.* **58** (1991) 571.
66. Chrones J. and Germain R.R., *Fuel Science & Technol. Int.* **7** (1989) 783; Eccles R.M., *Fuel Proc. Technol.* **35** (1993) 21.
67. Schüth F., *Stud. Surf. Sci. Catal.* **135** (2001) 1.
68. Lin Y., Zhang W. and Pinnavaia T.J., *J. Amer. Chem. Soc.* **122** (2000) 8791; *Angew. Chem.* **113** (2001) 1295.
69. Zongtao Zhang et al., *J. Amer. Chem. Soc.* **123** (2001) 5014; *Angew. Chem.* **113** (2001) 1298.
70. Bagshaw S.A. and Hayman A.R., *Adv, Mater.* **13** (2001) 1011.
71. Martens J.A. et al., *Stud. Surf. Sci. Catal.* **130** (2000) 293.
72. Maxwell I.E., *Catal. Today* **1** (1987) 385.

CHAPTER 7

# C4-C6 ALKANE ISOMERIZATION

F. SCHMIDT, E. KÖHLER,
*SÜD – CHEMIE AG, Waldheimer Str 13,*
*D-83052 Bruckmühl/*

## 7.1 Introduction

Fuel standards for the year 2000 were set by the European community, with plans for more stringent specifications on sulfur and aromatics for the year 2005 (Table 7.1).

**Table 7.1** Gasoline specifications for Europe and the United States

| | Europe 2000 | Europe 2005 | California (CARB Phase I) |
|---|---|---|---|
| RVP (Summer), bar | 6 | AO-II | 5 |
| Sulfur, ppmw | 150 | 50 | 30/80 |
| Benzene, vol.-% | 1.0 | 1.0 | 0.8/1.2 |
| Olefins, vol.-% | 18 | AO-II | 4.0/10.0 |
| Aromatics, vol.-% | 42 | 35 | 22/30 |
| Oxygen, wt.-% | 2.7 | AO-II | 1.8/2.2 |

Because of the very high volumes of fuels compared to chemicals, the excess aromatics caused by the reduction of aromatics in fuels (from 43 vol.-% to 34 vol.-%) cannot fully be used by the chemical industry. The entire market for solvents or aromatics based monomers for the polymer industry is too small. Moreover, the aromatics are high RON (research octane number) components. Therefore, a considerable amount of the aromatics needs to be converted to components with highest possible RON.

The increase of 1 unit of the RON corresponds to about 900.000 US$ per year for a 300.000 tpa hydroisomerization unit (1). In Figure 7.1, several major refinery processes to improve RON are shown: these include isomerization, reforming, addition of FCC-Naphtha, alkylation, addition of oxygenates or polygas or butanes. The effect of these options with respect to the new specifications is different for each particular process. Keeping in mind the Californian ban on MTBE and also the fact that the oxygenate content has to be reduced to half of the present value, the most favourable options are alkylation and isomerization.

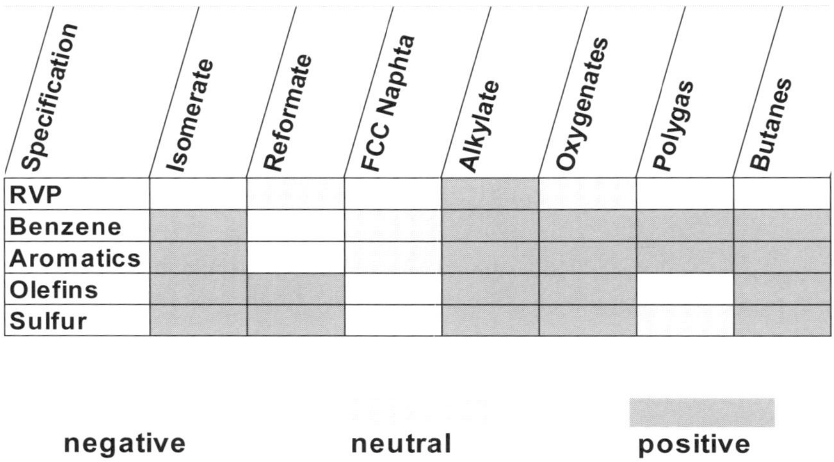


**Fig. 7.1** Options to react on the new European Fuels Specifications

## *7.1.1 Aromatics reduction & RON gap*

The need for benzene reduction is one of the determining factors in the way refiners will have to modify their process portfolio to meet future specifications. Apart from lowering the reformer severity, pre-fractionation and post-fractionation provide viable tools to reduce benzene in the gasoline pool. Pre-fractionation and subsequent hydrogenation of benzene is a typical solution. However, the products (cyclohexane and alkyl-cyclohexanes) are low in octane. Therefore, this option is only feasible if the refinery is not short in octane. The octane loss can be compensated for by the addition of oxygenates or preferably by the addition of alkylates. If more octane is needed, post-fractionation is one of the solutions.

However, the most economical and octane effective route to reduce benzene is to eliminate benzene and benzene precursors from the naphtha reformer feed by pre-fractionation in an up-stream splitter and then send these precursors to an isomerization unit. With this approach, benzene will be saturated and isomerised to a mixture of higher octane saturated cyclics. Depending on the feed benzene content, various solutions can be applied to handle the heat load from the saturation of benzene. These heat management solutions entail saturation of the benzene in the isomerization reactor by employing low feed temperatures, internal quenching, or feed dilution with recycle; or by the installation of an upstream hydrogenation catalyst, either nickel or noble metal based, in a separate reactor. The attractiveness of these options depends on the benzene content in the feed.

### *7.1.2 Sulfur*

Upcoming regulations will also call for lower sulfur in gasoline. Controlling sulfur in gasoline will most probably involve the hydrotreating of FCC gasoline. This method will not only result in reduced sulfur but also hydrogenate olefins and benzene/aromatics during the saturation process. In both cases, there is an octane loss from the hydrogenation of either benzene/aromatics or olefins during the saturation process, creating another need for octane compensation by isomerization and/or alkylation.

In the future, there will be a need for feed streams to isomerization to be composed of fractions from totally different sources, including those with very high benzene content. To make up for octane losses and debottleneck the gasoline production, it is particularly attractive to feed light naphtha to the isomerization unit without pre-treatment. The commercially proven CKS ISOM process, licensed exclusively by Kellogg Brown & Root and using SÜD-CHEMIE's HYSOPAR catalyst can be applied in any of the cases outlined above, including sulfur rich, untreated feeds (2, 3). Although the hydrogenation activity is dramatically inhibited with traditional isomerization catalysts when sulfur is present in the feedstock, even benzene saturation in the presence of about 50 ppmw sulfur can be considered with this sulfur tolerant catalyst as has been demonstrated (4). As the HYSOPAR catalyst also hydrogenates sulfur compounds, CKS ISOM can help control benzene and sulfur with respect to future specifications, thus relieving the load on upstream naphtha hydrotreaters and reformers and making a positive contribution to the overall refinery economics.

## 7.2 Proven isomerization catalysts and processes

Isomerization increases the octane value of light straight run naphtha by rearranging the low octane straight chain C5-C6 paraffins into their higher octane branched isomers as shown in Table 7.2.

Isomerization is controlled by thermodynamic equilibrium with the isomer balance being a function of temperature. This is demonstrated in Figures 7.2 and 7.3.

**Table 7.2** Research Octane Number (RON) of pentanes and hexanes

| | Compound | RON |
|---|---|---|
| NC5 | n-Pentane | 61.7 |
| IC5 | iso-Pentane | 92.3 |
| NC6 | n-Hexane | 24.8 |
| 3-MP | 3-Methylpentane | 73.4 |
| 2-MP | 2-Methylpentane | 74.5 |
| 2,2-DMB | 2,2-Dimethylbutane | 91.8 |
| 2,3-DMB | 2,3-Dimethylbutane | 102.0 |

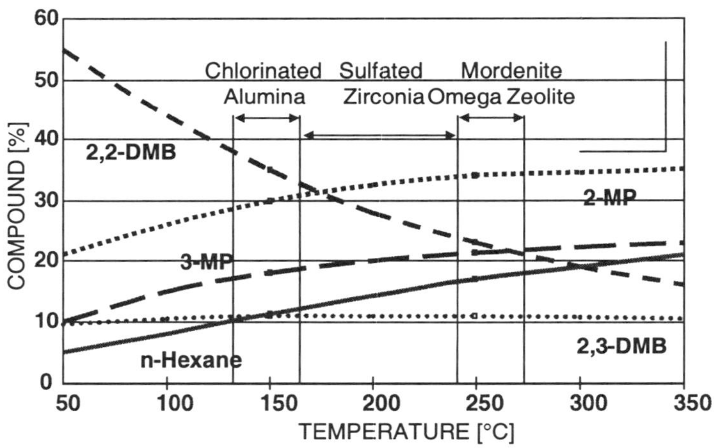


**Fig. 7.2** Equilibrium distribution of hexanes

Commercial hydroisomerization catalysts have both a noble metal based hydrogenating function and an acid function (Table 7.3). Traditionally, the acid component is provided by (i) a zeolite or by (ii) a chlorinated alumina substrate or by (iii) a sulfated zirconia carrier, the latter both being extremely intolerant of sulfur, water, and other feed contaminants. The zeolite is generally a mordenite and not a Y-zeolite. However, catalysts based on zeolite omega have been shown to be superior to mordenite-based catalysts, but no up-scale to commercial use has been reported for omega zeolite containing hydroisomerization catalysts (see below).

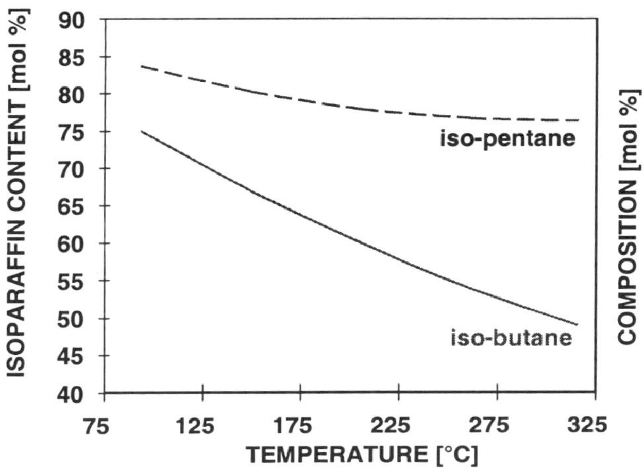


**Fig. 7.3** Equilibrium distribution of butanes and pentanes

**Table 7.3** Commercial isomerization processes and catalysts

| Process | Licensor | Zeolite based catalyst | Metal Oxide Based catalyst | Cl/Al2O3 based catalyst |
|---|---|---|---|---|
| Hysomer, TIP, Penex | UOP/ Union Carbide | HS-10, I-7 | LPI-100 ($ZrO_2/SO_4$) | I-8 I-80 |
| Ipsorb Hexorb | IFP/ Procatalyse (AXENS) | IS 632 | | IS 612A, IS 614A |
| BP/ Engelhard | BP/ Engelhard | | | IRD 291 |
| (no process) | AKZO Nobel/ TOTAL | | | AT-2G |
| CKS ISOM | CEPSA KBR Süd-Chemie | HYSOPAR | HYSOPAR-SA* (proprietary) | |

- not yet commercialised

The noble metal component may be either palladium or platinum: the effect of the concentration of both metals on methylpentane as well as on dimethylbutane selectivity in C6 hydroisomerization on lanthanum and ammonium Y-zeolite with Si/Al of 2.5 has been studied by M.A. Lanewala et al. (5). They found an optimum of metal content for that reaction between 0.1 and 0.4 wt.-%. The noble metal has several functions: (i) to increase the isomerization activity of the zeolite; (ii) to support the saturation of the coke precursors and hence prevent deactivation, as was shown by H.W. Kouvenhoven et al. (6) for platinum on hydrogen mordenite; (iii) to support the hydrodesulfurization activity of the catalysts in sulfur containing feedstocks.

From the thermodynamic point of view, the hydroisomerization reaction is not pressure sensitive. However, because the catalyst contains the acid function, hydrocarbon cracking is an unavoidable side reaction. The cracking reaction however should depend on the total pressure. Table 7.4 shows laboratory results obtained in a bench scale reactor at 250°C (482°F), $H_2/HC = 1$ with a synthetic feedstock containing 50wt% of $n\text{-}C_5$, 25 of $iC_5$, 20 of $n\text{-}C_6$, 5 of methylcyclopentane and no heptane (Feed 1). At a liquid hour space velocity (LHSV) of $2h^{-1}$ an increase of the total pressure from 20 bar to 30 bar reduces the cracking selectivity $S = \Sigma\ C4/\Sigma\ HC$ from 1.6 to 1.1 wt.-%, whereas at a LHSV of 1 $h^{-1}$ no effect can be observed.

With feed 2 which, like feed 1, contains 50, 25 and 20 wt% of $n\text{-}C_5$, $i\text{-}C_5$ and $n\text{-}C_6$ but 2.5 wt% of $n\text{-}C_7$ and 2.5 w% of methylcyclopentane (instead of 5) the gas production at 20 bar, 250°C and LHSV of 2 is higher (2.7wt%, Table 7.5) than with feed 1 (1.7 wt%, Table 7.4).This reflects the higher cracking affinity of $C_7$ hydrocarbons. Furthermore, Table 7.5 shows that the increase of the $H_2/HC$ ratio from 0.25 to 1 (hence of the hydrogen pressure from 4 to 10 bar) causes a significant decrease in the cracking selectivity. Therefore, the higher hydrogen pressure not only improves the hydrodesulfurization activity and reduces coking by hydrogenating the coke precursors but also reduces the gas make caused by higher hydrocarbons in the feedstock.

Small quantities of sulfur and water will severely reduce isomerization performance and at moderate levels will permanently deactivate chlorinated alumina (that cannot be regenerated) or sulfated zirconia based catalysts. Traditional zeolite based catalysts are also lacking sulfur tolerance, but to a lesser extent.

**Table 7.4** Effect of pressure on hydroisomerization activity of Pt-H-MOR
Conditions: T=250 °C, H2/HC molar ratio = 1, Feed 1

| Pressure, bar | 30 | 30 | 20 | 20 |
|---|---|---|---|---|
| LHSV, h-1 | 1 | 2 | 1 | 2 |
| C5 Activity ΣI-C5/ΣC5 wt.-% | 67 | 57 | 67 | 60 |
| C6 Activity A 2,2 = 2,2-DMB/ΣC6 wt.-% | 18 | 13 | 18 | 15 |
| C6 Activity A 2,3 = 2,3-DMB/ΣC6 wt.-% | 8 | 7 | 8 | 7 |
| C6 Activity A 2 = 2-MP/ΣC6 wt.-% | 29 | 28 | 29 | 28 |
| C6 Activity A 3 = 3-MP/ΣC6 wt.-% | 18 | 18 | 18 | 18 |
| Cracking Selectivity S = Σ-C4/Σ HC wt.-% | 1.7 | 1.1 | 1.7 | 1.6 |

**Table 7.5** Effect of pressure on cracking selectivity in hydroisomerization of synthetic feedstock on Pt-H-MOR Conditions: T=250 °C, Total pressure = 20 bar, LHSV = 2 $h^{-1}$, Feed 2

| $H_2$/HC Molar Ratio | Cracking Selectivity, wt.-% C4- |
|---|---|
| 0.25 | 4.3 |
| 0.43 | 3.5 |
| 1 | 2.7 |
| 1.5 | 2.6 |

The first commercial zeolite catalyst based process is known as Hysomer, a once-through process developed by Shell (7). The process operates at 27 to 30 bar hydrogen pressure at about 250°C. The octane gain is reported to be up to 10 points. Aromatics are hydrogenated to naphtenes and do not poison the catalyst. C7+ hydrocarbon compounds at concentrations lower than 4 wt.-% are acceptable without influencing activity and selectivity of the catalyst. Higher concentrations have a negative effect, because the higher hydrocarbons are cracked to gases, which are believed to hinder the effective mass transport of the other molecules. Sulfur was claimed to be tolerable up to 35 ppmw, although no commercial unit is known to operate at such conditions. According to the early publications (7), water up to a level of 50 ppmw does not weaken the performance. The catalyst could be regenerated in situ. A further improvement of the octane number could be achieved by combining the hydroisomerization with the UNION CARBIDE ISOSIVE product n/iso separation system. This process is known as the TIP (Total Isomerization Process), now offered by UOP.

Paraffin isomerization is a bifunctional process comprising: (i) hydrogenation and dehydrogenation steps catalyzed by platinum sites; (ii) isomerization of the olefinic intermediates on the acid sites of the catalyst; (iii) transport of the intermediates from the platinum to the acid sites and vice versa (8). It is a well-known phenomenon for standard noble metal catalysts that sulfur blocks this reaction mechanism and as a consequence substantial inhibition of their hydrogenation capabilities is experienced.

Isomerization processes employing any of these traditional catalysts, except HYSOPAR, require full hydrotreating of the feedstock. As a result, very few traditional isomerization units in the world operate today with more than 1 ppmw sulfur. Furthermore, a catalyst whose activity is heavily impacted by sulfur cannot cope with benzene under high sulfur conditions either. The HYSOPAR catalyst was particularly developed to show evidence of high tolerance to feedstock sulfur and to cope with benzene even at high sulfur conditions. This opens up new opportunities for the refiner to feed benzene to isomerization units.

## 7.3 Zeolite based catalyst systems

### *7.3.1 Mordenite based catalysts*

Commercial zeolite based hydroisomerization catalysts comprise alumina bound and platinum impregnated dealuminated mordenite. The activity and selectivity of the hydroisomerization of n-paraffins is strongly influenced by acid leaching. The influence of silica to alumina ratio has been studied for pentane isomerization over platinum mordenite many times since one of the first papers published (6). Table 7.6 shows an example reported by H.W. Kouwenhoven (6).

**Table 7.6** Influence of $SiO_2/Al_2O_3$ ratio on C5 hydroisomerization activity of Pt-H-MOR [6]. T=250°C, p=30 kg/cm2, $H_2/C_5$ molar ratio = 2.5

| $SiO_2/Al_2O_3$ | Relative Activity |
|---|---|
| 10 | 100 |
| 17 | 135 |
| 25 | 84 |

Because several other factors also influence the activity of the catalyst, different authors report slightly different optimum $SiO_2/Al_2O_3$ ratios for maximum activity.

Acid leaching is decreasing the number of Brönsted acid sites (9,10) and creates mesoporosity inside the zeolite crystallites (11). This mesoporosity has been visualized by 3D TEM images (12). In that paper it has been shown that increasing mesoporosity decreases diffusion limitations. This causes shorter residence time and facilitates desorption of products, whereby secondary reactions of the intermediates are depressed and selectivity is improved.

A decreasing number of acid sites and increasing porosity are acting in opposite directions with respect to catalyst activity. In case of unmodified mordenite, the total number of acid sites is high but only few are accessible due to mass transfer limitations. Decreasing the total number of acid sites but increasing the number of accessible sites by leaching results in an optimum catalyst formulation. Different optimum silica to alumina ratios are reported in the literature. This is simply caused by the sensitivity of the mordenite system towards synthesis and dealumination conditions. Therefore, applying the same dealumination procedure to various mordenites leads to different results as the dealumination behaviour depends on the mordenite synthesis procedure, even if the mordenites were prepared by using the same silica to alumina batch composition (13). Hence, for each mordenite in the 'as synthesised' form the subsequent leaching procedure has to be optimized. This statement is also true for different mordenites of the same overall chemical composition but produced by different mordenite suppliers.

Steaming usually produces extra framework aluminum, but even in the case of leaching some extra framework aluminum may be left on the mordenite. In some cases they can be removed by ammonium ion exchange, if desired, followed by calcination. However, depending on the amount and location of the extra framework aluminum, as the case may be, these species may increase or decrease the catalyst activity. An early review on the dealumination of mordenite was given by Karge and Weitkamp (14). In particular for hydroisomerization of paraffins, this effect has been studied by A. Corma et al. (15, 16).

Besides the intracrystalline porosity, the number and strength of the acid sites and the EFAL/FAL ratio, the porosity of the binder/zeolite composite have to be optimized, too.

### *7.3.2 Omega based catalysts*

Laboratory studies have shown that omega (MAZ structure type) based paraffin hydroisomerization catalyst shows higher activity than mordenite based catalyst and better selectivity, i.e. higher octane due to higher yield of di-branched paraffins compared to mordenite performance (17). The isomerization of a C5/C6 cut at 15 bar results in a final calculated RON of 80.4 for the alumina bound dealuminated PtH-MOR catalyst supplied by IFP with undisclosed (most likely similar) Si/Al ratio, measured at 265 °C compared to a RON value of 80.9 for an alumina bound dealuminated PtH-MAZ catalyst with bulk Si/Al = 16, measured at 250 °C. Both measurements were performed in a bench-scale tubular reactor with a volume of 50 $cm^3$ of 2 mm diameter extrudates with WHSV of 1.5 $h^{-1}$ and $H_2$/HC of 4. This higher yield is reported to be due to both the unique structural properties of the MAZ structure type and the higher acid strength of its Brönsted sites.

## 7.4 Metal oxide catalyst systems

In contrast to zeolite based catalysts, chlorinated alumina and sulphated zirconia are sensitive to feedstock impurities but show higher activity, and hence can be used at lower temperatures, where the thermodynamic conditions are more favourable (Figures 7.2 and 7.3). Although, with respect to the thermodynamic equilibrium, the chlorinated alumina is superior to the zirconia based catalyst, the latter has the big advantage of being regenerable. Figure 7.4 shows the results of micro reactor tests of the newly developed HYSOPAR-SA catalyst, a platinum promoted metal oxide catalyst, compared to the zeolitic HYSOPAR catalyst. These data demonstrate the higher activity of the metal oxide catalyst. But it is also obvious that during the conditioning phase the gas make of the metal oxide catalyst is much higher than in the case of the mordenite based catalyst. This disadvantage needs to be compensated for by a careful start-up procedure and proper temperature control. HYSOPAR-SA was developed so as to overcome the intrinsic disadvantages of MO catalysts , i.e. extreme sensitivity to feedstock poisons, in particular water. As a matter of fact, this catalyst can tolerate 50 ppm of water which is far superior to competitive catalysts.

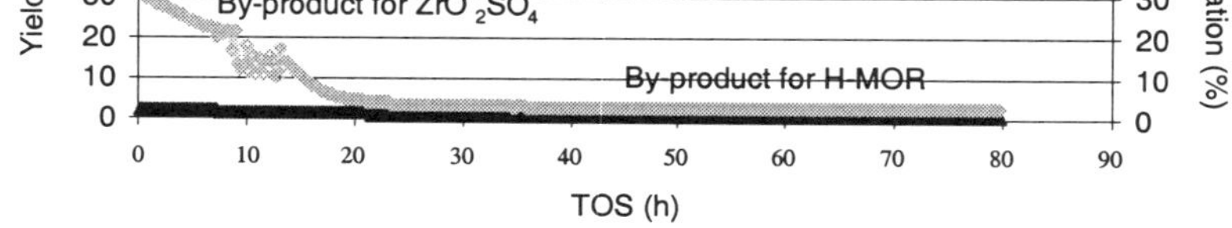


**Fig. 7.4** Comparison of Mordenite (250°C) vs. Sulfated Zirconia (200°C) based Isomerization Catalysts; Time on stream
20 bar; H2/HC = 1:1; feed = pure n-pentane; WHSV = 2 l/h (20.71 ml H2/min; 0.107 ml n-pentane/min); 2 g catalyst (sieve fraction of crushed material 0.5 – 1 mm)

## 7.5 CKS ISOM and HYSOPAR commercial performance at high sulfur conditions

The requirements for a new isomerization technology were identified to be (i) tolerance to feed impurities, (ii) robustness, (iii) low capital cost and low permanent operating costs. Reviewing the existing isomerization processes, based on chlorinated alumina catalysts, the following features of an improved catalyst were identified that could lead to a process meeting the company's objectives:

- Sulfur tolerance: eliminate or reduce feed desulfurization expense
- Water tolerance: eliminate feed and hydrogen dryers
- Higher tolerance for other feed impurities: system more forgiving to changes in feed quality
- Regenerable catalyst: reduce catalyst replacement costs and increase catalyst life
- Eliminate continuous catalyst promoter addition: lower cost of construction materials and elimination of effluent treatment systems

The HYSOPAR catalyst has been developed by CEPSA on a laboratory scale. The modification of the laboratory procedure to allow commercial scale production was successfully completed by SÜD-CHEMIE who exclusively manufacture this catalyst. HYSOPAR is a strongly acidic mordenite based catalyst. Its most striking feature is the extreme tolerance for sulfur and other feed contaminants.

The HYSOPAR catalyst shows steady and high performance even if sulfur is present in the feed up to 100 ppmw. In all cases with less than 100 ppmw sulfur in the feedstock sulfur pre-treatment is not required. For feeds having a sulfur content greater than 100 ppmw, a regenerable caustic extraction process can be employed, which is significantly lower in capital and operating cost than a hydrotreater. Moreover, the reduction of the catalyst activity due to high sulfur (up to 200 ppmw) in the feed is only temporary and can be recovered by simply passing a lower sulfur feed over the catalyst. Furthermore, activity reduction due to a more significant sulfur breakthrough can also be recovered by reversibly regenerating the catalyst.

HYSOPAR's superior performance under extreme conditions has been proven in CEPSA's 8000 BPSD isomerization unit at Algeciras, Spain, which is based on CKS ISOM C5 recycle process technology (2, 3).

## 7.6 Benzene saturation with a zeolitic hydroisomerization catalyst at high sulfur conditions

As mentioned in part 7.2, reporting on the Shell Hysomer Process, the aromatics in the feed are quantitatively hydrogenated to the corresponding naphtenes and present no problems to the catalyst selectivity, even in concentrations of up to 2 wt.-% (4). The commercial performance of HYSOPAR includes experience with up to 5 wt.-% of benzene in the isomerization feed (4).

The effect of even higher benzene levels has been studied on a demonstration plant scale: a feed of 15 wt.-% benzene and 85 wt.-% n-C5 was completely saturated at a liquid hourly space velocity of 4 $h^{-1}$ and a pressure of 30 bar at a weight average bed temperature (WABT) as low as 120°C (3).

To study the benzene saturation under high sulfur conditions, the same feed, but spiked with 30 ppmw of sulfur (as ethyl mercaptan), has been used (4). Sulfur in the effluent was analyzed to calculate the hydrodesulfurization activity.

The HDS reaction for a feed with the composition: 15% Bz + 85% n-C5 + 30 ppmw sulfur is already completed at about 170°C, as shown in Figure 7.5. This shows that HYSOPAR is an efficient hydrodesulfurization (HDS) catalyst under these conditions despite the effect of sulfur on benzene hydrogenation activity. To complete the benzene conversion the temperature needs to be increased from 225-235°C to 250-260°C with 30 ppmw S in the feed. At these high temperatures, thermodynamic equilibrium between $nC_5$ and iC5 was practically reached, indicating that the impact of S on isomerization is not pronounced. Some ring cleavage of cyclic compounds was also observed, accompanied by an increase in gas production. The results are summarized in table 7.7.

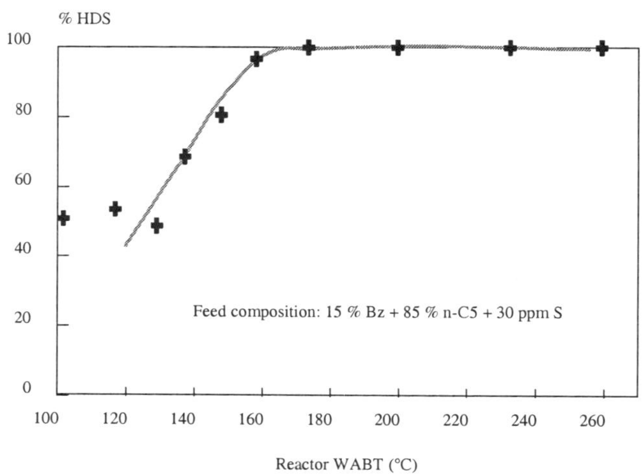


**Fig. 7.5** Hydrodesulfurization (HDS) Activity (18)

## 7.7 Conclusions

Hydroisomerization is one of the few major refinery processes that allow refineries to cope with the future fuel regulations on the one side and the necessity to supply premium fuel with the necessary octane on the other side. Due to the limited volume the chemical industry can cope with in addition to the present level, future reduction of the aromatics in fuels will force the refineries to convert as much of the aromatics as possible to fuel components. One possible option is to feed the one-ring aromatics such as benzene to an isomerization unit. A state of the art hydroisomerization catalyst such as HYSOPAR is very active for benzene hydrogenation at temperatures as low as 100°C, where 100% hydrogenation is achieved, and can cope with up to 15 wt.-% of benzene in the feed. When sulfur in the range of 50 ppm is present in the feed, a partial inhibition of

**Table 7.7** C5 Isomerization activity and Gas & iso-C6 Yield [18]
Conditions: T=250 °C, Total pressure = 30 bar, LHSV = 1 $h^{-1}$

| Feed Composition | Activity i-C5/C5 total | Yield Gas wt.-% | Yield Cyclic i-C6 wt.-% | Yield Paraffinic i-C6 wt.-% |
|---|---|---|---|---|
| n-C5 | 65+-3 | 1.7+-0.2 | | |
| n-C5 + 30 ppmw S | 65+-3 | 1.7+-0.2 | | |
| 85 wt.-% n-C5 +15 wt.-% Benzene | 55+-3 | 5+-2 | 6+-2 | 7+-2 |
| 85 wt.-% n-C5 +15 wt.-% Benzene + 30 ppmw S | 55+-3 | 4+-2 | 5+-2 | 6+-2 |

benzene hydrogenation may occur, which can be attributed to sulfur adsorption on the catalyst surface. This sulfur retention produces a reversible deactivation of the catalyst in regard to its hydrogenation function, while the isomerization reaction is affected to a lesser extent. Increasing the reaction temperature and/or processing cleaner feedstocks can compensate for both effects.

The new developments in catalyst improvements and in process optimization have brought new opportunities to directly process virgin naphtha from atmospheric distillation along with benzene and benzene precursors extracted from reformer feeds. This can eliminate the need for a separate naphtha hydrotreater as well as separate downstream processing of benzene such as hydrogenation with noble metal or nickel catalysts. This means considerable capital savings and lower operating costs. Furthermore, this can not only help to resolve the benzene problem but also provides an attractive tool in compensating for benzene losses from removing benzene from the gasoline pool, even under unfavourable conditions in terms of feedstock impurities like sulfur.

## References


1. Köhler E, private communication.
2. Floyd F.M., Perez Pascual M. and Köhler E., *PETROTECH 1998 Conference Proceedings*, **1, 35**, Bahrain.
3. Floyd F.M., Gilbert M.F., Pérez Pascual M. and Köhler E., *Hydrocarbon Engineering*, (September 1998) 1.
4. Köhler, E., Perez Pascual, M., Floyd, F.M. and Gilbert, M.F., AIChE 1999 Spring National Meeting, March 14-18, 1999, Houston, Session *Recent Advances in Isomerisation* (Group T5).
5. Lanewala M.A., Pickert P.E. and Bolton A.P., *J. Catal.* **9** (1967) 95.
6. Kouwenhoven H.W., *Adv.Chem.Series* **121**, (1973) 529.
7. Kouwenhoven H.W. and Van Zijll Langhout W.C., *Chem. Eng. Prog.* **67(4)** (1971) 65.
8. Guisnet M. and Fouche V., *Applied Catalysis*, **71** (1991) 307.
9. Corma A., Martinez A., in *Catalytic Activation and Functionalisation of Light alkanes: Advantages and Challenges*, Eds. Derouanne E.G. et al., (Kluwer Academic, Dortracht, 1988) 35.
10. Tromp M. et al., *J. Catal.* **190** (2000) 209.
11. Meima G. R. , *CATTECH* ***2*** **(1)** (1979) 411.
12. van Donk S. et al., to be published in *Stud. Surf. Sci. Catal* (2002)
13. Burgfels G. and Schmidt, F., in *Science and Technology in Catalysis 1994* (Kodansha Ltd, 1995) 465.
14. Karge H.G. and Weitkamp J., *Chem.-Ing.-Tech.* **58** (1986) 946.
15. Corma A., Frontela J., Labaro J. and Perez M., *Proceedings of the "Symposium on Alkylation, Aromatization, Oligomerization and Isomerisation of short Chain Hydrocarbons over Heterogeneous Catalysts"* , (Prepr. – Am. Chem. Soc., Div. Pet. Chem., 1991) 36 (4), 833.
16. Labaro J., Frontela J. and Corma A., *EP 0 409 303 A1* (1990).
17. McQueen D. et al., *J. Catalysis*, **161** (1996) 587.
18. Perez Pascual M. A., CEPSA, private communication

CHAPTER 8

# BASE OIL PRODUCTION AND PROCESSING

M. DAAGE
*ExxonMobil Process Research, P.O. Box 2226*
*Baton Rouge, LA 70810, USA*

## 8.1 Introduction

With the development of heavy machinery and internal combustion engines, the production and manufacturing of lubricant base oils became essential. In the early days of the petroleum industry, lubricant base oils were obtained by acid treating and wax settling of Pennsylvania crude oils. In response to increasing demand, many processes were developed to upgrade the quality of less desirable feedstocks: conventional processes include hydrogenation and solvent extraction. It is important to understand that today many different feedstocks are used and that their processing involves multiple consecutive steps, leading to a wide variety of process combinations, tuned for the production of a multiplicity of lubricant grades. This paper provides background information for a non-specialist and will review selected examples of base oils processing, and characteristics of the many available options currently implemented in the industry.

## 8.2 Base oils - characteristics and properties

### *8.2.1 Base oil terminology*

Lubricants are high value products, generally formulated for a broad variety of uses. Typically, lubricants are used in the automotive industry and other industrial applications, including medicinal uses. More specific uses include engine oils, Automatic Transmission Fluids (ATF), machine oils, greases, turbine oils, electrical oils, drilling fluids, food grade oils and white oils. Refineries produce the base oils or basestocks by refining crude oils or fractions thereof. Finished products are blends of basestocks with or without additives. The additives are added to correct or enhance key properties of the lubricant for a specific use. It may include addition of anti-oxidants, anti-wear components, detergents and the like. The additives packages can either be bought from a third party manufacturer or be a proprietary blend of additives. Basestocks may be called by various names and the most common include Neutral basestocks, Bright stocks and Grades. Generally, Neutral names are referred to by their viscosity at 100°C, such as 100N, 150N or 600N for example. Bright stocks are heavy lubes made by the processing of vacuum resids. Their names refer to their appearance and their typical viscosity is 2,500 SUS at 40°C. Grades refer either to their viscosity or trademarks: for example SAE 5, 10, 30 or ISO 22, 32,... A more comprehensive glossary of acronyms and terms specific to base oils has been published elsewhere (1).

### *8.2.2 Basestocks - properties and definitions*

The basestock molecular composition and its physical properties determine the performance of the finished products. Of particular importance are the Viscosity Index or VI, the saturate content and the wax content. High VI is important because it improves volatility, fuel economy and operating range of automobiles. Higher saturate content provides superior oxidation stability and better soot handling. Wax content is a critical parameter in the composition of a basestock because it impacts the low temperature performance of the finished lubricant: the lower the wax content, the lower the pour point. Typically, basestocks are categorized by references to their composition. Table 8.1 summarizes the categories according to the American Petroleum Institute (API).

**Table 8.1**

| API Group | Saturates | Sulfur | VI | Manufacturing process |
|---|---|---|---|---|
| I | < 90 wt% | > 0.03 wt% | 80 - 119 | Solvent extraction |
| II | > 90 wt% | < 0.03 wt% | 80 - 119 | Hydroprocessing |
| III | > 90 wt% | < 0.03 wt% | > 120 | Wax isomerization, H/C GTL |
| IV | n.a. | n.a. | | Polyalphaolefins (PAO) |
| V | | | | All other basestocks |

Due to an increasing demand in performance of lubricants, characterizing a basestock is often complex and requires the measurement of many chemical and physical properties. Viscosity, which measures the fluidity of the basestock ranges from 2 to 20 cSt @ 100°C for Neutrals and can be as high as 32cSt for Bright stocks. The most common method used is Brookfield viscosity. The viscosity Index or VI represents the inverse measure of the change of viscosity with temperature. The VI ranges from 85 to about 105 for conventional basestocks and can be as high as 150 for specialty grades such as Exxsyn, PAO or XHVI for example. The pour point and the cloud point are tied to the low temperature properties of the base oil. The pour point is defined as the temperature at which the basestock becomes nearly solid: it ranges typically from -9 to -24 °C. The cloud point is the temperature at which wax crystals appear. The cloud point is higher than the pour point and often the difference in temperature is referred to as the pour/cloud spread. Acceptable spreads are generally less than 15°C. When higher spreads are observed, haze problems are encountered. For example, Bright stocks are often further processed to meet the cloud point requirements. Volatility is another critical property, which needs to be met. Noack volatility measures the actual evaporation of the basestock at 250°C under flowing air for a period of 1 hour, typically 20-35 wt%. Alternatively, GCD (Gas Chromatographic Distillation) may be used to measure volatility by utilizing the front end of the boiling curve (e.g. 10% at 375°C). Other properties often measured include color and its changes by exposure to light, the residue left on ignition or Con Carbon. Finally, the molecular composition is often reported as the content of saturates (paraffins and naphthenes), aromatics and asphaltenes. In general, higher quality basestocks result from lower viscosity, higher VI, lower volatility, lower heteroatom content and higher saturate content.

### *8.2.3 Base oil feedstocks and upgrading chemistry*

Feedstocks, appropriate for base oil processing, are generally derived from crude oils by vacuum distillation (VPS) of the atmospheric resid. The vacuum resid contains the heaviest material from which Bright stocks are recovered. Often several distillate fractions are recovered from the VPS and further processed, most often in batch operations. Typically, the lube distillate fractions are characterized by their boiling range: 330 to 385°C for light lube, 385 to 440 °C for medium lube, 440 to 565 °C for heavy lube and above 565 °C for vacuum resid. If any crude may be used, preferred crudes have high paraffin content and high wax content (W. Texas, Arab Light for example) or high naphthene content and low wax content (S. Louisiana, for example).

Lube distillate fractions have complex molecular compositions, due to the high molecular weight and high boiling points of its components. Typically, lube distillates are complex mixtures of hydrocarbons having at least 20 carbons. Chemical compounds typically found in such distillates include paraffins, naphthenes, multi-ring aromatics and alkylaromatics as well as sulfur and nitrogen containing derivatives. A direct consequence of this complexity is the recognition that each lube distillate fraction will require specific upgrading schemes and strategies to maximize the yield and quality of the base oil. The process steps are devised to eliminate the condensed aromatics and heteroatom containing compounds, increase the hydrogen content, convert wax to isoparaffins and preserve isoparaffins.

### 8.2.4 Base oil chemistry and processing steps

Because of the wide range of molecular composition, many types of conversions may be used for upgrading the performance and the quality of the base oils, resulting in a large variety of process configurations. Typical processing stages are based on hydroprocessing or solvent extractions.

Deasphalting is an extractive-precipitation process generally used for treating the heavier fractions, such as atmospheric and vacuum residua. The purpose of the deasphalating process is to remove asphaltenes and contaminant metals such as vanadium and nickel. Deasphalting improves color, stability, viscosity and hydrogen content. It reduces Con Carbon, aromatic content and heteroatom content, hence facilitating further hydroprocessing. Commonly, propane and supercritical light hydrocarbons up to n-heptane are used for deasphalting. Modern deasphalating units comprise baffle towers and rotating discs coupled to a solvent recovery train (2, 3).

Solvent extractions are also extensively used for removing aromatics from lighter fractions. The importance of many solvents used in the past, such as liquid $SO_2$, aniline, phenol, nitrobenzene and Chlorex, has dwindled, since they cannot fulfill current quality, selectivity and environmental needs. Modern liquid-liquid extraction processes use furfural and N-MethylPyrrolidone (NMP). Typical NMP based processes include EXOL N Extraction (4) and Texaco MP Refining Process (5). Solvent extraction allows the recovery of a raffinate, the properties of which have been significantly improved: raffinates have lower aromatic content, lower heteroatom content, lower viscosity, higher wax content, higher hydrogen content and better color. Typical designs include rotating disc contactors and weir trayed towers (2, 6).

Hydrocracking is another process that can be used advantageously. It is a more severe process and is most often used for treating low quality crude oils. The purpose of hydrocracking is to remove heretoatoms and aromatics components by a wide variety of chemical reactions, which include hydrogenation, hydrodesul-furization, hydrodenitrogenation and dealkylation. The severity at which the process is operated is often driven by the VI improvement requirements. The catalysts are metal sulfides (CoMo, NiMo, NiW,...) supported on an acidic support such as amorphous silica-alumina or large pore zeolites (HY for example). Another benefit of hydrocracking is to increase the concentration of paraffins, which are formed by the acid dealkylation.

However, the high severity of hydrocracking is detrimental to the stability of the base oil and subsequent hydrofinishing is often needed to stabilize the hydrocracked base oil. Hydrocracking may also be used advantageously for treating waxes, for example slack wax (7) and Fischer-Tropsch wax. In this particular use, the purpose of the process is centered on controlling the molecular weight conversion while obtaining a high level of isomerization. Noble metal catalysts supported on amorphous acidic support may be used in these applications because of the very low concentration of heteoatoms such as sulfur.

Hydroprocessing is a central technology in base oil processing. In contrast to hydrocracking, the purpose of hydroprocessing is to increase the quality of the base oil by hydrogenating the aromatic compounds, while eliminating the heteroatoms such as sulfur and nitrogen. This approach is beneficial to high yield since boiling conversion is minimized. Typical processes utilize trickled bed reactors and the catalysts are sulfide metal, generally supported on low acidity support such as alumina. Depending upon the heteroatom content of the feed, noble metal catalysts such as Pt, Pd or Pt/Pd catalyst may be used in an hydrofinishing step, designed to stabilize the base oil by hydrogenation of olefin and residual aromatics. A typical application is the hydrogenation of white oils for which strict requirements are imposed on the aromatic content for medicinal use.

Hydroisomerization is also a key process. In this process, linear paraffins are converted to isoparaffins. This reaction greatly improves the pour point of the base oil, but results in a loss in VI. The catalyst is often noble metal supported on a controlled acidity support. The catalyst formulations are often proprietary and may utilize an amorphous silica-alumina or a modified molecular sieve.

Dewaxing is the most important process involved in the manufacturing of base oil. This step is critical for meeting the cold flow requirements. There are different options that may be used. Dewaxing might be achieved by solvent extraction/precipitation of the wax. The precipitation is generally affected at low temperature and the use of chilling equipment is common. The solvent may vary, but Methyl Ethyl Ketone (MEK) and Methyl IsoButyl Ketone (MIBK) are widely used. The Texaco Solvent Dewaxing Process and the Exxon Dillchill™ Dewaxing process are examples of solvent dewaxing (8). However, if solvent dewaxing is effective for improving the cold flow property, it does reduce the yield of finished base oil. More advanced processes, based on a catalytic conversion of n-paraffins, are more suitable to improved yield. The catalytic conversion of n-paraffins is controlled by the use of zeolites or molecular sieves, which pore sizes and structure are tuned for carrying shape selective cracking and hydroisomerization. Typical processes that have been successfully commercialized are the Mobil Lube Dewaxing Process (MLDW) (9), the Chevron Isodewaxing Process (10) and more recently Mobil Selective DeWaxing (11).

## 8.3 Base oil production

The production of base oils is highly dependent upon the quality requirements for the finished base oil. Group I base oils, which have limited requirements in hydrogen and heteroatom content, have traditionally been produced by a combination of solvent extraction and solvent dewaxing. Group II and III base oils require a more severe and selective processing and the 'state-of-the-art' lube plant is all catalytic. In between these two options, a large range of variations may be implemented, leading to hybrid lube plants. In this section we will review a conventional lube plant, a state-of-the-art plant and, finally, an example of an hybrid plant.

### *8.3.1 Conventional lube manufacturing plants*

As mentioned above, the production and finishing of base oils can be combined and coordinated in different ways. Traditionally, base oil manufacturing plants consist of distillation, deasphalting, solvent extraction and solvent dewaxing. A typical block scheme of such a plant is illustrated in Fig. 8.1. Alternatively, a hydrofinishing unit or a clay treatment may be added after the dewaxing to further improve the stability and color of the base oil. Not shown here for purpose of simplicity, are the blocks for solvent recovery and recycling of the plant. Another important aspect of a base oil manufacturing plant, is that the process is operated in blocked operations, i.e. relatively narrow distillation cuts (about 50 to 100°C) are processed individually. The blocked operation is preferred to the processing of a wide cut because it allows better control of the quality and yield of base oils.

The performance of solvent extraction, particularly the yield of base oil extracted, is highly dependent upon the design of the extractor used, as well as the type of solvent. For example, Mobil obtained good lube production rates by redesigning the internals of the extractors and by better control of the temperature profile (12). Solvent Dewaxing Units (SDU) are strongly affected by the solvent composition and more importantly by the solvent utilization. Many units are limited by solvent recovery. Typically, the raffinate from the aromatic extraction unit is mixed with a primary solvent and the resulting mixture is chilled by heat exchange with a refrigerant stream. The desired temperature is achieved by addition of cold solvent before separation by filtration in a rotary drum. The filtrate is then sent to a solvent recovery section, where the solvent is separated from the dewaxed oil. Both refrigeration and operating constraints affect the number of rotary filters required for production, particularly if flash towers are used in the solvent recovery section. Significant improvements have been achieved by the incorporation of a membrane separation system (Max-Dewax), which allows a partial bypassing of the solvent recovery system (13). The proprietary polyimide membrane removes the MEK solvent from the dewaxed oil filtrate at or near dewaxing temperatures. The solvent may be recycled to the dewaxing process without the need for additional cooling. The availability of this additional cold solvent allows a lower feed viscosity to the filter, thereby increasing the rate of filtration.

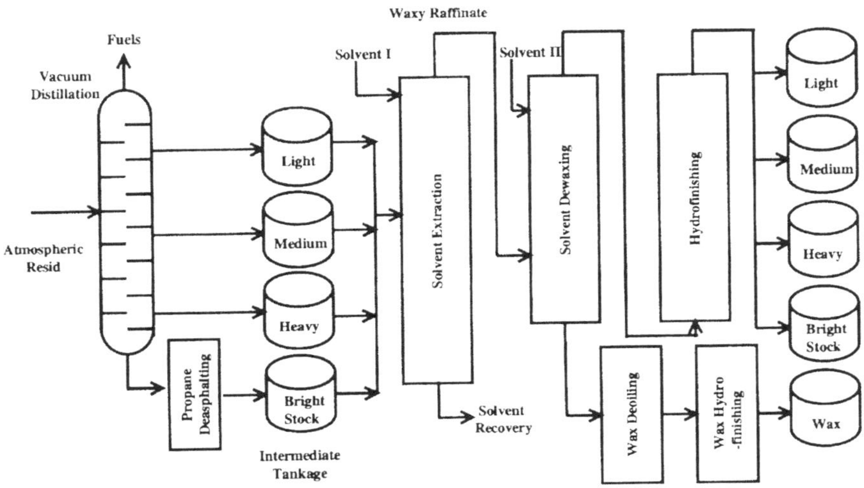


**Fig. 8.1** Conventional lube plant

Conventional lube manufacturing plants, based on solvent extraction and solvent dewaxing, are still widely used for producing Group I base oils. However, group II and III require further and more advanced upgrading processing steps to meet higher quality.

### *8.3.2 'All Catalytic' lube plant: state-of-the-art*

Faced with increasing pressures to improve base oil quality (groups II and III), new plant configurations are needed as an alternative to conventional solvent processing plants. Even though various hydroprocessing options may be used, catalytic hydrocracking provides a preferred mean for improving feedstocks and refined products.

The state of the art, today, is based on an 'all catalytic' lube plant, which does not rely on solvent processing. An example of such a plant is Mobil's Jurong plant located in Singapore (11). The configuration of the plant relies on the use of a lube hydrocracker, coupled with a selective catalytic dewaxing unit (Fig. 8.2).

The hydrocracking unit (LHDC) is designed to convert vacuum gas oils (VGO). The hydrocracking section provides an enhanced flexibility to crude selection and converts undesirable aromatics into higher VI compounds with improved oxidation stability. The proprietary hydrocracking catalyst is operated at high partial pressure of hydrogen and useful byproducts include naphtha and high quality kerosene and distillate.

The waxy hydrocracked bottoms, which have a high wax content, are then processed in an isomerization/dewaxing unit to convert high molecular weight, low pour point paraffins into low pour point high VI isoparaffins. Mobil Selective DeWaxing (MSDW) utilizes a proprietary zeolite formulation, which was first commercially applied in Jurong. The MSDW reaction chemistry is based on classic bifunctional catalysis. Linear paraffins are dehydrogenated on the metal function and isomerized on the acid function. The skeletal isomerization is controlled by the pore geometry of the zeolitic component. The yield of lighter hydrocarbons generated by strong acid functions has been reduced and some aromatic hydrogenation is also obtained on the strong metal function. The catalyst has a slow deactivation and a high tolerance to sulfur and nitrogen. Controlling the MSDW catalyst for making on-specification base oil is achieved with minor temperature adjustments

Mobil selected the "all catalytic" plant because of its low capital and manufacturing costs relative to a conventional solvent processing plant. The plant flexibility to produce multiple grades of base oil is achieved by controlling the VI of the hydrocrackate and by using the MSDW catalyst which can operate in the presence of sulfur and nitrogen contaminants. Group II and group III base oils have been produced and their quality exceeds that of base oils produced by solvent processing. Further information on the interdependence between the hydrocracking section and the dewaxing section can be found in the literature (11, 14).

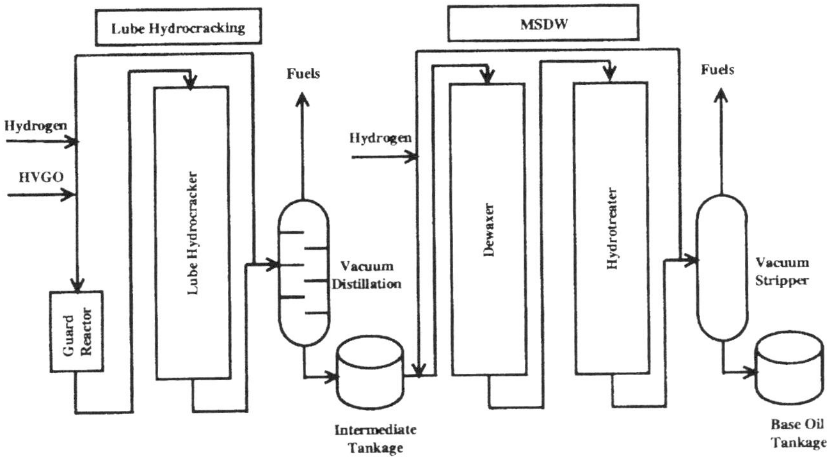


**Fig. 8.2** 'All Catalytic' lube plant

### *8.3.3 Hybrid lube plants*

Depending upon the specific needs, conventional plants may be converted to hybrid plants to improve capacity and/or product quality. An example of a hybrid configuration is used by Shell to manufacture their XHVI. In this particular case, the plant couples a lube hydrocracker with a solvent dewaxing unit. There are several modifications or improvements of known technologies that can be used as a 'Mix and Match to Needs' to upgrade existing plants.

Technologies or technological improvements that are available today include:

- Integration with a fuels hydrocracker
- Improved solvent dewaxing: use of solvent mixtures
- Membrane assisted solvent processing (13)
- Raffinate hydroconversion
- Lube hydroisomerization
- Advanced catalytic dewaxing (14)
- Wax hydrofining
- Improved solvent extraction

Overall, the wide multiplicity of plants, designs and configurations is characteristic of the variety of feedstocks, products and to the complexity of the sequential processing need to produce base oils.

## 8.4 Raffinate HydroConversion (RHC)

Raffinate HydroConversion (RHC) corresponds to a severe hydroprocessing, which is applied only to selected stocks, as opposed to the treatment of the entire slate. The purpose of the hydroconversion, which is carried on a proprietary sulfided catalyst, is to increase the VI while preserving the yield of lubes and wax. RHC is very effective at removing heteroatoms such as sulfur and nitrogen and at hydrogenating aromatics.

Thermal diffusion separation can be used to characterize the differences in composition and quality obtained by RHC. Fig. 8.3 is a schematic description of the thermal diffusion separation apparatus. In thermal diffusion separation, two concentric tubular walls are separated by a small gap and each wall is maintained at a different temperature. The difference in temperature creates thermal convection current, which allows the separation of the liquid consistent by their density. At equilibrium, the low density molecules are concentrated at the top of the separation device and the high density molecules accumulate at the bottom. Using a collection of 10 ports distributed along the length of the external tube, samples are collected and analyzed for VI and composition. As expected, low density molecules are essentially paraffinic whereas high density molecules are more aromatic in nature.

As illustrated in Fig. 8.4, RHC significantly increases the VI, particularly for ports 7 to 10. This is characteristic of a lower concentration of multi-ring aromatic components, at least a portion of which have been converted by hydrogenation to naphthenes. Furthermore, in addition to the increase in VI, the RHC treated oil has a lower Noack volatility and a similar pour point.

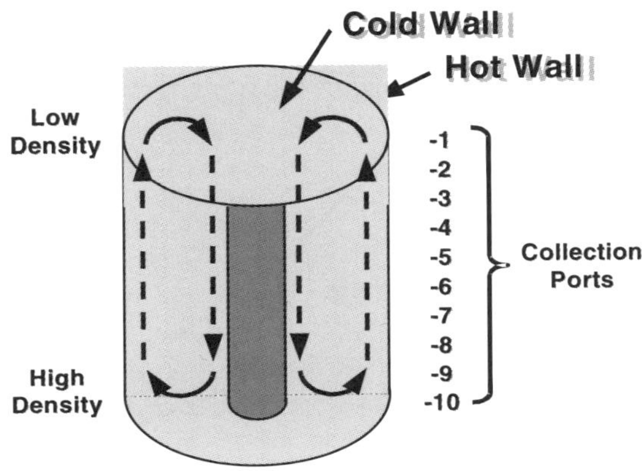


**Figure 8.3** Thermal diffusion separation

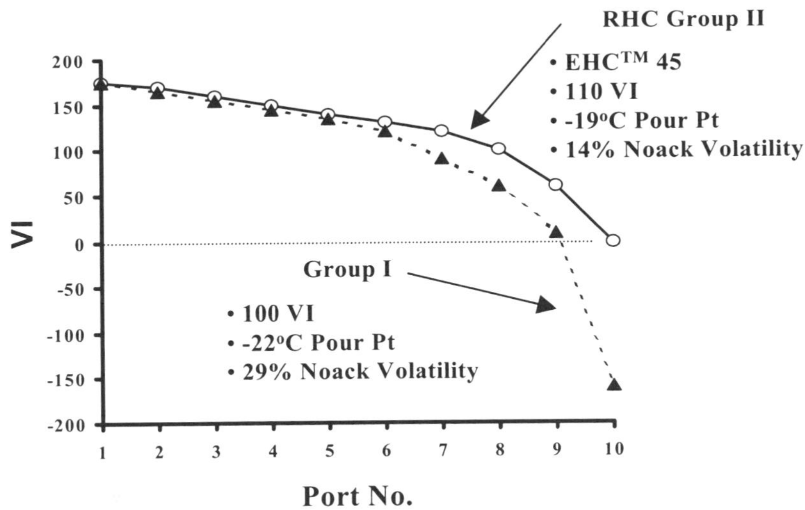


**Fig 8.4** Comparison of RHC Group II with a conventional Group I

Another advantage of RHC, is that it provides the flexibility to tailor the product by adjusting the severity of the processing. Fig. 8.5 illustrates the ability of RHC to decrease both Noack volatility and viscosity for a given feedstock.

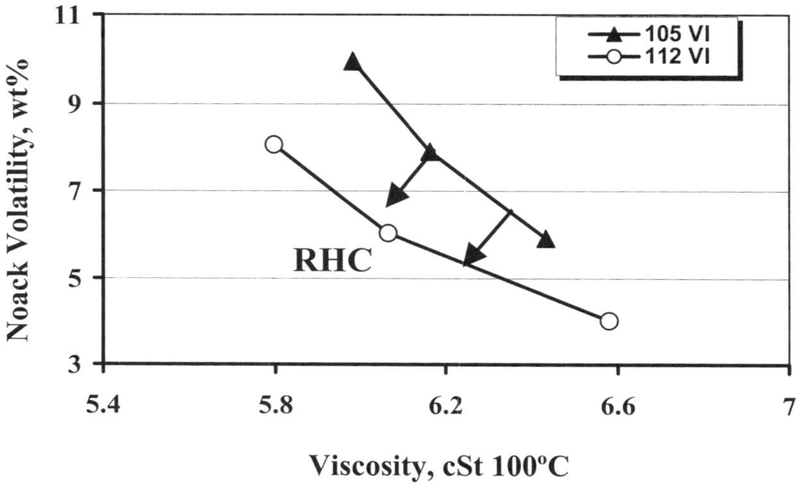


**Fig. 8.5** Impact of RHC on viscosity and noack volatility

## 8.5 Selective catalytic dewaxing

As mentioned earlier, base oils are required to meet stringent cold flow properties, such as pour point and cloud point. This is achieved by dewaxing of the base oil, i.e. by removing the high molecular weight linear paraffins as done by solvent dewaxing. More recent approaches have been centered on the use of catalysts, particularly the use of zeolitic catalytic materials.

### *8.5.1 Shape selectivity in catalytic dewaxing*

Catalytic dewaxing has been historically tied to the discovery of medium pore zeolites, more specifically 10 ring zeolites. Examples of 10 ring zeolites include ZSM-5, ZSM-11, ZSM-23. Other microporous materials such as SAPO-11 have also been used. The formulation of dewaxing catalysts is complex and relies on the control of at least three functions. One key function is provided by a catalytic metal, the role of which is to provide an hydrogenation/dehydrogenation function for paraffins activation. Once an olefin is formed, the conversion occurs by means of an acidic function, provided by the zeolite component. The strength of the acid function is controlled by the structural properties of the zeolite, while the acid site concentration is adjusted by dealumination or by substitution of the framework aluminum. A critical aspect of dewaxing is the need to selectively convert linear paraffins while preserving the isoparaffins, which have good VI and low pour point. High selectivity for dewaxing is achieved by controlling the diffusion and adsorption in microcrystalline materials and has led to the development of shape selective catalysts. Figure 8.6 gives a schematic representation of the shape selectivity concept.

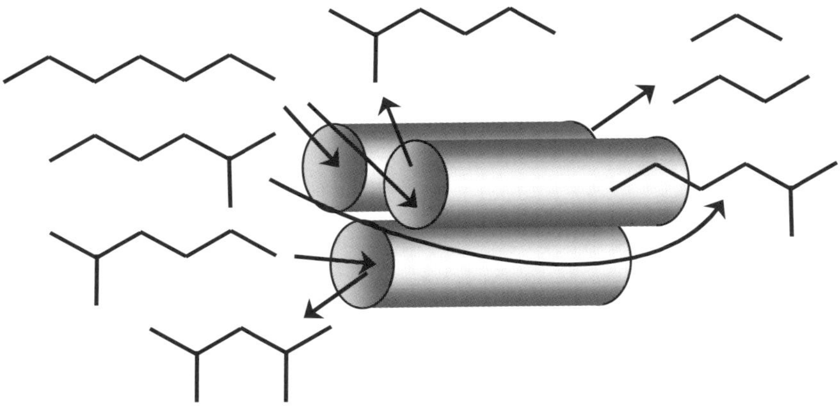

**Fig. 8.6** Shape selective reactions of paraffins

The concept of shape selectivity relies on the control of the adsorption and diffusion of the molecule in the pore of the catalytic material. Linear paraffins have the smallest critical diameter among hydrocarbons and can therefore penetrate smaller pore structures. In the figure, for example, the n-heptane molecule penetrates the tubular pore. Upon reaction on an active site, it may desorb as an isomer, such as 2-methylhexane, or crack to form propane and butane. In contrast with the n-heptane, the 2-methylhexane molecule may not penetrate the pore when the linear branch of the molecule does not line-up with the pore, and therefore may not react, due to the larger critical diameter of the molecule near the tertiary carbon. If the linear branch of the molecule is long enough, it may line up with the pore, and react to form a dibranched molecule. Once both ends of the hydrocarbons are branched, the molecule cannot penetrate the pore and remains unchanged. There are many variations and interpretations of such molecular 'traffic control' by use of specific pore structures. For longer paraffins, a mechanism based on a 'key-lock' adsorption involving neighboring pores has also been proposed (15).

### *8.5.2 Using zeolites for selective dewaxing*

The first evidence of a shape selective catalyst was discovered in the selective cracking of linear paraffins from a mixture of hydrocarbons and led to the commercialization of the Selectoforming process (16). Erionite was an effective catalyst for the conversion of small paraffins, but its use was limited to $C_{11}$ and lower paraffins. Unlike erionite, the rate of cracking on medium pore zeolites increases with the carbon number of the molecule and decreases with the bulkiness of the molecule (17). On large pore zeolites, the hydrocracking of long chain paraffins becomes independent of the chain length and multibranched paraffins are produced using an Y zeolite (18, 19, 20). Medium pore zeolites are characterized by their 10 ring structure as opposed to small pore zeolites (8 ring) and large pore zeolites (12 ring). Small pore zeolites, such as erionite and zeolite A, are used for the adsorption of paraffins or the conversion of small paraffins. Large pore zeolites, such as faujasite, zeolite L and mordenite for example, are used for the conversion of heavy aromatics. Medium pore zeolites are the zeolites of choice for the selective conversion of linear paraffins and include ZSM-5 and ferrierite for example. Other molecular sieves having similar pore dimensions, such as SAPO-11, may also be used. The shape and diameter of the pore in medium zeolite can vary depending on the framework structure, but generally has a dimension comprised of between 4.5 and 6 angstroms. The structure of the zeolite pores may be unidimensional (1-D) to intersecting three-dimensional (3-D) and even have dual pore systems. Table 8.2 and Fig. 8.7 give typical examples of different medium pore zeolite structures of interest.

**Table 8.2** Typical structure of medium pore zeolites useful for dewaxing

| Pore System | 1-Dimensional | 2-Dimensional | 3-Dimensional |
|---|---|---|---|
| 10-R | ZSM-22 (4.4x5.5)<br>ZSM-23 (4.5x5.2)<br>SAPO-11 (3.9x6.3) | NU-87 (4.7x6.0) | ZSM-5 (5.3x5.6)<br>ZSM-11 (5.3x5.4) |
| 10-R x 8-R | | Ferrierite (4.2x5.4)<br>ZSM-57 (5.1x5.4) | Wenkite (2.6x4.9) |

Pore dimensions in Angstrom in parentheses (21)

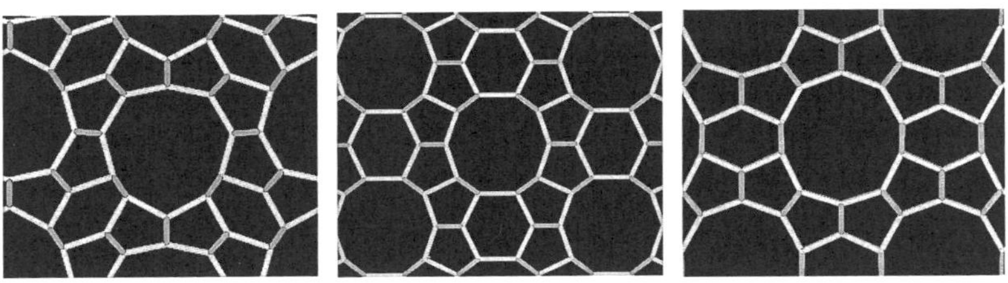

**ZSM-23 (MTT)** **ZSM-35 (FER)** **ZSM-5 (MFI)**

**Fig. 8.7** Examples of pore geometry in medium pore zeolites

The shape of the 10-R pore may be round, oval or even 'teardrop'. Subtle effects on the shape selectivity catalysis could be associated with the uniformity of the pore, its shape and its connectivity to other pores. The later effect, in particular, is important for two- and three-dimensional zeolites, where the intersection of the pore leads to the formation of pockets or cages of slightly larger dimension.

### *8.5.3 Identification and testing of selective dewaxing catalysts*

The first generation of dewaxing catalyst relied on the selective cracking of linear paraffins (Fig. 8.8).

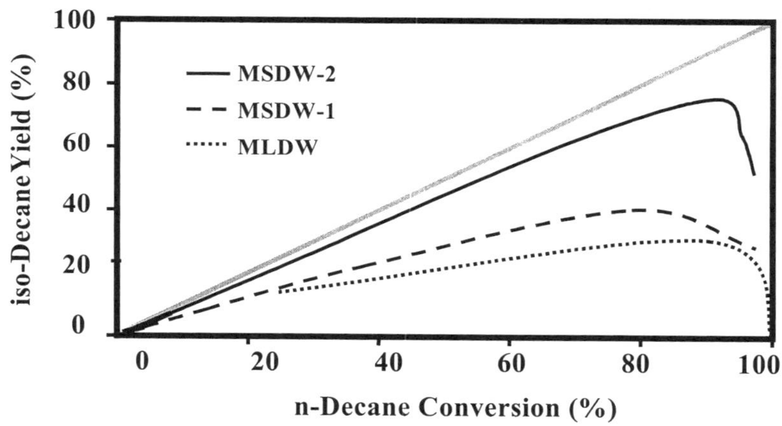


**Fig. 8.8** Selective Hydroisomerization of n-Decane (22)

The MLDW catalyst is characteristic of the early generation of dewaxing catalysts. As seen in Fig 8.8, only a fraction of the n-decane is converted to isodecanes. More importantly, the yield of isodecane continuously increases with the conversion of n-decane and passes through a maximum at about 90% conversion, corresponding to a maximum yield of about 30%. At higher conversion, secondary reactions convert the isodecanes into cracked products, rapidly resulting in a significant loss of $C_{10}$ hydrocarbons. The loss of isodecanes also results in a loss of VI. MLDW is the most widely used catalytic process for the dewaxing of lube base stocks and utilizes a proprietary catalyst based on ZSM-5 (9). There are several variations of similar catalysts and processes that have been commercialized, for example BP Catalytic Dewaxing Process using mordenite (23), and the Chevron Catalytic Dewaxing Process (24).

Given the high VI of isoparaffins, it is clear that it would be advantageous to selectively convert linear paraffins to their corresponding isoparaffins instead of cracking them to fuel range molecules. Fig 8.8 shows two examples of new generation ExxonMobil catalysts which exhibit such selectivity. MSDW-1(Mobil Selective DeWaxing) already represented a significant improvement in performance, allowing the preservation of about 40% of the $C_{10}$ molecules as isodecanes. MSDW-2 represents the 'state-of-the-art' in the development of such catalysts and more than 70% of the decane is converted to isodecanes (25).

Hydroisomerization is not only critical to the yield of base oil, but its control significantly impacts on both the pour point and the VI. Table 8.3 shows the impact of branching on the melting point of decanes and nonanes (26, 27).

**Table 8.3** Impact of branching on Melting Point (°C) of Hydrocarbons

| | Number of Branches | | | | |
|---|---|---|---|---|---|
| Hydrocarbons | 0 | 1 | | 2 | |
| Nonanes | -53.5 | 2Me | -80.4 | 2,2DiMe | -113 |
| | | 3Me | -107.0 | 2,6DiMe | -103 |
| | | 4Me | -113.0 | 2,3DiMe | -116 |
| Decanes | -29.6 | 2Me | -74.5 | 2,7DiMe | -54.0 |
| | | 3Me | -84.8 | | |
| | | 4Me | -98.7 | | |
| | | 5Me | -87.7 | | |

The melting point and the pour point are essentially the same. The addition of branches as methyl groups decreases the melting point by at least 30°C and up to 60°C. A closer look at the table shows that it is not the number of methyl or branches that appear to impact the melting point, but the presence of at least one branch. However, there is a significant difference when the branch is located near the end of the alkyl chain, i.e. in position 2: the melting point only decreases by about 30°C or half the diminution observed for the other branching positions. It is therefore preferable to have the methyl located toward the center of the alkyl chain to minimize the pour point. The conversion of linear paraffins to isoparaffins also impacts the VI (Fig. 8.9).

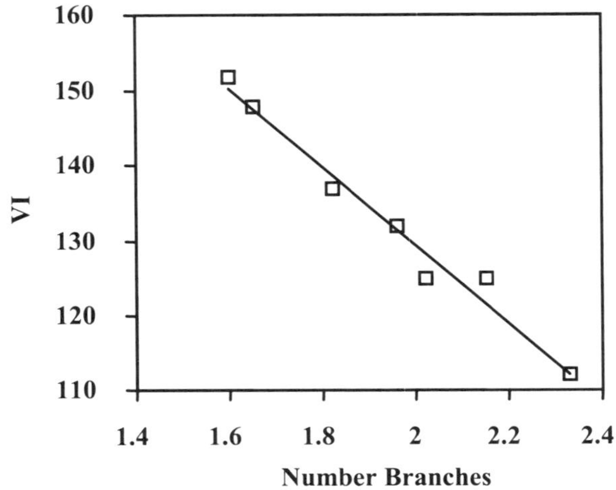


**Fig. 8.9** Impact of branching on VI for $C_{24}$ hydrocarbons (28)

The hydroisomerization of n-C24 on the Isodewaxing catalyst reported by Miller, demonstrates the relation between the degree of branching, as measured by $^{13}C$ NMR, and the change in VI of the isomerized fluid. Based on this work, Miller (28) proposes that the conversion of the linear paraffin to its 2-methyl substituted analog lowers the pour point, and that further isomerization by migration of the methyl substituant toward the center of the molecule enhanced the reduction of the pour point. The addition of the second methyl then leads to a loss of VI and eventually to a loss of yield by acid cracking. When a high selectivity toward 2-methyl paraffins is observed in the conversion of linear paraffins, the high selectivity is often attributed to a 'pore mouth' reaction where only part of the molecule penetrates the pore of the zeolite (28, 29, 30). Another interpretation is based on the 'key/lock' adsorption concept (15).

It is the complex relationship between the yield of base stock, its VI and its cold flow properties (pour point and/or cloud point) that leads to the differentiation and optimization of selective dewaxing catalysts and processes, where 'shape selective' catalysis is of critical importance. For example, Figure 8.10 shows the advantages of MSDW-2 over earlier technologies such as solvent dewaxing and MSDW-1 (25).

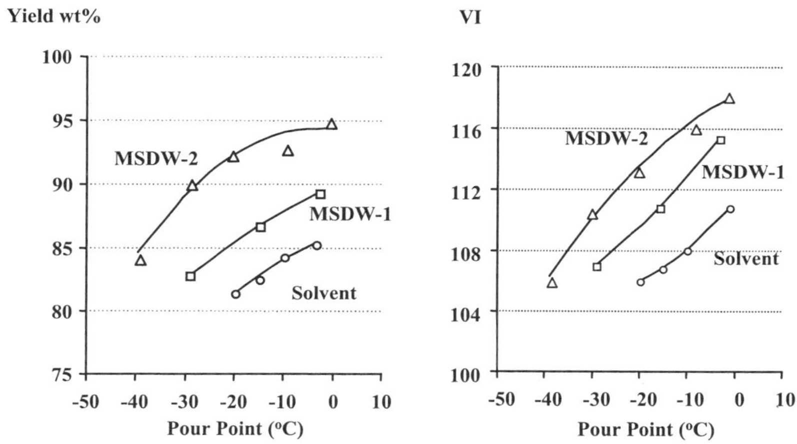


**Fig. 8.10** Selective dewaxing of hydrocracked light neutral feedstock (14, 25)

Finally, Figure 8.11 provides an example of the effect of the feedstock type on the relative performance of such processes.

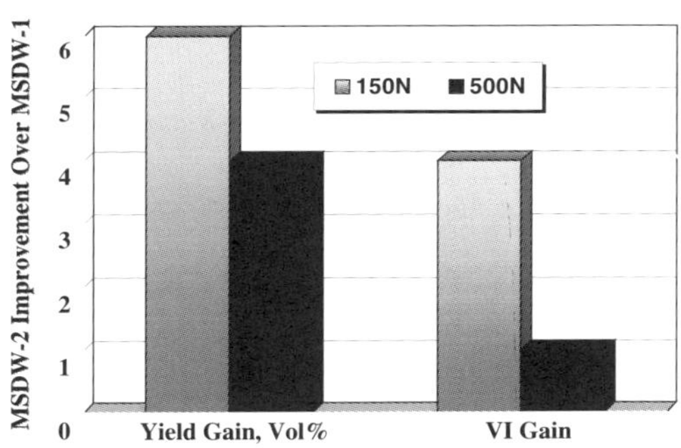


**Fig. 8.11** MSDW-2 performance gains for two grades of basestocks

## 8.6 Conclusion

Base oils processing and production relies on a wide multiplicity of non-catalytic and catalytic processes. Meeting stringent quality and performance requirements has led to the development of new processes and plant configurations, including the modification of existing conventional solvent based plant to the construction of the state-of-the-art 'all catalytic' lube plant. Base oils are complex fluids, which require specific upgrading processes, such as solvent extraction or hydroprocessing, coupled with dewaxing processes to improve their low temperature properties. One of the most recent technologies implemented in that field is based on the use of medium pore molecular sieves and shape selective catalysis. The progress made in selective dewaxing leads to the ability to produce high performance base oils such as Group II and III lubricants.

## Acknowlegments

The author's thanks go to my colleagues D.O. Marler, J. Beck, D. Boate, W. Murphy and T. Degnan for their helpful contributions and comments.

## References

1. Sequeira A., *Lubricant Base Oil and Wax Processing* (Marcel Dekker, Inc, New York 1994) 262-278.
2. *The LEDA Process for Low Energy Solvent Deasphalating*, (Foster Wheeler Technical Publication, Foster Wheeler, July, 1983).
3. Chang C.P. and Murphy J.R., *Deasphalating, Encyclopedia of Chemical Processing and Design* (Marcel Dekker, Inc), **14** (1983) 140-165.
4. Sankey B.M. et al., *Proceedings of the Tenth World Petroleum Congress*, **4** (1979) 407-414.
5. Sequeira A. et al., *NPRA Annual Meeting, San Antonio, Texas*, Paper No. AM-79-20, (1979) 27-29.
6. Bushnell J.D. and Fiocco R.J., *1980 Proceedings-Refining Department, American Petroleum Institute*, **59** (1980) 159-167.
7. Bijwaard H.M.J. et al., *The Shell Hybrid Process, an Optimized Route for HVI (High Viscosity) Lube oil Manufacture,* Petroleum Refining Conference of the Japan Petroleum Institute, Tokyo, (1986).

8. Scholten G.G., *Solvent Dewaxing, Encyclopedia of Chemical Processing and Design* (Marcel Dekker, inc, New York), **15** (1983) 353-370.
9. Smith K.W. et al., "New Process Dewaxes Lube Base Stocks", *Oil & Gas J.,* **78** (1980) 75-85.
10. Miller S.J. et al., *Advances in Lube Base Oil Manufacture by Catalytic Hydroprocessing,* National Fuels and Lubricants Meeting of the NPRA, Houston, Texas, Paper FL-92-109, (1992).
11. Wuest R.G. et al., *Oil & Gas J.*, **July** (1999) 70-73.
12. Banta F. et al., *Oil & Gas J.*, **July** (1998) 70-74.
13. Bhore, N.A. et al., *Oil & Gas J.*, **November** (1999) 67-74.
14. Helton T.E. et al., *Oil & Gas J.*, **July** (1998) 58-67.
15. Souverijns W. et al., *Progress in Zeolite and Microporous Materials, Studies in Surface Science and Catalysis*, **105** (1997) 1285.
16. Chen N.Y. et al., *Oil & Gas J.,* **66** (1968) 154.
17. Chen N.Y. et al., *J. Catal.*, **52** (1978) 453.
18. Jacobs P.A. et al., *Chem. Soc. Faraday Disc.*, **72** (1981) 353.
19. Martens J.A., *PhD thesis*, (Catholic University Leuven, Belgium, 1985).
20. Weitkamp J. et al., *Appl. Catal*, **8** (1983) 123.
21. Meier W.M., *Atlas of Zeolite Structure Types*, (Butterworths, 2nd edition, 1987)
22. Beck J., *New Developments in Shape Selective Catalysis*, Mobil Catalysts Corporation of Japan, 15th anniversary Symposium, Tokyo, Japan, November, (2000).
23. Hargrove J.D., *Encyclopedia of Chemical Processing and Design* (Marcel Dekker, New York), **15** (1983) 346-352.
24. Zakarian, J.A. et al., *All Hydroprocessing Route for High VI lubes,* presentation at the AIChe Spring Meeting, New Orleans, (1986).
25. Tabak, S.A. et al., *ExxonMobil Advances in Catalyst and Process Technology for Lubes*, Mobil Catalysts Corporation of Japan, 15th anniversary Symposium, Tokyo, Japan, November, (2000).
26. Rossini F.D., *Selected Values of Physical and Thermodynamical Properties of Hydrocarbons and Related Compounds*, API Research Project (Carnegie Press, 1953).
27. *Handbook of Chemistry and Physics*, 81st edition (CRC Press, 2000).
28. Miller S.J., *Wax Isomerization for Improved Lube Oil Quality*, AIChE Spring Meeting, New Orleans, LA, March 8-12, (1998).
29. Ernst S., *Angew. Chem. Int. Ed. Engl.*, **35** (1996) 63.
30. Martens J.A. et al., *Angew. Chem, Int. Ed. Engl.*, **34** (1995) 2528.

# PARA-XYLENE MANUFACTURING: CATALYTIC REACTIONS AND PROCESSES

F. ALARIO
*Institut Français du Pétrole, 1 et 4 Avenue de Bois Préau,*
*F 92852 Rueil Malmaison cedex, France,*

M. GUISNET
Laboratoire de Catalyse en Chimie Organique, UFR Sciences , 40, Avenue du Recteur Pineau, 86022 Poitiers Cedex, France

## 9.1 Overview (1-7)

The most valuable isomer among the four C8 aromatics (ethylbenzene –EB-, para-xylene –pX-, ortho-xylene –oX- and meta-xylene –mX-) is pX, which is a chemical intermediate for the production of polyethyleneterephtalate (PET) fibers, resins and films. pX production is then driven by PET fiber and resin markets, which are growing at an estimated 4-6% per year for the next decade. The highest growth rates are observed in emerging countries, mainly in Asia, where most of the latest large capacity plants have been constructed.

The main source of pX is the 'mixed xylenes' cut, containing the four C8 aromatics including EB. The mixed xylenes are produced by catalytic reforming (82%), steam cracking (9%) and toluene disproportionation or toluene-heavy aromatics transalkylation (9%). The xylenes distribution is generally close to thermodynamic equilibrium values, while EB content depends on the origin of the cut (Table 9.1).

**Table 9.1** Typical composition of mixed xylenes

| % | Steam Cracking | | Catalytic reforming | | Thermodynamics (500°C) | |
|---|---|---|---|---|---|---|
| EB | 50 | --- | 17 | --- | 10 | --- |
| pX | 10 | 20 | 19 | 23 | 22 | 24 |
| mX | 25 | 50 | 47 | 57 | 46 | 52 |
| oX | 15 | 30 | 17 | 20 | 22 | 24 |
| Σ | 100 | 100 | 100 | 100 | 100 | 100 |

The boiling point difference between pX and mX being only 0.8°C (Table 9.2), pX must be separated from the other three isomers by other means than distillation (crystallization or adsorption). The remaining raffinate (Figure 9.1), containing mainly EB, mX and oX, is then processed into an isomerization unit where pX is 'reformed'. The obtained C8 aromatics cut, containing the xylenes in proportion close to their thermodynamic equilibrium values, is recycled back to the 'xylenes column' (or 'rerun column') for further pX separation. A loop is then created which is called the 'aromatics loop'.

**Table 9.2** Main physical properties of BTX

| | | | 'Mixed Xylenes' | | | |
|---|---|---|---|---|---|---|
| | B | T | EB | pX | mX | oX |
| MW (g/mol) | 78.12 | 92.15 | | 106.17 | | |
| Spec. Grav. (20/4) | 0.8765 | 0.8669 | 0.867 | 0.8611 | 0.8642 | 0.8802 |
| Melting point (°C) | 5.5 | -59.4 | -94.9 | 13.3 | -47.9 | -25.2 |
| Boiling point (°C) | 80.1 | 110.6 | 136.1 | 138.3 | 139.1 | 144.4 |

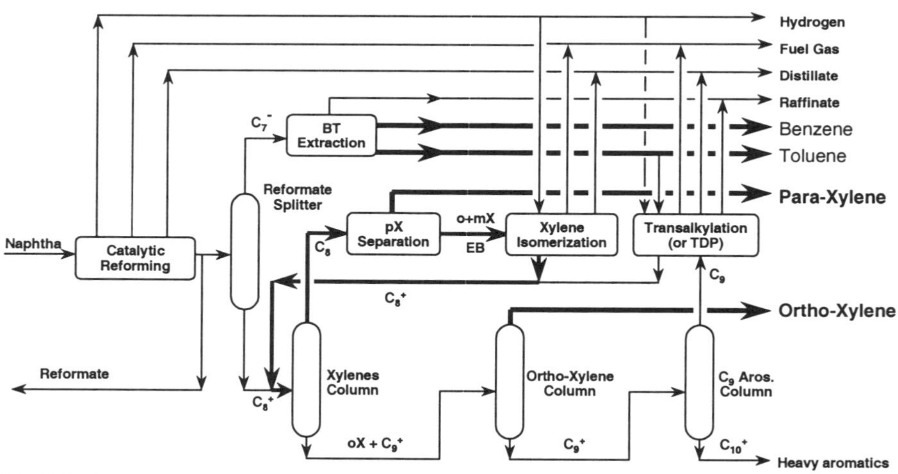


**Fig. 9.1** The aromatics complex

There are two types of isomerization processes depending on EB transformation (xylene isomerization occurs in every industrial process):

(a) EB reforming type where EB is isomerized into xylenes
(b) EB dealkylation type where EB is deethylated into benzene and ethane

Type (a) maximizes the overall pX production of the complex, while type (b) co-produces additional high purity benzene. Also, type (b) may be a smart solution for debottleneckings of type (a) complexes since it allows a higher feed space velocity than type (a) (see below).

Both processes use bifunctional catalysts generally comprising well-dispersed platinum and at least one acid zeolite generally with the MOR structure (Figure 9.2a) for type (a) and with the MFI structure (Figure 9.2b) for type (b).

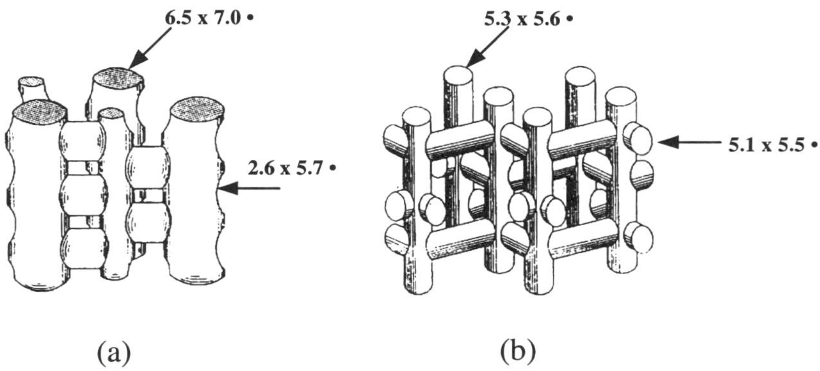


**Fig. 9.2** Pore structure of MOR (a) and MFI (b) zeolites

An interesting way to increase pX production of an aromatics complex is to transform less valuable aromatics such as toluene and $C9_+$ aromatics into xylenes. This is achieved either by toluene disproportionation (TDP) or by toluene $C_9^+$ aromatics transalkylation. There are two types of disproportionation processes: Normal TDP produces a xylene cut where pX content is at thermodynamic equilibrium, while with STDP pX content is much higher (para-selectivity). This is possible when a selectivated zeolite (generally MFI) is used. For normal TDP as well as for transalkylation processes, mordenites doped with a hydrogenating metal to limit coking are used as catalysts.

The catalytic reactions occurring in C8 aromatics isomerization and toluene disproportionation or transalkylation with heavy aromatics are given in the following paragraphs. The catalysts used for these reactions and the processes are also described.

## 9.2 Mixed xylenes isomerization

### *9.2.1 Xylene transformations*

Xylene inter-isomerization is a well-known acid catalyzed reaction. Over acid catalysts disproportionation is generally observed as a side reaction. Over the bifunctional catalysts used in the commercial processes xylenes can also undergo two other primary reactions:

- hydrogenation into dimethylcyclohexane followed by isomerization,
- isomerization into EB, this reaction being significantly limited by thermodynamic equilibrium (approximately 8% of EB in C8 aromatic cuts at 410°C).

Only the acidic reactions are examined here, the other reactions will be presented with ethylbenzene transformation.

A recent review paper (8) has been published on the mechanisms of xylene isomerization over acid solid catalysts. Two mechanisms have been identified:

- the well-known intramolecular mechanism involving 1,2 methyl shifts in benzenium ion intermediates (Figure 9.3a),
- an intermolecular mechanism involving successive xylene disproportionation and fast transalkylation between trimethylbenzenes and xylenes (Figure 9.3b).

The first mechanism is largely predominant with average pore size zeolites such as MFI, because steric constraints in the vicinity of the acid sites inhibit the formation of the bulky disproportionation and transalkylation intermediates. This is also the case with large pore tridimensional zeolites such as FAU zeolites. The second mechanism was shown to be predominant with monodimensional mesoporous aluminosilicates such as MCM41. In the long, regular, non-interconnected channels of these molecular sieves, xylene molecules undergo, before desorption, successive reactions of disproportionation and transalkylation (9). This mechanism could also play a significant role in monodimensional pore zeolites (10).

With bifunctional Pt acid catalysts, a third mechanism can participate in xylene isomerization, involving the same intermediates as ethylbenzene isomerization (2,11).

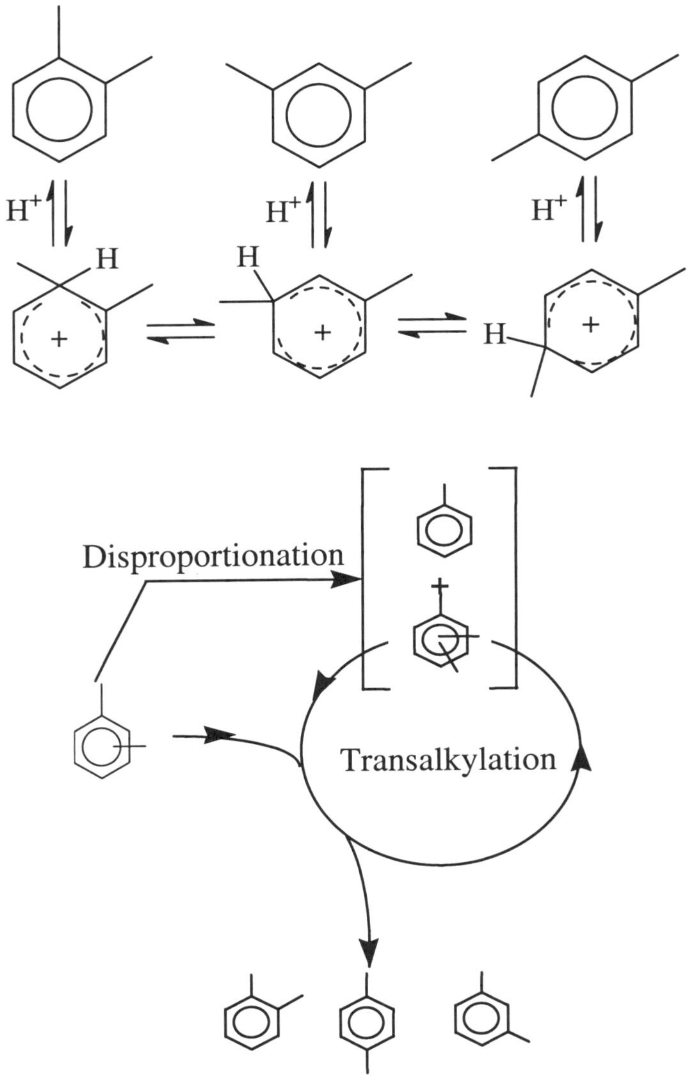


**Fig. 9.3** Mechanisms of xylene isomerization- a: Intramolecular mechanism (1,2 methyl shifts)
b: Intermolecular mechanism via disproportionation and transalkylation steps

The participation of protonic acid sites in xylene isomerization is clearly demonstrated by correlations between the isomerization rate and the concentration of protonic sites of silica alumina with various alumina contents (13), alkaline-earth and rare earth FAU zeolites (14, 15), MFI zeolites (16), etc. Evidence is also provided by the fact that protonic sites participate in alkylbenzene disproportionation. On the other hand, it seems most unlikely that Lewis acid sites play a direct role in xylene isomerization and disproportionation (8).

Large differences exist between the xylene disproportionation/isomerization ratios (D/I) found with acid catalysts. With zeolites the size of the space available near the acid sites was shown to play a determining role (2). The smaller the size of the intracrystalline zeolite cavities, the lower the ratio between the rate constants of disproportionation and isomerization: 0.05 at 316°C with a FAU zeolite (diameter of the supercage of 1.3 nm), 0.014 and 0.01 with MOR and MAZ (0.08 nm). Steric constraints which affect the formation of the bulky bimolecular transition states and intermediates of disproportionation (Figure 9.4) would be responsible for this observation. However, the very low value of D/I (0.001) obtained with MFI (2), the channel intersection of which has a size of 0.85 nm, is also due to other causes: limitations in the desorption of the bulky trimethylbenzene products of disproportionation from the narrow pores of the zeolite (~ 0.6 nm) and most likely the low acid site density of the used sample (Si/Al=70 instead of 5-15 with the large pore zeolites).

Indeed, the rate of disproportionation (but not the rate of isomerization) could depend on the density of the protonic acid sites, or of their proximity (17). This seems quite possible when the complex mechanism (six successive steps) of disproportionation (Figure 9.4) is considered. Indeed, two different acid sites could be involved in steps 1, 2 and 3 and in steps 4, 5 and 6. Recent results with NaHMOR catalysts (18) confirm this decrease in D/I with the density of the protonic sites (Figure 9.5), and hence are in favour of this demanding character of xylene disproportionation. However, with large series of HFAU zeolites differing by their Si/Al (19, 20) ratio the situation was not so simple; this can be related to rapid deactivation effects as well as to heterogeneity in the distribution of hydroxyl groups (hence of protonic sites) in dealuminated samples (20). Furthermore, no clear conclusion can be drawn on the dependence of the D/I ratio with the acid strength. Indeed, according to certain authors, disproportionation would demand stronger acid sites than isomerization (21) whereas the reverse is proposed by other authors (22).

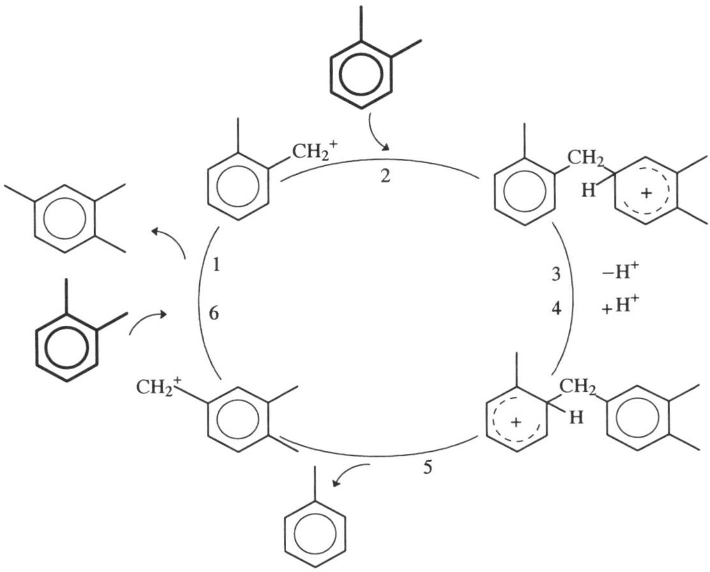


**Fig. 9.4** Mechanism of xylene disproportionation

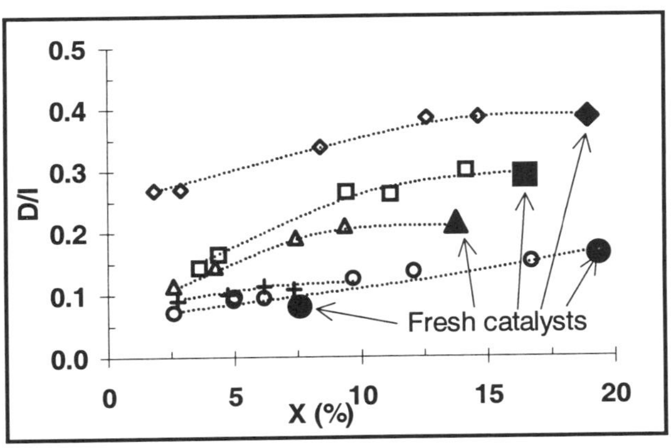


**Fig. 9.5** Transformation of m-xylene over HMOR10 and NaHMOR10 samples. Disproportionation / Isomerisation rate ratio (D/I) versus m-xylene conversion (X%) on fresh and on deactivated catalysts; HMOR10 (◊), 14NaHMOR (□), 28NaHMOR (Δ), 45NaHMOR (+), 63NaHMOR (○)

### *9.2.2 Ethylbenzene transformations*

Under hydrogen flow, various reactions can be observed during ethylbenzene transformation over bifunctional Pt/acid catalysts. Some of them occur through bifunctional catalysis (reactions 1, 2, 6), the other through acid (reactions 3,4) or metal catalysis (reaction 5).

1- Isomerization into xylenes
2- Hydrodealkylation into benzene and ethane
3- Disproportionation into benzene and diethylbenzene
4- Transalkylation between ethylbenzene and xylene products (i.e. a secondary reaction)
5- Hydrogenation of aromatics into C8 naphthenes
6- Secondary cracking of naphthenes

#### 9.2.2.1 Ethylbenzene isomerization

The mechanism of ethylbenzene isomerization reported in Figure 9.6 was firstly proposed by Weisz (23).

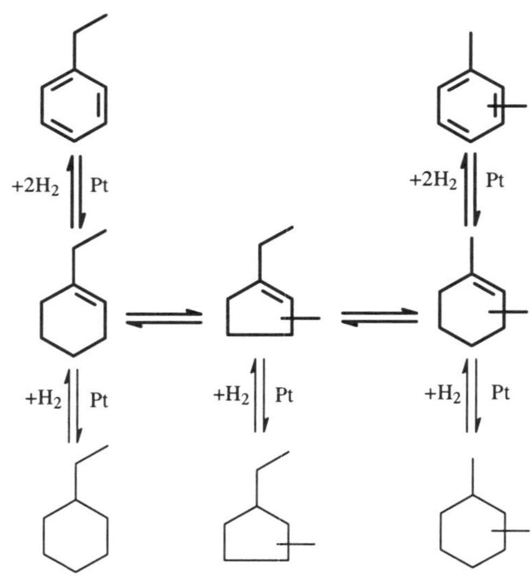


**Fig. 9.6** Mechanism of ethylbenzene isomerization over bifunctional catalysts

This mechanism was confirmed by various experiments carried out over a series of Pt/fluorinated alumina catalysts (24). In particular, the comparison of the reactivities of ethylbenzene and of ethylcyclohexane, the values of the kinetic orders with respect to ethylbenzene and to hydrogen are in favour of ethyl and dimethylcyclohexene intermediates rather than ethyl and dimethylcyclohexane which are, however, formed in much more significant amounts.

As it is generally the case with bifunctional catalysis processes, the balance between hydrogenating and acid functions determines for a large part the catalyst activity. This was quantitatively shown for series of bifunctional catalysts constituted by mechanical mixtures of a well dispersed Pt/Alumina catalyst and of mordenite samples differing by their acidity and their porosity (25). The balance between hydrogenating and acid functions was taken as nPt/nH+ the ratio between the number of accessible platinum atoms and the number of protonic sites determined by pyridine adsorption.

Whatever the mordenite component, the turnover frequency TOF (per protonic site) for isomerization firstly increases with nPt/nH+, which indicates that for low values of this nPt/nH+ ratio, the hydrogenation steps are the limiting steps of isomerization (Figure 9.7).

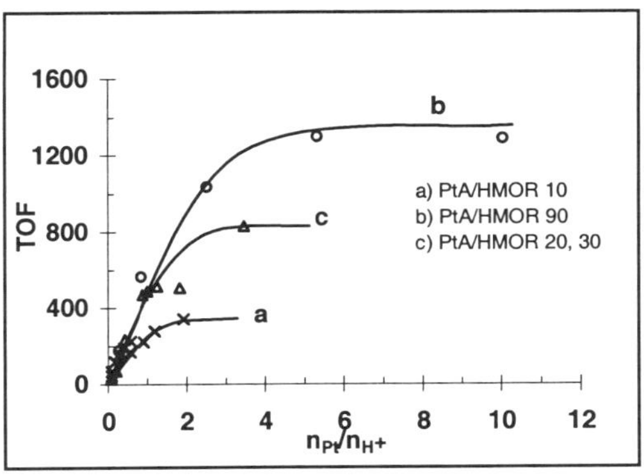


**Fig. 9.7** Turnover frequency values TOF ($h^{-1}$) for ethylbenzene isomerization of the protonic sites of various bifunctional PtA/HMOR as a function of $n_{Pt}/n_{H}+$ the ratio between the concentrations of accessible platinum sites and of protonic sites able to retain pyridine adsorbed at 423 K

A plateau is obtained above a certain value of nPt/nH+, the limiting step being then the isomerization of ethylcyclohexene intermediates on the protonic sites. Curiously, the value of TOF at the plateau depends very much on the mordenite sample. This is not due to differences in acid strength but to diffusion limitations.

These diffusion limitations are particularly pronounced with HMOR10 (Si/Al ratio of 10) catalysts, which can be explained by the quasi absence of mesopores (26, 27). With the other samples and especially with HMOR90, mesopores created by dealumination allow a quasi tridimensional diffusion of organic molecules in the pore system, decreasing or suppressing diffusion limitations. These diffusion limitations are also responsible for the abnormally low value of TOF per protonic site found with HMOR10 catalysts for hydrodealkylation and for disproportionation, i.e. for the other primary reactions involving acid catalytic sites.

The isomerization of ethylcyclohexenes into dimethylcyclohexenes which occurs through successive six-ring contraction and five ring expansion (24, 28) via protonated cyclopropane intermediates is a relatively facile reaction, hence does not require strong protonic sites. Furthermore, the selectivity of ethylbenzene transformation on bifunctional catalysts depends a great deal on the zeolite pore structure. A high selectivity to xylenes can be obtained by using large pore zeolites, especially mordenites, whereas practically no xylenes are formed on MFI catalysts. This could be related to the high selectivity of PtMFI catalysts for hydrocracking. Indeed, while with PtMOR catalysts the hydrocracking of C8 naphthenes is limited, with PtMFI catalysts nearly all the C8 naphthenes are transformed into light products. It can be suggested that, due to their slow migration in the narrow pores of MFI, olefinic intermediates underwent isomerization and cracking steps during their migration between two platinum crystallites (29).

### 9.2.2.2 Ethylbenzene disproportionation

The mechanism of ethylbenzene disproportionation depends on the zeolite pore structure (3). With large pore zeolites, this reaction occurs mainly through the carbocation chain mechanism proposed for xylene disproportionation (Figure 9.4) which involves benzylic carbocations and diarylmethane intermediates. With MFI zeolites in the pores of which steric constraints limit the formation of the bulky diarylmethane intermediates, ethylbenzene disproportionation occurs mainly through a successive dealkylation-alkylation process:

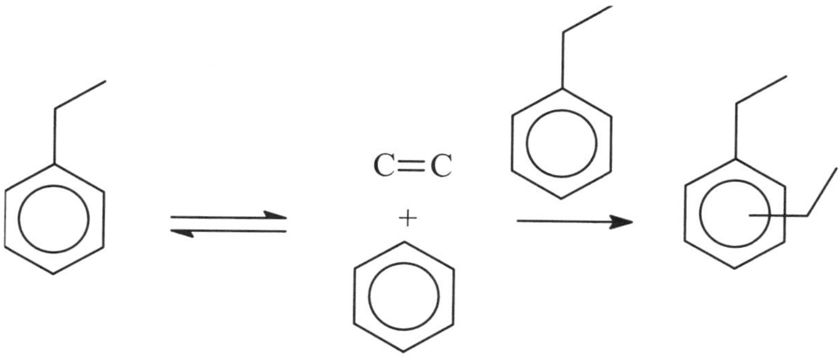


This difference in mechanism is clearly demonstrated by substituting bifunctional catalysts for acidic catalysts (29). The introduction of platinum in MFI catalysts leads to a large decrease in the rate of ethylbenzene disproportionation (divided by 6), which is due to a large consumption of ethylene by hydrogenation as shown by the large increase in the rate of dealkylation. On the other hand, the introduction of platinum in MOR catalysts leads to a limited change in the rates of disproportionation and dealkylation.

With mordenite catalysts on which disproportionation occurs through benzylic carbocations and diarylmethane intermediates, the rate of this bimolecular reaction was shown to be dependent on the acid site density: the turnover frequency for disproportionation is roughly proportional to the square of the concentration of acid sites (Figure 9.8 (25)). As suggested for xylene disproportionation, this probably means that xylene disproportionation requires two protonic sites for its catalysis (the first one for steps 1, 2, 3 and the second for steps 4, 5 and 6 of Figure 9.4).

### 9.2.2.3 Ethylbenzene dealkylation

Acid dealkylation of EB into benzene and ethylene occurs through the following mechanism:

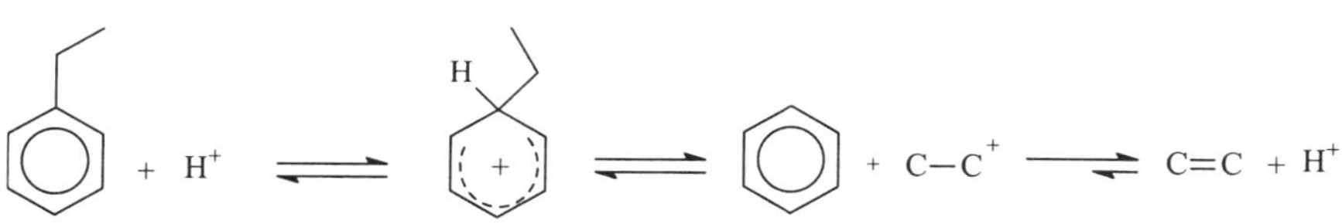

involving unstable ethylcarbenium ions with a consequently high energy barrier (4). Therefore this reaction should be sensitive to the acid strength of the zeolite. The higher the strength, the greater should be the activity of the acid sites. This is apparently the case with mordenite catalysts (25, 27). Thus the greatest turnover frequency (per protonic site) of HMOR20 bifunctional catalysts for hydrodealkylation can be related to the presence of very strong protonic sites (able to retain pyridine at 752K) on this zeolite only. Furthermore, the pore structure also seems to determine for a large part the dealkylation activity. Thus at 410°C, the HMFI(27) zeolite was found to be 7 times more active for dealkylation than HMOR(10) (29). One possible explanation could be the longer contact time between ethylbenzene molecules and protonic sites in the narrower pores of HMFI.

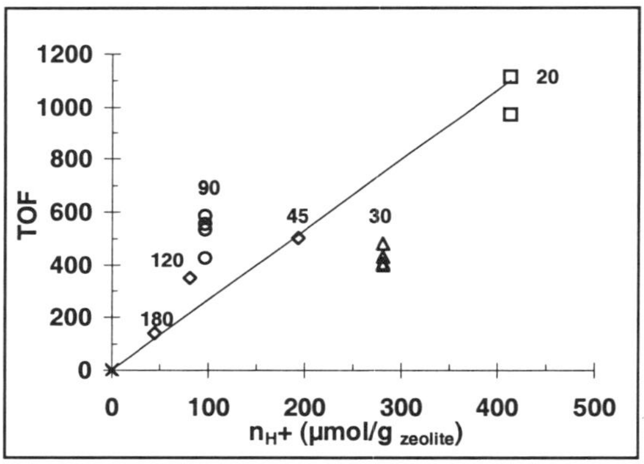


**Fig. 9.8** Turnover frequency values TOF ($h^{-1}$) of the protonic sites for ethylbenzene disproportionation over various PtA/HMOR catalysts with different proportions of PtA and HMOR and different framework Si/Al ratios of the zeolite versus the concentration of protonic sites $n_{H^+}$ per gram of zeolite

## *9.2.3 Catalysts and processes for xylene and ethylbenzene isomerization*

There were two main kinds of catalytic systems used in the 60s, both containing platinum: chlorinated aluminas and steamed silica-aluminas. Their selectivity and coke-stability were acceptable, but they were not active enough. In the mid-70s, Engelhard came up with the first industrial mordenite-based catalyst that was a major improvement. The great majority of today's processes use Pt-mordenite systems.

The main undesired reactions occurring with these catalysts are:

- Xylene and EB disproportionation/transalkylation,
- EB dealkylation,
- Naphthene ring opening leading to C8-paraffins.

According to a recent study, the highest values of isomerization selectivity can be obtained with bifunctional Hmordenite catalysts with high values of the ratio between the concentrations of Pt and protonic sites accessible by reactant molecules, i.e. with catalysts on which the acid step is the limiting step of ethylbenzene isomerization. Furthermore, because of the reduction of the disproportionation activity, Na exchange of HMOR has a positive effect on the selectivity.

The typical operating conditions of xylene and EB isomerization processes are shown in Table 9.3. These conditions minimize the above side reactions. Pressure, temperature and H2/HC ratio are key parameters that define the partial pressure of $C8^{=}$ naphthenes intermediates for EB isomerization. Naphthene cracking and disproportionation/transalkylation are responsible for the C8 aromatics net losses that affect the overall pX yield. The C8 recycled stream from the isomerization unit to the separation unit is three times higher than the fresh feed stream (since there cannot be more than ~ 24% of pX in the C8 aromatic cut after isomerization). This means that each percent of loss in the isomerization unit will decrease the pX yield by 3%. For example, when standard mordenite-based catalysts lead to 4% of net losses, the overall pX yield is roughly 88%.

**Table 9.3** Typical operating conditions of Xylene and EB isomerization processes

| | | |
|---|---|---|
| - Fixed-bed reactor | | |
| - Pressure | 7-15 bar | $C_8^{=}$ naphthenes partial pressure (Pressure, Temperature, $H_2/HC$) |
| - Temperature | 370-430°C | |
| - $H_2/HC$ | 3-10 mol/mol | |
| - WHSV | $3\text{-}5h^{-1}$ | |
| - EB conversion | 25-35% | |
| - $pX_{ATE}$ = | $\dfrac{\frac{[pX]}{[X]\text{ feed}} - \frac{[pX]}{[X]\text{ product}}}{\frac{[pX]}{[X]\text{ feed}} - \frac{[pX]}{[X]\text{ th. equilibrium}}}$ | >92% |
| - $C_8$ losses | 4-7% (cracking and disproportionation only) | |
| - Cycle length | 1-3 years | |
| - Some processes | Isomar(UOP), Octafining II (Engelhard-IFP), Aris (VEB Leuna), Isolene II (Toray), 'RIPP' | |

New generations of catalysts that allow higher pX yields are now commercially available. Recently, two new catalysts were announced by UOP (31) and IFP (32). UOP claimed that the new molecular sieve-based catalyst I-210 could achieve a pX yield higher than 90%, while IFP's Oparis, containing a new zeolite, could reach 92%+ yields.

### *9.2.4 Catalysts and processes for xylene isomerization with ethylbenzene dealkylation*

MFI zeolites seem to be the most efficient for EB dealkylation, in terms of activity, selectivity and stability. In the 70s, on metal-free MFI catalysts, EB was disproportionated into benzene and diethylbenzenes. As indicated above, with MFI catalysts, ethylbenzene disproportionation occurs through a deethylation-ethylation mechanism, with ethylene as desorbed intermediate. The addition of a metal (carried out early 80s) allows a rapid and irreversible conversion of ethylene into ethane with a consequent shift of ethylbenzene transformation from disproportionation to hydrodealkylation. The selectivity is highly sensitive to temperature that must be in the range 380°C-460°C to limit both alkylation and naphthene cracking.

The typical operating conditions of xylene isomerization with EB dealkylation processes are shown in Table 9.4. An interesting point is the high WHSV(weight hourly space velocity) these processes can reach. For an existing complex debottlenecking, switching from an EB reforming isomerization (WHSV < 5 h-1) into an EB dealkylation unit can be a smart solution to avoid changing the reactor volume. High purity benzene can be produced, but the economical interest depends on the benzene market to which a given refiner has access.

**Table 9.4** Typical operating conditions for Xylene Isomerization and EB dealkylation processes

| | |
|---|---|
| - Fixed-bed reactor (dual-bed catalyst in Mobil's process) | |
| - Pressure | 5-15 bar |
| - Temperature | 370-430°C |
| - $H_2$/HC | 2-4 mol/mol |
| - WHSV | 3-12 $h^{-1}$ |
| - EB conversion | 30-80% |
| - pX ATE | > 95% |
| - Xylene losses | 0.5-10% |
| - Cycle length | 1-8 years |
| - Some processes | Advanced MHAI (ExxonMobil), Isomar (UOP), Aris (KataLeuna), 'RIPP', Xylofining (IPCL) |

## 9.3 Toluene disproportionation (TDP) and transalkylation with $C_9$+ aromatics

These processes present several advantages:

- EB-poor C8 aromatics cuts are produced which favors further pX separation
- Low-value toluene and C9+ aromatics (that are generally sent to the gasoline pool because of their high octane number) are transformed into high value pX
- Toluene hydrodealkylation units can be easily revamped into TDP or transalkylation units

### *9.3.1 Toluene disproportionation (TDP)*

There are two kinds of TDP processes: normal TDP which leads to thermodynamic mixture of xylenes, and selective TDP (STDP) which leads selectively to pX (>80%). Typical operating conditions of these processes are given in Table 9.5. With STDP, toluene conversion is low (<30%) whereas with normal TDP, this conversion (45-50%) is close to the maximum thermodynamic equilibrium (60%). Mordenite associated with a hydrogenating metal in order to limit coking is used for normal TDP and a selectivated MFI zeolite for STDP.

**Table 9.5** Typical operating conditions of TDP processes (all fixed-bed reactors)

| | Normal TDP | STDP |
|---|---|---|
| - Pressure | 20-40 bar | 20-40 bar |
| - Temperature | 380-500°C | 420-480°C |
| - $H_2$/HC | 1-6 mol/mol | 1-2 mol/mol |
| - WHSV | 1-6 $h^{-1}$ | 3-5 h-1 |
| - Tol. conversion | 45-50% | < 30% |
| - pX/X | ~ 24% | > 80% |
| - Cycle length | > 1 year | > 1 year |
| - Some processes | MTDP(ExxonMobil)<br>Tatoray(UOP-Toray)<br>T2BX(TotalFinaElf) | MSTDP, MTPX,<br>PxMax(ExxonMobil)<br>PxPlus(UOP) |

Toluene disproportionation (TDP) is a well-known acid reaction, occurring through the same mechanism as xylene disproportionation (Figure 9.4). Like this latter reaction, toluene disproportionation requires most likely two protonic sites for it catalysis, hence the density of protonic sites has a very positive effect on the catalyst activity. Furthermore, the bimolecular intermediates (methyldiphenylmethane and the corresponding carbocations) of toluene disproportionation being bulky, steric constraints can exist in the pores of 10 membered ring zeolites. However, above 450°C, MFI is an effective catalyst for TDP, which means that, at least at high temperatures, methyldiphenylmethane intermediates can be easily formed at the channel interactions of this zeolite. With a REX zeolite, pX is favored at low conversion, indicating a low ratio between the rates of toluene disproportionation and xylene isomerization ($k_I/k_D<10$). With ZSM5 zeolites, because of steric constraints on disproportionation, $k_I/k_D$ was found very high ( 7000 at 482°C) and xylenes were apparently formed in their thermodynamic equilibrium ratio. However, para selectivity can be observed after a zeolite post-treatment, such as a pre-coking at high temperature or a specific deposition of silicon containing compounds on the zeolite cristallite surface, and at low toluene conversion (generally<30%). Both treatments are responsible for a much faster diffusion of pX from the zeolite pores compared to oX and mX.

The main undesired reactions, which occur essentially with normal TDP, are toluene hydrodealkylation, toluene-xylenes transalkylation and ring cracking.

### *9.3.2 Toluene –C9+ aromatics transalkylation*

The desired reactions are shown in Figure 9.9. Deethylation of ethyl+ groups is a desired reaction because these ethyl+aromatics cannot lead to xylenes via transalkylation. The side reactions to avoid are demethylation, xylenes transalkylation with other aromatics and ring cracking.

The catalysts contain large pore zeolites and among them, mordenite seems to be used worldwide. With these large pore zeolites, transalkylation reactions occur through the mechanism proposed for xylene disproportionation (Figure 9.4), i.e. through bimolecular intermediates. It is most likely that hydrogenating metals are present to improve catalyst aging and to hydrogenate light olefins C2+ (from dealkylation) avoiding further alkylation. Group VIIB and VIII metals seem to be the most suitable, allowing a very limited ring cracking. Such metals are necessary when a few percent of C10+ aromatics are processed along with C9 aromatics for acceptable cycle lengths.

-Toluene/Benzene-$C_9^+$ A Transalkylation

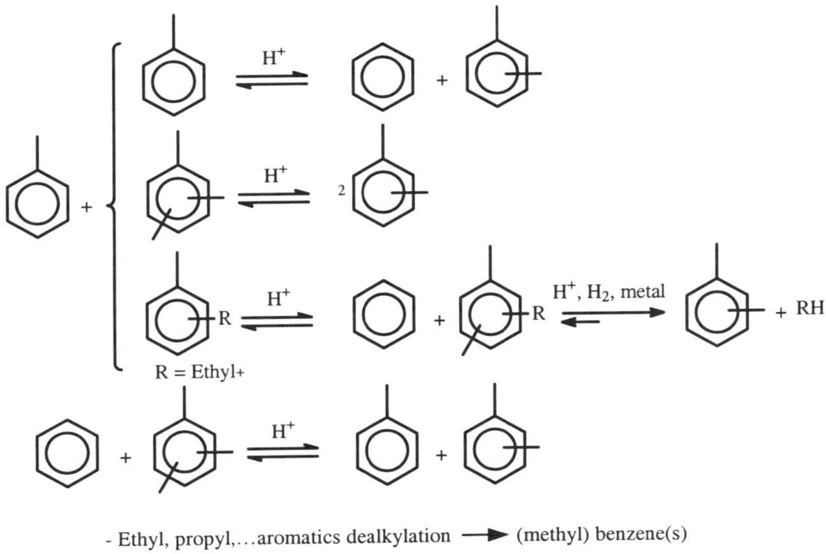


**Fig. 9.9** Toluene-$C_9^+$A transalkylation – desired reactions

Transalkylation processes operating conditions are given in Table 9.6. These conditions are similar to those of normal TDP except for the Xylenes-Plus process from ARCO-Lyondell which operates with a moving bed in absence of hydrogen.

**Table 9.6** Typical operating conditions of Toluene-C9+ aromatics transalkylation

| | |
|---|---|
| - Fixed-bed reactor (except Xylenes-Plus process: moving bed) | |
| - Pressure | 1*-40 bar |
| - Temperature | 380-520°C |
| - $H_2$/HC | 0*-6mol/mol |
| - WHSV | 1-4 $h^{-1}$ |
| - pX/X | ~ 24% |
| - Cycle length | > 1 year |
| - Some processes | *Xylenes-Plus(ARCO-Lyondell), Tatoray(UOP-Toray) TAC-9(Toray), Transplus(ExxonMobil-CPC) |

## 9.4 Conclusions

Zeolite catalysts play a significant role in the manufacture of pure paraxylene from the aromatic products of reforming and steam cracking. There are two types of processes for the isomerization of the $C_8$ aromatic cut, either with isomerization of ethylbenzene into xylenes or with hydrodealkylation into benzene and ethane. In both processes, bifunctional catalysts are used, the zeolite component being mordenites in the processes with ethylbenzene isomerization and MFI zeolites in the processes with hydrodealkylation. The paraxylene production of aromatic complexes can be increased by toluene disproportionation or toluene transalkylation with $C_9^+$ aromatics. A high paraxylene yield can be obtained when toluene disproportionation is carried out on modified MFI zeolites (e.g. by coking at high temperatures). Mordenites doped with hydrogenating metals are used for normal toluene disproportionation and transalkylation processes.

It should, however, be emphasized that new catalysts with zeolites other than MOR or MFI which give higher paraxylene yields were recently developed for the isomerization of the $C_8$ aromatic cut. Moreover, adsorption on FAU zeolites is now the main technique used for paraxylene separation (Chapter 10).

## References

1. Beck J. S., Haag W. O., *Handbook of the Heterogeneous Catalysis,* Ed. Ertl G., Knözinger H. and Weitkamp J., (VCH, Weinheim, **5** 1997) 2136.
2. Olson D.H., Haag W.O., *ACS Symposium Series,*. (American Chemical Society, Washington, DC, **284,** 1984) 275.
3. Guisnet M., *L'Actualité chimique*, (Avril 1998) 9.
4. Sie S.T. et al., *Erdöl Erdgas Kohle*, (November 1996) 463.
5. Baraqué M., *l'Actualité Chimique*, (Octobre 1997) 18.
6. Alario F., Barraqué M. and Marcilly C., *Génie des Procédés*, (Techniques de l'Ingénieur, Paris, Tome **J3** 1996) J 5920.
7. Guisnet M. and Gnep N.S., *Zeolites: Science and Technology,* Ed. Ribeiro F.R., Rodrigues A.E., Rollmann L.D., Naccache C., (NATO ASI Series, The Hague, **80** 1984) 571.
8. Guisnet M., Gnep N.S., Morin S., *Microporous and Mesoporous Materials*. **35-36** (2000) 47.
9. Guisnet M., Morin S.,Gnep N.S., *Shape Selective Catalysis*, Ed. Song C., Garces J.M. and Suji Y., (ACS Symposium Series, American Chemical Society, Washington **738** 2000) 334.
10. Jones C.W., Zones S.I., Davis M.E., *Appl. Catal.* **181** (1999) 289.

11. Silva J.M. et al., *Appl. Catal.* **125** (1995) 1.
12. Gnep N.S. and Guisnet M., *React. Kinet. Catal. Lett.* **22** (1983) 237.
13. Hansford R.C. and Ward J.W., *J. Catal.* **13** (1969) 316.
14. Ward J.W., *Zeolite Chemistry and Catalysis* Ed. Rabo J.A., (ACS Monograph 171, American Chemical Society, Washington, DC, 1976) 118.
15. Ward J.W., *J. Catal.* **13** (1969) 321.
16. Babu G.P., Hedge S.G., Kulkarini S.B. and Ratnasamy P., *J. Catal.* **81** (1983) 471.
17. Guisnet M., *Acc. Chem. Res.*, **23** (1990) 392.
18. Moreau, F. et al., accepted for publication in *Microporous and Mesoporous Materials*.
19. Corma A. et al., (American Chemical Society, *Symp. Ser.* **368** 1988) 555.
20. Morin S., Ayrault P., Gnep N.S. and Guisnet M., *Appl. Catal. A*: **166** (1998) 281.
21. Ratnasamy P., Sivasankar S. and Vishnoi S., *J. Catal.* **69** (1981) 428.
22. Morita Y., Kimura T., Kato F. and Tamagawa M., *Bull. Jpn. Petrol. Inst.* **14** (1972) 192.
23. Weisz P.B, *Adv. Catal.* **13** (1962) 137.
24. Gnep N.S. and Guisnet M., *Bull. Soc. Chim. Fr.*, (1977) 429 and 435.
25. Moreau F. et al., accepted for publication in *Ind. Eng. Chem. Research*.
26. Fernandes L.D. et al., *J. Catal.* **177** (1998) 363.
27. Moreau F. et al., *J. Catal.* **202** (2001) 402.
28. Sosa R.C., Nitta M., Beyer H.K. and Jacobs P.A., *Procedings of the 6 th International Zeolite Conference* Ed. Olson D. and Bisio A., (Butterworths, Guilford, 1984) 508.
29. Silva J.M. et al., *Appl. Catal.* **125** (1995) 15.
30. Terlouw T. and Gislon J.P., *Shell Internationale Research*, (Maatschappij B.V., EP N°**458378** (1991).
31. *European Chemical News*, 10-16 March, (1997).
32. Dupraz C., Alario F. and Renard P., *DGMK Conference*, Erlangen (Germany), Oct 13-15, (1999).

CHAPTER 10

# SEPARATION OF PARAXYLENE BY ADSORPTION

A. MĒTHIVIER
*Institut Français du Pétrole, 1-4 Av. de Bois Préau*
*92852 Rueil Malmaison Cedex, France*

## 10.1 Introduction

The paraxylene market has enjoyed high growth rates over the years and it is expected that this trend will continue in the future. Xylenes are produced from catalytic reforming of naphta. Isomerization combined with separation is allowed to produce high purity paraxylene in agreement with the market demand. In this respect, the separation of paraxylene is a key operation which requires high purity, high yield, and large capacity. Typical values for these figures are 99.9% for the purity, 96% for the recovery and 450,000 ton/year for the capacity (even more). Another trend for the market is the recent orientation towards co-production of metaxylene.

Crystallization and adsorption are both widely used to perform the separation; distillation is not used (except for orthoxylene separation) because of too small differences between the boiling points (Table 10.1). Despite the still high importance of crystallization, adsorption becomes the most widely used technique because of its high efficiency. The adsorbents which are used for selective adsorption of paraxylene are X or Y zeolites exchanged with adequate cations. Liquid phase Simulated Counter Current adsorption, which is the most efficient process, is generally used (1). In addition to the complexity of this process, the choice of an adsorbent selective for paraxylene is the critical point.

**Table 10.1** Physical properties of xylenes

| | p-Xylene | m-Xylene | o-Xylene | Ethylbenzene |
|---|---|---|---|---|
| Formula | $C_8H_{10}$ | $C_8H_{10}$ | $C_8H_{10}$ | $C_8H_{10}$ |
| Mol. weight | 106 | 106 | 106 | 106 |
| bp (°C) | 138.35 | 139.1 | 144.4 | 136.2 |
| mp (°C) | 13.26 | -47.87 | -25.18 | -94.97 |
| Diameter (nm) | 0.67 | 0.71 | 0.74 | 0.67 |

In this paper, after a general presentation of separation processes, we propose some key features to seize the world of paraxylene separation, the emphasis being placed on the adsorbent selectivity and on the separation process.

## 10.2 The adsorbent for paraxylene separation

### *10.2.1 The notion of selectivity*

The primary requirement for an economic separation process is an adsorbent with high selectivity and capacity. The selectivity may depend upon differences in either kinetics or thermodynamic equilibrium of adsorption. Differences in diffusion rates between molecules, due to steric effects, can be large enough to provide transient selectivity. The separation factor is the ratio between the diffusion coefficients of the molecules.

However, in most cases, selectivity based on differences in adsorption of the components at thermodynamic equilibrium is preferred, the separation factor being then:

$$\alpha_{AB}=\frac{Y_A/Y_B}{X_A/X_B}$$

where $Y_A$, $Y_B$, and $X_A$, $X_B$ are the mole fractions of components A and B in adsorbed phase and in fluid phase at thermodynamic equilibrium, respectively. This separation factor is directly related to the multicomponent adsorption isotherm and is generally dependent on the temperature and on the composition of the fluid phase. However, for an ideal Langmuir isotherm, the separation factor is independent on composition and equal to the ratio of the Henry's constants for A and B components.

For paraxylene separation, both kinds of selectivity can be observed. In the MFI structure, the aperture of the pores is sufficiently close to the dimensions of the molecules to make shape selectivity appear. However, the kinetic diameters of paraxylene and of ethylbenzene are identical, so that the selectivity is not effective for these two components. Moreover, the capacity of MFI zeolites is weak compared to other structures. More open structures which provide the opportunity to use equilibrium selectivity are preferred. The problem is that the selectivity is mainly due to interactions between the zeolite and the aromatic ring which are identical for all the xylenes. It will be shown in the following sections that this problem can be solved by using chosen FAU zeolites.

*10.2.2 Structure of X and Y zeolites*

X and Y zeolites belong to the faujasite family (FAU structure). X zeolites are characterized by a Si/Al ratio in the range 1 to 1.5 and Y zeolites refer to higher Si/Al ratios. The framework structure has a cubic symmetry. It is built of sodalite cages (also called $\beta$-cages) linked together by hexagonal prisms, creating a large cavity called a supercage (Figure 10.1). Each supercage is connected to four sodalite cages through a 6-ring window (free aperture diameter of 0.22 nm) and to four other supercages through a 12-ring window (free aperture diameter of 0.74 nm). There are eight supercages and eight sodalite cages per unit cell (2).

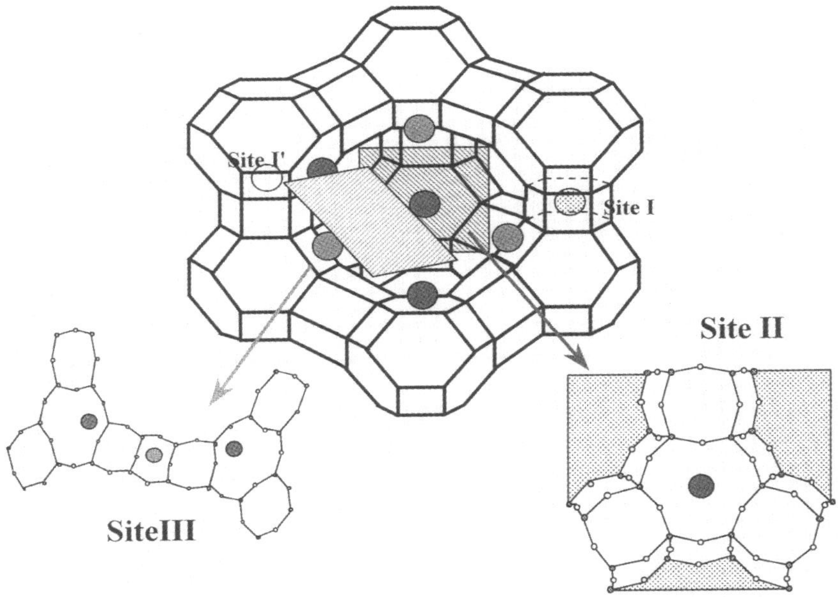


**Fig. 10.1** Faujasite structure

The compensation cations can be located in various positions. For example, in the BaX zeolite (3-5), barium cations are located in three crystallographic sites (Figure 10.1): Ba(1) is located in the center of the hexagonal prism (site I), Ba(1') in the sodalite cage in front of the 6-ring window connected with the hexagonal prism (site I'); the last site, Ba(2), is in the supercage (site II), and lies in front of the 6-ring window connected with the sodalite cage. Whereas large aromatic molecules like xylene molecules enter only supercages, small molecules such as water can be adsorbed in both supercages and sodalite cages. The adsorption sites are mainly located near the cations which neutralize the negative charges of the framework.

### *10.2.3 Factors determining the separation selectivity*

The selectivity of X and Y zeolites for paraxylene is a complex phenomenon. The molecules to be separated are very similar, as is shown in Table 10.1. Consequently, the selectivities are low and very sensitive to various parameters. The first one is the nature of the cation, which is able to reverse the selectivity. For example, NaY is selective for metaxylene and KY selective for paraxylene. This is clearly illustrated by the values given in patent literature (6) (some of the figures are reported in Table 10.2). They clearly demonstrate that, depending on the cation, the molecule which is preferentially adsorbed is either paraxylene, metaxylene or ethylbenzene.

Iwayama et al. (7) showed that paraxylene selectivity of alkaline Y zeolites exhibits a maximum with K (Table 10.2). This demonstrates that there is an optimum value in the cation properties (size, electronegativity and position). As was demonstrated by Seko et al. (8), the Si/Al ratio is also important: paraxylene selectivity increases with increasing Si/Al in the range of 1.2 to 3.

Researchers attempted to find correlations between the composition of the unit cell and the selectivity for paraxylene. In this respect, D. Barthomeuf (9) proposed an approach based on Sanderson's intermediate electronegativity which allows us to estimate the basicity of zeolite oxygens, and hence the strength of the acid-base interaction between xylene molecules and zeolites. It should be noted that these calculations provide an insight into the interactions between the zeolite structure and the molecules at low loading only, i.e. when the interactions between adsorbed molecules are negligible.

**Table 10.2** Selectivities of various X and Y exchanged zeolites

| Zeolite | Cation | Selectivity Px/Mx | Selectivity Px/Eb | Selectivity Mx/Eb |
|---|---|---|---|---|
| Y | Li | 0.72 | 1.52 | 2.13 |
| Y | Na | 0.75 | 1.32 | 1.74 |
| Y | K | 1.83 | 1.16 | 0.64 |
| Y | Rb | 0.96 | 1.51 | 0.64 |
| Y | Ba | 1.27 | 1.85 | 1.45 |
| Y | KBa | 3.75 | 2.10 | 0.56 |
| X | Na | 1.02 | 1.15 | 1.14 |
| X | BaK | 2.49 | 2.03 | 0.81 |

More recently, it was shown (10, 11) for a series of FAU samples that the zero loading adsorption enthalpies were very similar for paraxylene and metaxylene (Table10.3). This indicates that at low loading, very few differences can be expected for the adsorption of paraxylene and metaxylene.

**Table 10.3** Adsorption heats of paraxylene and metaxylene on X and Y zeolites (10, 11)

| Zeolite | Adsorption heat (kJ/mol) | |
|---|---|---|
| | paraxylene | metaxylene |
| NaX | 86.5 | 85 |
| NaY | 76 | 78 |
| KY | 76 | 76 |
| BaX | 115 | 115 |
| BaY | 107 | 115 |

The situation is dramatically different at high loading, i.e. at saturation of the supercages. Cottier et al. (12) demonstrated that loading is a key parameter which can strongly influence the selectivity. For example, with KY, selectivity for paraxylene is only obtained at high loading of the zeolite (over two molecules per cavity). This means that additional phenomena occurring at high loading are driving the selectivity. The structural work performed by Mellot (4, 13, 14), Descours (5) and Pichon et al. (15, 16) provide a good insight into this phenomena. Thanks to neutron diffraction combined with the Rietveld method, the authors located xylene molecules in the zeolite cavities as a function of loading. They demonstrated that in BaX zeolites, the first molecules adsorb close to site II and that the increase of the loading leads to strong rearrangements and minimization of the interactions between the aromatic ring and the cation. This is associated with increasing interactions between the methyl groups and the network oxygens, which is clearly illustrated in Figure 10.2. In this respect paraxylene selectivity is only obtained because of a favorable packing in the supercage. Infrared studies were also performed by Mellot and Descours (4, 5). They showed that infrared bands ($\delta$ C=C, for example) for the first adsorbed molecules were strongly shifted (compared to the liquid). At high loading, the bands went back to more or less a liquid behavior.

(a)

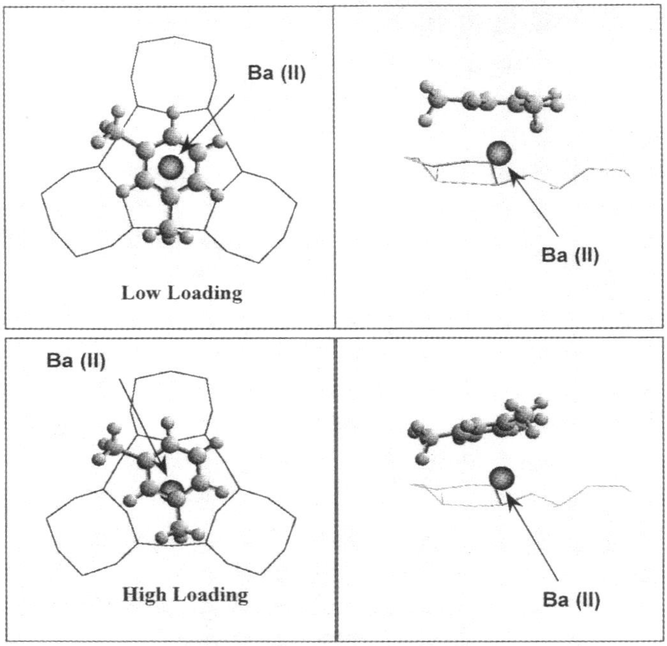


(b)

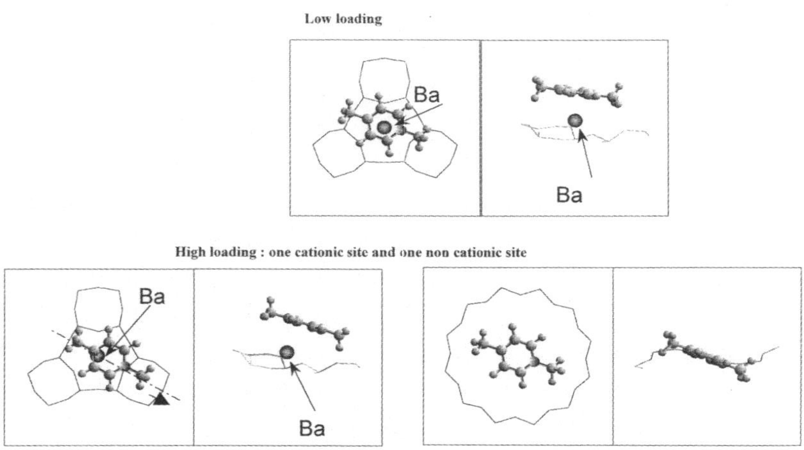


**Fig. 10.2** Location of metaxylene (a) and paraxylene (b) molecules as a function of loading in a prehydrated BaX zeolite

The last key parameter is the preadsorption of small polar molecules on the zeolite adsorbent. It is now well established that preadsorbed water can modify the selectivity of BaX zeolites. This was illustrated by Furlan et al. (17) who found a maximum in the curve of selectivity versus water content. Pichon et al. (15, 16) demonstrated that the water molecules were strongly interacting with the cations, consequently weakening the interaction between the aromatic ring and the cation. This is in favour of the paraxylene selectivity as it allows the occurrence of an additional non-cationic site for paraxylene. Again, the packing of paraxylene is more efficient than the packing of metaxylene. However, the drawback of water adding is a decrease in the adsorption capacity. Hence water content must be optimized.

The complexity of xylene adsorption over zeolites is too high to predict the selectivity from the chemical properties of the zeolite only (electronegativity of the cations, charge of the framework oxygens). The interactions between xylenes and the zeolite must necessarily be considered, which explains the important development of molecular simulation methods. This is supported by the work of V. Lachet et al. (18) who succeeded in reproducing the inversion of selectivity between KY and NaY with Grand Canonical Monte Carlo Simulations.

### *10.2.4 Conclusion*

Paraxylene selectivity is a complex phenomenon which can only be observed at high loading and for bulky and weakly charged cations. This phenomenon is related to entropy effects which allow paraxylene to be more efficiently packed in the zeolite micropores. This leads to paraxylene/metaxylene selectivity over three and paraxylene/ethylbenzene over two, which is enough for a separation process. It is now established that steric effects are very important to make the adsorbent selective for paraxylene. These effects are driven by the size, the number and the position of the compensation cations. The charge of the cation also plays an important role by controlling these effects.

Paraxylene selectivity can only occur by minimizing the electrostatic interactions between cations and aromatic ring, i.e. at high loading and with water molecules.

The prediction of the selectivity (in tendency and in value) is impossible, without using methods which simulate the complex interactions between molecules and zeolite adsorbents.

## 10.3 Separation processes

### *10.3.1 General principles*

Adsorption separation processes are all based on chromatographic processes which are discontinuous. However, for efficiency reasons, large scale processes need the continuous production of one or several pure products, which means that the adsorbent must be alternatively saturated and regenerated. The processes which can operate in liquid or gas phases can be divided into two classes:

- Cyclic systems in which the adsorbent bed is alternatively saturated and regenerated
- Continuous flow systems involving countercurrent contact between feed and adsorbent

Between these two classes, the processes can be sub-classified according to the methods of adsorbent regeneration.

For cyclic systems, four basic methods are generally used alone or in combination:

- Thermal swing operation, in which the bed is regenerated by heating with a stream of hot gas or hot liquid
- Pressure swing, in which the regeneration is obtained by reducing the pressure at constant temperature
- Purge gas stripping. The bed is regenerated by purging with a non adsorbing inert gas
- Displacement desorption. The method is essentially the same as for purge gas stripping but the regeneration stream is adsorbed and displaces the adsorbed species

The choice between these various regeneration methods depends on technical considerations as well as on economic factors. Thermal swing processes are the most common systems. They are efficient for strongly adsorbed species, but not suitable for rapid cycling, which means that the adsorbent cannot be used with maximum efficiency. On the other hand, pressure swing processes are preferable for weakly adsorbed species, but mechanical energy is more expensive than heat. Lastly, displacement desorption is well adapted for strongly adsorbed species and moreover reduces the ageing of the adsorbent compared to the other methods. The choice of the desorbent is critical in the system.

Only displacement desorption is used for paraxylene separation processes, which is carried out in continuous countercurrent systems.

### *10.3.2 Continuous countercurrent processes*

#### 10.3.2.1 Introduction

Countercurrent contact maximizes the driving force for mass transfer and therefore provides, in principle, a more efficient use of the adsorbent than in a cyclic process. However, this system requires one either to circulate the adsorbent, or, by an appropriate design of the fluid flow system, to simulate adsorbent circulation (as we will see later). Therefore, the technology required for operating these systems is highly critical. Once again, the balance between economic and technical criteria must be considered to choose the process. As a rough guide, this technology is generally used for difficult separations with highly valuable products.

#### 10.3.2.2 Actual countercurrent adsorption

To understand the concept of simulated moving bed chromatography, we will first investigate the simpler system called true moving bed chromatography which is described in Figure 10.3. In this system, a single chromatographic bed is separated into four distinct zones. This bed has two liquid inlets and two liquid outlets. In the true moving bed experiment, the chromatographic bed is moving from bottom to top. The mobile phase (the desorbent) is circulating in the opposite direction. It enters the column at zone I. In this system, the solid exits the column at zone I and is recycled to the bottom of the column at zone IV. The eluent does just the opposite and is recycled from zone IV back to Zone I. The feed containing A and B is continuously injected in the middle of the column. Provided that the affinities of A and B for the solid are different (A being the most strongly adsorbed), one can choose four different flow rates that will cause the displacement of A in the upper direction and the displacement of B in the opposite direction. A and B can then be removed through the extract and raffinate lines.

In this system, the continuous injection of the feed and the continuous collection of A and B from the extract and the raffinate is an ideal system which provides the maximum efficiency of the adsorbent. A and B are recovered diluted in the desorbent but can be obtained pure after distillation. From a practical point of view, the true moving bed concept would be extremely difficult to implement. Although recycling the liquid would be feasible, the circulation of a solid packing from the bottom to the top of the fixed column presents a mechanical challenge. This technological obstacle is bypassed by using the simulated moving bed.

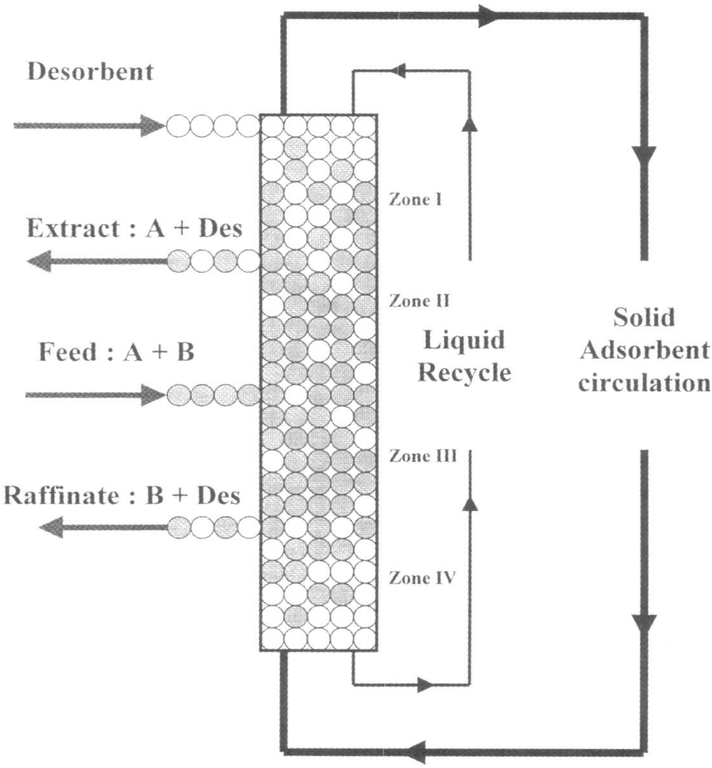


**Fig. 10.3** Principle of actual counter current adsorption.

10.3.2.3 Simulated moving bed adsorption

The concept of the simulated moving bed is based on the periodical change of the inlet and outlet along the fixed beds as it is shown in Figure 10.4. For practical reasons, the fixed bed is separated in a series of interconnected smaller beds. The inlets and outlets are moved periodically from one column connection to another via a complex valve system. The relative solid's movement is related to the shift period of the inlet-outlet lines. The monitoring of the system consists of maintaining a quasi stationary state where the profiles for A and B are fixed all along the bed and where A and B are separated (pure A and S withdrawn in the extract and pure B and S withdrawn in the raffinate). This is achieved by adequately choosing the four internal flow rates and the switching time. The number of beds must be high enough to achieve a quasi-continuous solid circulation. The system works efficiently, provided a high selectivity and capacity of the adsorbent, high diffusion rates and solvent efficiency can be obtained. This last point is a key factor as the solvent S must be able to be replaced in the solid by the less strongly adsorbed component B and must be able to replace the most strongly adsorbed component A. It is important here to notice that a global optimization of the system (process + solvent + solid) is necessary to achieve good performance. In particular, the ternary phase thermodynamic and diffusion behaviour must be controlled in order to be used in simulations.

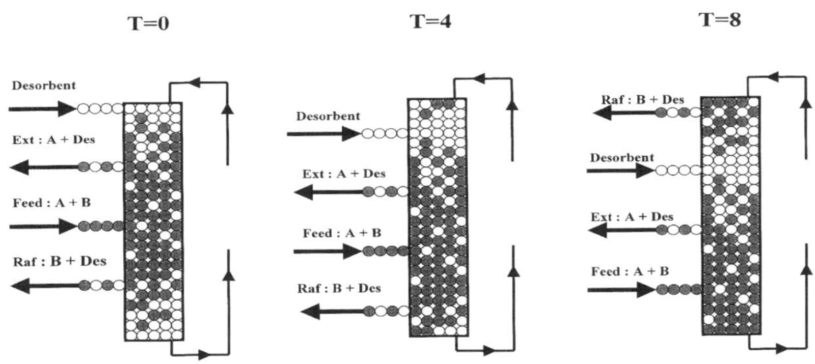


**Fig. 10.4** Principle of simulated counter current adsorption

### 10.3.2.4 Simulated counter current adsorption for paraxylene separation

Simulated moving bed is particularly well adapted to the production of high purity paraxylene. Two versions of the process exist:

- The stand alone process allows the direct production of high purity paraxylene. It works similarly to the process described in section 10.3.2.3. Generally two adsorbent towers of 12 beds each are required to achieve high performance
- The hybrid process combines the best features of adsorption and crystallization. It is particularly well adapted to retrofitting of existing units. It consists of a simplified adsorption section with only higher productivity, fewer stages and smaller adsorbent inventory. Low purity paraxylene is produced (90 to 95%) with a very high rate of recovery. The paraxylene coming from the adsorption section is then further purified in a small single stage crystallizer and the filtrate is recycled

back to the adsorption section. Ultra high purity paraxylene can be produced easily and economically with this technology.

The sieves which are used are BaX or KBaY zeolites. The zeolite cannot be used without appropriate shaping in order to avoid too high a pressure drop. Beads or granulates with a diameter smaller than 1 mm allow maximum packing, maximum diffusion rates and good mechanical properties.

The solvents which are chosen are aromatic compounds heavier or lighter than the $C_8$ aromatics, in order to allow easy separation by distillation. Paradiethylbenzene, toluene or indane are frequently cited in patents or papers. The adsorption strength of the solvent has, however, to be adjusted to the process (standalone or hybrid) in order to achieve direct high purity or high productivity and yield. The temperature has to be chosen for optimizing adsorption selectivity (which decreases with increasing temperature), while minimizing the mass transfer resistances and lowering the pressure drops in the columns. In order to take into account these constraints, the temperature is always between 150°C and 200°C.

## 10.4 Conclusion

X and Y zeolites exchanged with potassium and barium cations and associated with an adequate solvent are efficient for paraxylene separation by adsorption. The design of the adsorbent as well as of the process are both critical to reach economical production of paraxylene. The complete system, adsorbent plus process with the chosen configuration – stand alone or hybrid (adsorption + crystallization) - has to be optimized.

## References


1. Neuzil R.W., U.S. Patent **3,558,730**, (1971).
2. Meier W.M., Olson D.H. and Baerlocher Ch., *Atlas of Zeolite Structure Types*, 4th ed., (Elsevier London).
3. Barrer R.M. in *Zeolites and Clay Minerals as Sorbent and Molecular Sieves*, (Academic Press, New-York, 1978).
4. Mellot C., Thesis, *Université de Paris VI, France*, (1993).
5. Descours A., Thesis, *Université de Bourgogne, Dijon, France*, (1997).
6. Stine L.O. and Broughton D.B., US Patent **3636121**, UOP (1969).
7. Iwayama K., Suzuki M., *Studies in Surface Science and Catalysis* **83** (1994), 243.

8. Seko M., Miyake T. and Inada K., *Hydrocarbon Processing*, (1980), 133.
9. Barthomeuf D., *J. Phys. Chem.* **88** (1984) 42.
10. Bellat J. P., Simonot M. H. and Jullian S., *Zeolites* **15** (1995) 124.
11. Bellat J.P.and Simonot-Grange M.H., *Zeolites* **15** (1995) 219.
12. Cottier V., Bellat J.P., Simonot-Grange M.H. and Méthivier A., *J. Phys. Chem. B* **101** (1997) 4798.
13. Mellot C. et al., *Catal. Letters* **27** (1994) 159.
14. Mellot C. et al, *Langmuir* **11** (1995) 1726.
15. Pichon C., Méthivier A., Simonot Grange M.H. and Baerlocher C., *J. Phys. Chem.* **103** (1999) 10197.
16. Pichon C. Méthivier A. and Simonot Grange M.H., *Langmuir* **16** (2000) 1931.
17. Furlan L.T., Chaves B.C. and Santana C.C., *Ind. Eng. Chem. Res.* **31** (1992) 1784.
18. Lachet V., *Thesis Université de Paris Sud Orsay* (1998)
19. Lachet V., Buttefey S., Boutin A. and Fuchs A., *Phys. Chem. Chem. Phys.*, **3** (2001), 80.
20. Larson A.C. and Von Dreele R.B., *GSAS General Structure Analysis System*, LAUR 86-748, Los Alamos National Laboratory; (1994).
21. Rietveld H.M., *J. Appl. Cryst.* **2** (1969) 165.

CHAPTER 11

# AROMATIC ALKYLATION : TOWARDS CLEANER PROCESSES

J.S. BECK, A.B. DANDEKAR, T.F. DEGNAN
*ExxonMobil Refining and Supply Company, Process Research Laboratories*
*1545 Route 22 East Annandale, New Jersey 08801, USA*

## 11.1 Introduction

Alkylation of aromatic compounds is practiced commercially on a large scale throughout the world. Some of the major products produced via alkylation include: ethylbenzene, subsequentally converted to styrene for polymer use; isopropylbenzene, a precursor to important solvents and chemical intermediates such as acetone and phenol; alkali benzene sulfonates, used extensively in detergent building; and alkylnapthalenes, precursors to advanced high strength polymers. Alkylation of aromatics is achieved with a variety of alkylating agents, including alcohols, olefins, ethers, acids, esters, ketones and halogenated derivatives of these.

Catalysts for aromatics alkylation range from common Brönsted acids such as sulfuric, phosphoric and hydrofluoric acids, to Friedel-Crafts agents, e.g., Lewis Acids such as aluminum chloride and other metal halides, and range further to solid acids such as silica aluminas and alumina itself, which may contain both Brönsted and Lewis sites. Catalysts for selective production of desired alkylated species include the wide variety of microporous aluminosilicate zeolites, which, by virtue of their discrete pore size and geometry and Brönsted acidity, are capable of highly specific production of target alkylated aromatic compounds, even so far as to direct the production of a single isomer of such a species. Several excellent references on aromatics alkylation have covered both synthetic and mechanistic issues concerned with the production of these materials (1-3). Some of the most important alkylaromatic compounds, their applications and general mechanistic trends are reviewed in the following subsections.

## 11.2 Industrially important alkylaromatics

### *11.2.1 Ethylbenzene (EB)*

Ethylbenzene (EB) is the precursor molecule for production of styrene monomer. It may be synthesized via alkylation of benzene and is also produced as a component of the C8 aromatics fractions obtained from catalytic crackers and reforming units (4). EB is converted to styrene via dehydrogenative or oxidative routes. As much as

90% of the world's EB production capacity is produced via the alkylation of benzene with ethylene. The alkylation of benzene with ethylene can be carried out in several ways, and the primary catalyst system used employs Friedel-Crafts alkylation with $AlCl_3$ as catalyst. One great drawback of these processes is the highly corrosive nature of the aluminum halide catalyst. With this main drawback as incentive, the development of a heterogeneous based zeolite catalyst for this transformation was explored. The Mobil/Badger vapor phase EB process, first commercialized in the 1970s, employs a highly selective ZSM-5 catalyst and produces higher EB yields than any other process now in commercial use. Successful alternatives to the Mobil/Badger vapor phase EB technology involve the liquid phase alkylation of benzene with ethylene. These types of processes also utilize zeolitic catalyst systems, and several are in commercial development or recently licensed, including ExxonMobil's EBMAX. The liquid phase operations are favorable for their decreased by-product yields.

### *11.2.2 Isopropylbenzene (IPB)*

The major industrial uses of isopropylbenzene (IPB) or cumene involve its conversion to phenol and acetone via oxidative routes (5). IPB is produced via the alkylation of benzene with propylene. Standard operating technology involves passing benzene and a propylene/propane stream over supported phosphoric acid. Earlier technology includes the use of sulfuric acid and aluminum chloride as catalysts. In the 1990s, liquid phase processes utilizing some state-of-the-art zeolite catalyst systems were developed and commercialized. Most of these technologies also have the attractive feature of transalkylation chemistry to dispose of unwanted, heavy diaalkylates. The advantages of such zeolite based processes rely on the ability to effect selective chemistries, the non-corrosive nature of the material, and reduction of catalyst costs as the materials may be regenerated. ExxonMobil has recently introduced a new commercial catalyst for cumene production. The new liquid phase alkylation process employs a new proprietary catalyst system, producing high yields of cumene in high purity at low operating cost.

### *11.2.3 Para-Ethyltoluene (PET)*

Dehydrogenation of para-ethyltoluene (PET) gives para-methylstyrene, a very useful polymer precursor molecule yielding polymers which have improved properties, such as higher flash-point and glass transition temperature, than those obtained via styrene polymerization. Early work in this area conducted at Mobil has demonstrated that modified HZSM-5 zeolite can result in an improved selectivity to the desired para-isomer. Characterization studies on these modified materials showed that the effect of the modifiers was to decrease the strength of Brönsted acid sites and increase the number of Lewis acid sites (6). Exxonmobil's PET catalyst system has been in commercial use for over 3 years now.

### *11.2.4 Cymene (CY)*

Cymene or isopropyltoluene is produced via alkylation of toluene with propylene. Cymene is an important intermediate in the production of cresol, and it is also used as an industrial solvent. Again, for both environmental and economic reasons, the use of zeolitic materials for this conversion has been studied. For example, Flockhart et al. have used zeolite Y to effect this reaction (7). They observed that the state of the zeolite, including its degree of ion-exchange and the temperature at which it was calcined, strongly affected the distribution of cymene isomers obtained. In order to enhance the selectivity to para-cymene, the direct precursor to para-cresol, various studies have focused on the use of surface modified zeolites, for example, ZSM-5 materials, including those produced by chemical vapor deposition (CVD) of silicate esters. These species serve to reduce surface acidity and change limit diffusion within the crystal.

### *11.2.5 Diisopropylbenzene (DIIPB)*

Diisopropylbenzenes are valued precursors to subsequent oxidation products. For example, 1,3-Diisopropyl- and 1,4-diisopropylbenzenes are used via oxidation chemistry to produce the materials resorcinol and hydroquinone, respectively.

### *11.2.6 Alkylnapthalenes [AN]*

Alkylnapthalenes are important monomers for the production of advanced aromatic polymer materials and are used in fine chemical synthesis as chemical intermediates. This area has recently been reviewed by Song et al. (8-9). For example, commercially produced vitamin K is prepared via a 2-methylnapthalene intermediate, and 2,6-dialkylnapthalene is an important precursor for the production of polyethylene napthalate (PEN), polybutylene napthalate (PBN) and liquid crystalline polymers. Use of heterogeneous catalysts to synthesize alkylnapthalenes has attracted recent attention (10-12). The great difficulty in preparing alkylnapthalenes in high selectivity is manifested in the great number of isomer positions which are possible. Compared to the dialkylbenzenes in which 3 isomers, ortho, para and meta are possible, there are ten possible isomers for a dialkylnapthalene system. Therefore, a catalyst chosen for this chemistry must be tuned in terms of activity and pore size to selectively produce the desired product.

### *11.2.7 Alkylphenols (AP)*

Alkylphenols may be produced via the alkylation of phenol with methanol. This reaction produces predominantly anisole and o-cresol with methylanisole and xylenol also being obtained. Strong Brönsted acids are not required to effect these transformations, as both amorphous aluminas and silica/aluminas are active catalysts. Often o-cresol can be produced in 100% isomeric selectivity, particularly when the reaction is run over amorphous alumina based catalysts. When zeolites are used isomer selectivities are changed.

## 11.3 General mechanistic trends

Alkylation of aromatics occurs principally via electrophilic substitution of the aromatic ring. This process involves the attack of an electrophile, E+, on the aromatic ring, e.g. ArH, which serves as a nucleophile. The result of this attack is the formation of the alkylated aromatic product and a proton as shown below:

$$E^+ + ArH \longrightarrow Ar\text{-}E + H^+$$

In the above process, C-H bond rupture, attachment of the electrophilic fragment to the ring and proton ejection restore ring aromaticity.

In the alkylation of arene systems with olefins, the above chemistry proceeds via a carbenium ion mechanism. Alkylation of an aromatic compound with an olefin occurs by the interaction of a Brönsted acid site of any of the catalysts with a participating olefin, creating a carbenium ion, via protonation of the double bond, and thus a polarized complex is formed, as shown below:

$$C{=}C + HA \rightleftharpoons C{-}C^+ A^- \overset{+ArH}{\rightleftharpoons} [C{-}C{-}ArH]^+ A^- \rightleftharpoons C{-}C{-}Ar + HA$$

The formation of this polarized complex intermediate may be achieved with any of the acid type species discussed above. However, it should be noted that in the case of the metal halides, alkylation is only achieved when a proton donor source such as water or a mineral acid is present.

The synthesis of aromatics alkylated in the desired positions (selective production of positional isomers) for commercial use has been studied extensively. The extent of alkylation, e.g. monoalkylates to polyalkylates, the position of the isomers and the yields obtained are influenced by a number of parameters, including reactivity of the alkylating agent and aromatic to be alkylated and steric constraints of the resulting alkylated products. On the catalyst side, issues such as type of catalyst, e,.g. its acid strength, reaction temperature, pressure and throughput or contact time over the catalyst are all important considerations. Needless to say, much research in this area involves tuning the above parameters to produce the highest value products, with a minimum number of side-products that can be easily separated from the compound of interest.

## 11.4 History and evolution of present technologies

### *11.4.1 Ethylbenzene*

The worldwide capacity for ethylbenzene is currently estimated to be 23 million metric tons/year (1 metric ton = 1000 kg) with an annual growth rate projected to be approximately 4%. Over 90% of the world's production of ethylbenzene is used in the manufacture of styrene. Other applications include paint solvents and pharmaceuticals.

Nearly 40% of the world's ethylbenzene capacity still utilizes the $AlCl_3$ -based Friedel-Crafts alkylation process first introduced in the 1950s. This process operates at low benzene/ethylene (B/E) ratios (2 to 3.5) and reasonable temperatures (250°C). However, over the past twenty years the operating costs associated with the corrosivity of $AlCl_3$, and other problems associated with its safe handling and disposal have induced most manufacturers to move toward zeolite catalyzed processes. Zeolite catalyzed processes are licensed by Mobil-Badger, Lummus - UOP, CDTech, and Dow. Over the last several years, the industry has moved to more selective ethylbenzene dehydrogenation catalysts, making the purity of ethylbenzene of even greater importance. The selectivity to the monoethylated product is very important in ethylbenzene synthesis, whereas ethylene oligomerization is normally a secondary consideration.

First introduced in 1980, the Mobil-Badger vapor phase process is still the most widely used zeolite catalyzed ethylbenzene manufacturing process. Since its initial commercial application in 1981, over 35 units have been licensed with a total annual capacity of nearly eight million metric tons. In the Mobil-Badger process, benzene is alkylated in a fixed bed reactor with ethylene in the gas phase using a ZSM-5 based catalyst (13, 14). Fresh and recycled benzene are combined with a diethylbenzene-rich recycle stream and fed, together with fresh ethylene, to an alkylation reactor. The process operates with a benzene to ethylene ratio of 5 to 20 (molar) at temperatures ranging from 370 to 420°C and pressures ranging from 0.69 to 2.76 Mpa (6.8 to 27.2 bar). Weight hourly space velocities, based on fresh feed, range from 300 to 400 $hr^{-1}$. The process has an overall ethylbenzene yield of a least 99.0%.

Together with the Badger Technology Center of Raytheon Engineers & Constructors, Mobil introduced second and third generation processes in 1986 and 1991. These next generation processes provided improved cycle lengths and higher yields (15). Cycle lengths improved from 60 days to more than one year, and yields improved to greater than 99.5%. Selectivity losses to both heavy aromatics and xylenes dropped by more than 50%. The energy efficiency of the Mobil-Badger process is reported to be high. Essentially all of the exothermic heat of reaction is recovered.

The Mobil-Badger process has also been modified to use dilute ethylene from FCC off-gas streams as a feedstock. The dilute ethylene process was first commercialized at Shell's Stanlow U.K. refinery in 1991 and has been in continuous use ever since.

EBMax is a liquid phase ethylbenzene process that uses Mobil's proprietary MCM-22 zeolite as the catalyst. This process was first commercialized at the Chiba Styrene Monomer Co. in Chiba, Japan in 1995 (16-18). The MCM-22-based catalyst is very stable. Cycle lengths in excess of three years have been achieved commercially. The MCM-22 zeolite catalyst is more monoalkylate selective than large pore zeolites including zeolites beta and Y. This allows the process to use low feed ratios of benzene to ethylene. Typical benzene to ethylene ratios are in the range of 3 to 5. The lower benzene to ethylene ratios reduce the benzene circulation rate which, in turn, improves the efficiency and reduces the throughput of the benzene recovery column. Because the process operates with a reduced benzene circulation rate, plant capacity can be improved without adding distillation capacity. This is an important consideration, since distillation column capacity is a bottleneck in most ethylbenzene process units. The EBMax process operates at low temperatures, and therefore the level of xylenes in the ethylbenzene product is very low, typically less than 10 ppm.

In the EBMax process, benzene is fed to the bottom of the liquid-filled multi-bed reactor. Ethylene is co-fed with the benzene and also between the catalyst beds. Polyethylbenzenes, which are almost exclusively diethylbenzenes, undergo transalkylation with benzene in a second reactor. Mobil-Badger offers both liquid phase and vapor phase transalkylation processes. The vapor phase process removes benzene feed coboilers such as cyclohexane and methylcyclopentane as well as propyl and butylbenzenes. Because the EBMax process produces very low levels of propyl and butylbenzenes, for most applications, the more energy efficient liquid phase process is preferred. Worldwide, there are currently ten licensed EBMax units with a cumulative ethylbenzene production capacity of five million metric tons per year.

### *11.4.2 Cumene*

Cumene capacity topped 9.5 million metric tons in 1998 and is projected to reach 10.4 million metric tons by the end of 2003 (19). Like ethylbenzene, cumene is used almost exclusively as a chemical intermediate. Its primary use is in the co-production of phenol and acetone through cumene peroxidation. Phenolic resins and bisphenol A are the main end uses for phenol. Bisphenol A, which is produced from phenol and acetone, has been the main driver behind increased phenol demand. Its end use applications are in polycarbonate and epoxy resins. The growth rate of cumene is closely related to that of phenol and is expected to be approximately 5.1% per year worldwide over the next five years. Process technologies for both chemicals have been moving away from conventional aluminum chloride and phosphoric acid catalyzed Friedel-Crafts alkylation of benzene, toward zeolite-based processes.

Prior to 1992, virtually all cumene was produced by propylene alkylation of benzene, using either solid phosphoric acid (SPA) or aluminum chloride as catalysts. The SPA process, currently licensed by UOP, was developed in the 1940s primarily to produce cumene for aviation fuels (20, 21). More than 40 SPA plants have been licensed worldwide. The SPA catalyst consists of a complex mixture of orthosiliconphosphate, pyrosiliconphosphate, and polyphosphoric acid supported on kieselguhr. To maintain the desired level of activity, small amounts of water are continuously fed to the reactor. The water continually liberates $H_3PO_4$, causing some downstream corrosion. SPA process conditions include pressures that range from 3.0 to 4.1 MPa, (29 to 40 bar), temperatures that range from 180 to 230°C, benzene:propylene ratios from 5 to 7, and weight hourly space velocities (WHSV's) that are between 1 and 2. Approximately 4 to 5 wt% of the product consists of di- and triisopropylbenzenes.

In the early 1980s Monsanto introduced an $AlCl_3$ process based on the same chemistry used in the ethylbenzene process. This process can be operated at lower benzene/propylene ratios than the SPA process because $AlCl_3$ can transalkylate the polyalkylated benzenes back to cumene. The process also operates at temperatures lower than the SPA process because the more highly acidic anhydrous $AlCl_3$ tends to produce significantly more undesired n-propylbenzene at equivalent temperatures. This technology is currently used in five plants.

Over the past seven years, cumene producers have begun to convert to the more environmentally friendly and more efficient zeolite-based processes. Principal among these are processes licensed by Dow, CDTech, Mobil-Badger, Enichem, and UOP. The zeolite based processes produce higher cumene yields than the conventional SPA process because most of the diisopropylbenzene byproduct is converted to cumene in separate transalkylation processes. Operating and maintenance costs are reduced because there is no corrosion associated with the zeolite catalysts. Finally, environmental concerns associated with the disposal of substantial amounts of phosphoric acid laden kieselghur or $AlCl_3$ are eliminated by the use of zeolites because they are regenerable and can be safely disposed of through digestion or landfilling.

First commercialized at Georgia Gulf's Pasadena, TX plant in 1994, the Mobil-Badger Cumene process consists of a fixed-bed alkylator, a fixed-bed transalkylator and a separation section (22, 23). Fresh and recycle benzene are combined with liquid propylene in the alkylation reactor where the propylene is completely reacted. Recycled polyisopropylbenzenes are mixed with benzene and sent to the transalkylation unit to produce additional cumene. Trace impurities are removed in the depropanizer column. Byproduct streams consist of LPG (mainly propane contained in the propylene feedstock) and a small residue stream, which can be used as fuel oil.

The Mobil-Badger process produces very pure cumene, 99.97 wt% at 99.7 wt% yield. Ethylbenzene, propylbenzene, and butylbenzene levels are an order of magnitude lower than those obtained with the SPA catalyzed process. The high cumene purity is primarily attributable to the high monoalkylation selectivity of the MCM-22 catalyst. This catalyst is unique in its ability to minimize propylene oligomerization while still exhibiting very high activity for benzene alkylation. Commercially, the catalyst has demonstrated cycle lengths in excess of two years. Ultimate catalyst life is in excess of five years. The Mobil-Badger process is used in seven commercial units worldwide. The seven plants have a combined cumene capacity of 3.1 million metric tons per year, or approximately 50% of the worldwide cumene capacity. An additional three plants are scheduled to start up with this process by 2001. This will increase total licensed capacity to 4.1 million metric tons.

### *11.4.3 MCM-22 catalyst technology*

Historically, major improvements in mature chemical process technologies have always coincided with advances in catalysis. This has certainly been true in ethylbenzene technology, where significant improvements in the process yield, product purity, and process economics have been realized through the development of improved zeolite-based catalysts. Among the most critical properties in a desirable alkylation catalyst is the mono-selectivity to ethylbenzene. Although all modern ethylbenzene technologies provide for a secondary transalkylation step that converts the polyethylbenzenes formed in the primary alkylation step back to ethylbenzene, the quantities of impurities and heavy byproducts per unit of ethylbenzene formed are often higher in the secondary step than in the alkylation step. Application of an improved alkylation catalyst will lead to the formation of relatively reduced quantities of polyethylbenzenes and final heavy byproducts, resulting in improved product purity as well as yield, and the accompanying reduction in capital cost.

Compared to other technologies, the Mobil/Raytheon EBMax$^{TM}$ process is much more selective towards the monoalkylation of benzene to ethylbenzene, owing to the use of Mobil's novel zeolite molecular sieve, MCM-22 (Mobil Composition of Matter 22). Like ZSM-5, MCM-22 is considered to be a medium pore zeolite; However, it possesses unique structural features that have not been seen before in molecular sieves. The structure of MCM-22 was solved by Mobil scientists using high resolution microscopy and X-ray diffraction analysis (24, 25). It consists of two-independent non-intersecting channel systems, each accessible through 10-member ring apertures. One of these pore systems is defined by two-dimensional, sinusoidal channels of slightly elliptical cross sections. The other is comprised of supercages whose inner free diameter is defined by 12-member rings. Structural analysis also indicates that the surface of MCM-22 crystal is covered with 12-ring pockets, each of which is half of a super cage (Figure 1).

MCM-22 crystallizes in thin sheets or plates. Within these sheets, there is evidence of the existence of a buried T-site (a Si or Al atom) in the framework structure that is not accessible to a channel wall. This buried T-site appears to be unique to MCM-22 and gives the zeolite structural stability even under severe conditions that would destroy other less-stable zeolites. MCM-22 has been actively studied and over 50 US patents have been issued on its use in varied areas, including hydroprocessing, aromatic/olefin alkylation, paraffin/olefin alkylation and dispro-portionation processes, as well as specialty chemical applications.

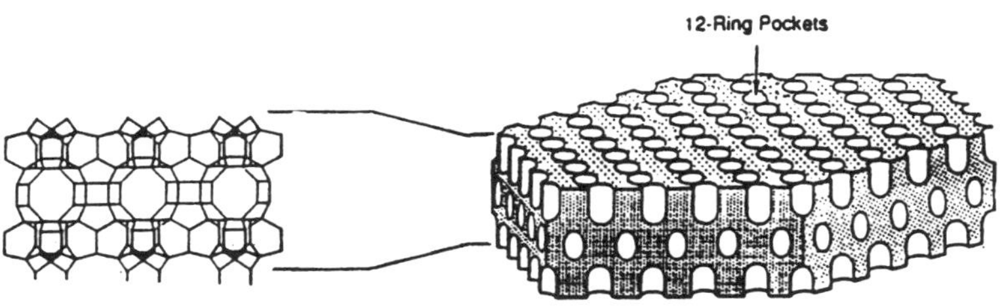


**Fig. 11.1** Schematic illustration of a single-layer hexagonal crystallite of MCM-22 showing surface pockets

It has been shown that single ring aromatic alkylation reactions such as benzene to ethylbenzene take place primarily within the 12- ring (12-MR) system, and that the 10-ring (10-MR) system contributes little to the ethylbenzene reaction. A key feature of MCM-22 is its ability to operate stably at low benzene-to-ethylene ratios with minimal production of polyethylbenzenes (PEBs) or ethylene oligomers. The excellent ethylbenzene selectivity of the MCM-22 catalyst is likely due to confinement effects within this 12-MR pore system and to the very facile desorption of ethylbenzene from the surface pockets of MCM-22. MCM-22's monoalkylation activity surpasses that of any other commercially demonstrated zeolite alkylation catalyst by a wide margin. Under EBMax$^{TM}$ operating conditions, ethylene oligomerization is also entirely suppressed. MCM-22 also exhibits substantially higher activity than other zeolites, resulting in catalyst loads that are a fraction of those required for the competing catalysts.

MCM-22 is a commercial catalyst with a production history of over seven years, and operation history of over five years. Catalyst performance in these applications has been outstanding. In MCM-22's first ethylbenzene commercial application at Denka's Chiba Styrene Monomer Company Ltd., a cycle length of over three years has been demonstrated without any significant aging of the catalyst or change in yields. Like ZSM-5, MCM-22 is regenerable, and is environmentally inert.

Alkylation chemistry over MCM-22 is characterized by essentially complete ethylene conversion with very high selectivity to ethylbenzene over a wide range of operating conditions. The reaction

| | | | | |
|---|---|---|---|---|
| $C_6H_6$ | + | $C_2H_4$ | ==> | $C_6H_5\text{-}C_2H_5$ |
| Benzene | | Ethylene | | Ethylbenzene |

is highly favored thermodynamically and is irreversible for practical purposes. However, alkylation of benzene does not terminate at monoalkylbenzene. Part of the ethylbenzene formed is further reacted to diethylbenzenes, triethylbenzenes, tetraethylbenzenes, etc., collectively referred to as polyethylbenzenes (PEBs). For example,

| | | | | |
|---|---|---|---|---|
| $C_6H_5\text{-}C_2H_5$ | + | $C_2H_4$ | ==> | $C_6H_4\text{-}(C_2H_5)_2$ |
| Ethylbenzene | | Ethylene | | Diethylbenzenes |

The monoalkylation selectivity of the alkylation step refers to the fraction of ethylene that reacts to form ethylbenzene, as opposed to forming polyethylated species. To suppress the formation of PEBs, benzene must be fed to the alkylation reactor in considerable excess (frequently five to seven times the stoichiometric requirement). Equipment in the alkylation reaction and benzene recovery systems must therefore be sized to accommodate the flow of excess benzene, and energy must be expended to recover the excess benzene from the reactor effluent. However, the superior monoalkylation selectivity and stability of MCM-22 permits operation with reduced benzene feed rates - in the range of two to four times the stoichiometric requirement - without excessive PEB formation (see Table 11.1).

**Table 11.1** Comparison of Product Selectivities for MCM-22, Beta, and USY Catalysts (220°C, 3.44 MPa, and 4:1 benzene/ethylene molar ratio)

| | MCM-22 | Beta | USY-1 | USY-2 |
|---|---|---|---|---|
| Ethylene Conversion, wt% | 97.1 | 97.2 | 96.3 | 32.8 |
| Ethylene WHSV, $hr^{-1}$ | 1.0 | 2.0 | 1.0 | 1.0 |
| Time on Stream, days | 3.8 | 5.2 | 0.8 | 1.8 |
| Product Selectivity, wt% | | | | |
| Ethylbenzene (EB) | 89.834 | 81.949 | 63.952 | 72.287 |
| Diethylbenzenes (DEB) | 9.470 | 15.517 | 12.729 | 13.003 |
| Triethylbenzenes (TEB) | 0.584 | 2.121 | 3.986 | 3.848 |
| Tetraethylbenzenes (TetraEB) | 0.044 | 0.127 | 1.319 | 1.043 |
| Ethylation Selectivity) | 99.93 | 99.71 | 81.99 | 93.18 |
| Aliphatics | 0.000 | 0.000 | 1.873 | 0.070 |
| Xylenes | 0.000 | 0.000 | 0.024 | 0.000 |
| Styrene | 0.000 | 0.000 | 0.008 | 0.000 |
| Cumene, n-PropylBz, Et-Tol | 0.000 | 0.000 | 1.161 | 0.743 |
| Butylbenzene | 0.061 | 0.158 | 2.960 | 1.041 |
| Butylethylbenzene | 0.000 | 0.016 | 1.157 | 0.735 |
| Diphenylethane | 0.008 | 0.095 | 4.578 | 1.473 |
| Others | 0.000 | 0.016 | 6.255 | 2.755 |
| (By-Product Selectivity) | 0.069 | 0.285 | 18.016 | 6.819 |
| DEB/EB, wt% | 10.54 | 18.94 | 19.90 | 17.27 |
| TEB/EB, wt% | 0.65 | 2.59 | 6.23 | 6.23 |
| TetraEB/EB, wt% | 0.05 | 0.16 | 2.06 | 2.06 |

A common side reaction in alkylation is ethylene oligomerization. Oligomers are undesirable by-products by themselves; they are also precursors to non-aromatic saturates, higher alkylbenzenes (cumene, n-propylbenzene, butylbenzenes, etc.) and heavies. Some of these byproducts - mainly $C_7$-$C_9$ non aromatics, cumene, and n-propylbenzene - boil close to ethylbenzene and may end up as impurities in the product. Moreover, since some of these oligomers are benzene co-boilers, a drag stream may be needed to purge these from the system. Non-aromatics are particularly undesirable impurities in ethylbenzene because in the styrene plant they become non-aromatic saturates and unsaturates which will appear in the styrene product and the toluene byproduct. In the ethylation of benzene over MCM-22, however, the rate of formation of ethylene oligomers is extremely low. Very small quantities of heavies, primarily dicyclic compounds such as diphenylethane, are formed in the alkylator, along with even smaller quantities of butylbenzenes, the latter of which are destroyed in the transalkylation reactor. The alkylation reaction is otherwise very clean. There are no detectable levels of by-products originating from isomerization reactions (e.g. xylenes) or cracking (e.g. propylbenzenes) in the alkylation reactor effluent.

## 11.5 Future of alkylation technologies

### *11.5.1 Ethylbenzene*

The approximately 40% of the world's ethylbenzene capacity that still uses $AlCl_3$ is testimony to the efficiency and economy of this process. To capture this segment of the industry, zeolite catalysts that operate at close to the same very low benzene to ethylene ratios that make the $AlCl_3$ process economically attractive will have to be developed. Heat management in fixed bed reactors becomes a design concern at the low benzene to ethylene ratios that characterize the $AlCl_3$ process. Hence, process and catalyst innovations will have to evolve concurrently to achieve the goal of low benzene to ethylene ratios.

Another opportunity for advancement in ethylbenzene synthesis is in the development of liquid phase processes that can handle low cost feedstocks, including dilute ethylene such as ethane/ethylene mixtures. The use of dilute ethylene has become increasingly attractive since it has the potential to debottleneck ethylene crackers. Currently higher temperature, vapor phase technologies can tolerate contaminants that enter with the dilute ethylene feed from FCC units. However, these same contaminants can accelerate catalyst aging in lower temperature, liquid phase operations because they are more strongly adsorbed at the lower temperatures. Acid catalysts that tolerate elevated levels of contaminants would facilitate the development of dilute ethylene-based processes. These same catalysts could be useful in applications where lower cost or lower quality benzene feeds are all that are available.

Future ethylbenzene alkylation catalyst development efforts will also undoubtedly focus on systems that convert less conventional feedstocks including ethane and ethanol. The two-step Dow ethane based process for ethylbenzene production is believed to be uneconomical because of its high capital investment requirement (26). However, it is a very attractive concept, and could be implemented if more efficient catalysts or improved process designs could be developed.

There are similar opportunities for improved transalkylation catalysts and processes. Heavies make with the current zeolite-based transalkylation catalysts are already very low, however improved catalysts could lead to even lower byproduct makes and yield losses due to residue formation. The majority of the reaction byproducts in the zeolite-catalyzed processes are produced in the transalkylator. There are opportunities to develop more active transalkylation catalysts that allow operation at lower temperatures where byproduct formation can be minimized.

Process development efforts will likely focus on increased integration of the alkylation and transalkylation stages. The two processes frequently operate at different temperatures. The challenge here is to match activities and aging rates. Other efforts will likely focus on improved process heat integration.

### *11.5.2 Cumene*

It is virtually inevitable that zeolite catalyzed, liquid phase production of cumene will eventually displace the SPA and $AlCl_3$ based processes. The situation is not analogous to ethylbenzene, since the predominant competing technology that zeolite processes face is the far less efficient and environmentally undesirable SPA process. The current zeolite-based processes allow for significant debottlenecking of SPA plants with little additional capital. More importantly, the zeolite processes have been able to produce cumene with purities that cannot be approached by either the SPA or the $AlCl_3$ process. Tighter product specifications, brought about by the more selective zeolite processes, will drive all manufacturers to select one of the zeolite technologies, possibly in the very near term.

Additional research will undoubtedly be directed toward identifying improved alkylation and transalkylation catalysts. Alkylation catalysts that operate at benzene to propylene ratios of three or possibly even lower are desirable because they greatly reduce the need for costly separation capacity and because they can debottleneck existing units. Transalkylation and alkylation catalysts that produce even lower byproduct concentrations at low benzene to propylene levels are likely to be the focus of future research and development efforts. The challenge of combining both lower byproduct make and lower benzene to propylene feed ratios involves overcoming higher aging rates implicit in using higher propylene feed concentrations and identifying more monoalkylation selective catalysts with activity similar to today's zeolite catalysts. Materials that surpass even MCM-22's superior ability to operate at high propylene concentrations without catalyzing excessive oligomerization could lead to important advances in cumene synthesis.

Current zeolite catalysts already operate at process temperatures that require minimal external heat addition. Heat integration and heat management will be of increasing concern at the lower benzene to propylene ratios because the cumene synthesis reaction is highly exothermic ($\Delta H_f$= -98 kJ/mole). Recycle, particularly in the alkylation reactor, is likely to become increasingly important as a heat management strategy. The key will be how to limit the build-up of byproducts and feed impurities in these recycle loops, particularly as manufacturers seek cheaper and consequently lower quality feedstocks. As in the case of ethylbenzene, process and catalyst innovations will have to develop concurrently.

Recently, there have been several process developments that allow for the production of phenol without co-production of acetone. These would tend to compete directly with the need to produce cumene. Developments include the AlphOx process, a joint development of Solutia and the Boreskov Institute of

Catalysis (27-30) which allows for the direct oxidation of benzene to produce phenol. Economic analyses have shown that these are attractive only in specific instances where, for example, a cheap source of $N_2O$ is available. Nevertheless, these developments have shown that direct oxidation is possible and further innovations in this area should probably be expected. The demands for acetone and phenol have generally tended to follow each other. However, as bisphenol A becomes an even more important end use for phenol and acetone, there will be a need for a separate source of phenol. The synthesis of bisphenol A requires two moles of phenol for every one mole of acetone, while the peroxidation of cumene produces one mole of each. Still, processes such as the direct oxidation of benzene are unlikely to have a major impact on cumene demand in the short term since there are competing processes such as Mitsui's for converting acetone back to propylene.

There is also the prospect of increased demand for some of the cumene by-products such as diisopropylbezene (DIPB). The production of diphenols from DIPB is important for synthesis of resorcinol (from meta-DIPB) and hydroquinone (from para-DIPB). It is likely that the market may soon see the introduction of cumene based processes that specifically produce these isomers with high degrees of selectivity.

## References

1. Olah G., ed., *Friedel-Crafts and Related Reactions*, (Interscience Publishers, New York, 1964).
2. Pines H., *The Chemistry Of Catalytic Hydrocarbon Conversions,* (Academic Press Inc., New York, 1981) 59-79.
3. Venuto P.B., *Organic Catalysis Over Zeolites: A Perspective On Reaction Paths Within Micropores*, Review presented at Symposium On New Catalytic Chemistry Utilizing Molecular Sieves, ACS National Meeting, Chicago, 1993 And Symposium Proceedings *Microporous Materials.* **2** (1994) 297-411.
4. McKetta J.J. and Cunningham W.A., eds., *Encyclopedia of Chemical Processing and Design*, **Vol. 20** (1989) 77-78.
5. *Ullmann's Encyclopedia Of Industrial Chemistry,* Vol. A 13, (1989) 257-258.
6. Kaeding W.W. et al., *J. Catal.*, **67** (1981) 159.
7. Flockhardt B.D., Liew K.Y., and Pink R.C., *J. Catal.* **72** (1981) 314.
8. Song C. and Shobert H.H., *Fuel Processing Technol.*, **34** (1993) 157.
9. Song C. and Kirby S., Symposium On New Catalytic Chemistry Utilizing Molecular Sieves, 206th National American Chemical Society Meeting, Chicago IL, (August 1993).
10. Fraenkel D., Cherniavsky M., Ittah B. and Levy M., *J. Catal.* **101** (1986) 273.
11. Katayama A. et al., *Chem Soc. Chem Comm.*,(1991) 39.

12. Moreau P., Finiels A., Geneste P. and Solofo J., .*J. Catal.* **136**, (1992), 487.
13. Dwyer F.G., in *Catalysis of Organic Reactions: Chemical Industries,* Ed. Moser W.R., (Marcel Dekker, New York 1981) 39.
14. Chen N.Y., Garwood W.A. and Dwyer F.G., *Shape Selective Catalysis in Industrial Applications*, (Marcel Dekker, New York, 1996) 208.
15. Dwyer F.G., Ram S., AIChE Spring National Meeting, Houston, TX, (April 7-11, 1991).
16. B. R. Maerz, C. R. Venkat, S. S. Chen, D. N. Mazzone, DeWitt (1996) *Petrochemical Review*, Houston, (March 3-21 1996).
17. D. N. Mazzone, P. J. Lewis, C. R. Venkat, B. R. Maerz, *Hydrocarbon Asia*, **7(3)** (1997) 56.
18. J. C. Cheng, T. F. Degnan, J. S. Beck, Y. Y. Huang, M. Kalyanaraman, J. A. Kowalski, C. A. Loehr, D. N. Mazzone, *Stud. Surf. Sci. Catal*. **105** (1998) 56.
19. M. A. Caesar, SRI Consulting PEP Report No. 219 (1999) 2.
20. V. N. Ipatieff, U. S. Patent 2,382,318 (1945)
21. P. R. Pujado, J. R. Salazar, C. V. Berger, *Hydrocarbon Process*., **55 (3)** (1976) 91.
22. Chemical Week, (June 14, 1995), 16.
23. *Hydrocarbon Processing*, (October 1996), 40.
24. J. B. Higgins, R. B. LaPierre, J. L. Schlenker, A. C. Rohrman, J. D. Wood, G. T. Kerr, and W. J. Rohrbaugh, *Zeolites* **8** (1988) 447.
25. M. E. Leonowicz, J. A. Lawton, S. L. Lawton, and M. K. Rubin, *Science*, **264** (1994) 1910.
26. S. Naqvi, Process Economics Program (PEP) Review 97-3, SRI Consulting, Menlo Park, CA, (1998).
27. G. I. Panov, *Appl. Catal. A: Gen.*, **98** (1993) 1.
28. A. K. Uiarte, 3rd World Congress on Oxidation Catalysis, San Diego, CA, (Sept. 1997).
29. K. Walsh, Chem. Week, (March 18, 1998) 10.
30. Chem. Market. Rep., (Dec. 30, 1996) 250.

CHAPTER 12

# METHANOL TO OLEFINS (MTO) AND BEYOND

PAUL BARGER

*UOP LLC, Des Plaines, IL, USA*

## 12.1 Introduction

The conversion of methanol (and/or dimethylether) to light olefins (MTO) using molecular sieves is a technology that can be a key for the economical utilization of natural gas resources, particularly in remote locations in the world. This paper reviews some of the requirements for an economically attractive natural gas conversion process and the desirability of producing high-valued light olefins (ethylene and propylene) and the corresponding olefin polymers. Molecular sieves have been known to be effective catalysts for converting methanol and other oxygenates to olefin since the early 1970s (1, 2). Much of the early work was an offshoot of the methanol-to-gasoline (MTG) process development that used catalysts based on the medium-pore sized ZSM-5 zeolite (3, 4). In the 1980s, it was found that small-pore sized non-zeolitic molecular sieves, such as silicoaluminophosphates (SAPOs), provided performance advantages for this process (5-7). The key performance characteristics of these classes of catalysts will be compared as well as the current understanding of the reaction mechanism occurring within the pores of the molecular sieves. UOP and Norsk Hydro have jointly developed the UOP/HYDRO MTO process for the conversion of methanol to ethylene and propylene that uses the MTO-100™ catalyst, based on the SAPO-34 molecular sieve (8). Some of the key aspects of the development of this process will be reviewed, including reactor design, olefin product purity and long-term catalyst stability. Finally, molecular sieve catalysts are used commercially for a variety of other olefin inter-conversion reactions, such as skeletal isomerization and cracking. A few examples of these processes will be highlighted.

## 12.2 Natural gas conversion

Worldwide natural gas reserves have been increasing steadily over the last 10 years and are comparable in size to the reserves of crude oil. However, much of this gas is located in remote sites without access to pipelines, where transportation costs to get the gas to established markets are prohibitive.

The development of processes for the conversion of natural gas, in particular methane, to higher valued products that can be easily transported is one of the key challenges for the use of this natural resource. At the present time, only the direct use as a fuel for heating or power generation and indirect chemical production, through synthesis gas, can be considered as economically viable options for methane utilization.

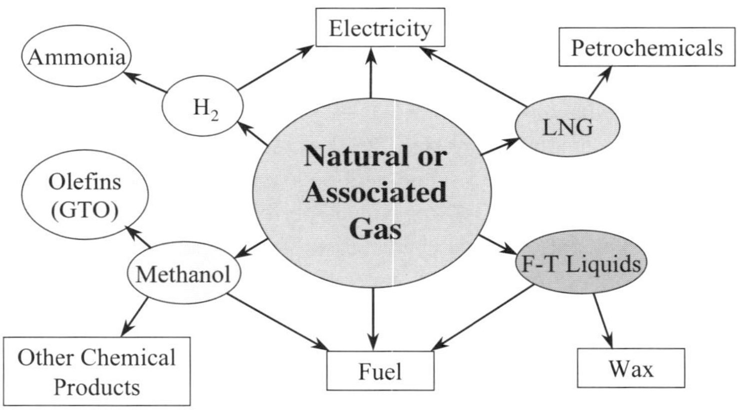


**Fig. 12.1** Options for natural gas utilization

The currently available processes for natural gas utilization are shown in Figure 12.1. While conversion of gas to methanol and ammonia has been commercially practiced for decades, the demand for these products is small compared with the overall supply of remote gas. The liquefaction of natural gas to LNG can allow the transport of gas over long distances, but the capital costs for both the on-shore liquefaction and re-vaporization facilities and the cryogenic transportation vessels are very high. The conversion of natural gas to hydrocarbons for fuels and petrochemicals using Fischer-Tropsch technology is currently being practiced in South Africa (Sasol, Mossgas) and Malaysia (Shell) where process economics are influenced by the political environment. The production of ethylene and propylene via methanol has been demonstrated in a 1 barrel/day demonstration unit (UOP/Norsk Hydro) and is currently available for commercial licensing (8).

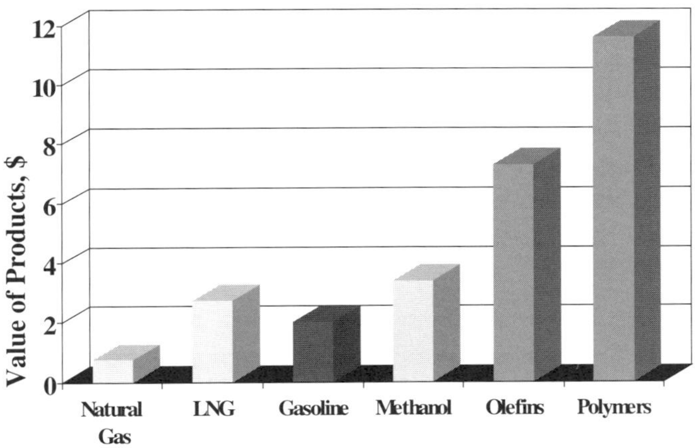


**Fig. 12.2** Value of products made from 1MM BTU natural gas by existing natural gas conversion processes

The value of the final product produced has a major influence on the economic attractiveness of natural gas conversion technologies is. Figure 12.2 shows the values of the major chemical products that can be produced from 1 MMBTU of natural gas via existing technology. Light olefins and olefin polymers produced via the MTO process have a substantially higher value than LNG, gasoline or methanol. The impact of these product values can also be seen in an economic comparison of the world-scale LNG, Fischer-Tropsch based gas-to-liquids (GTL) and MTO-based gas-to-olefins (GTO) plants that is shown in Table 12.1 (9). The MTO plant affords a significantly higher ROI (25%) compared with the other processes. In addition, it requires less natural gas feed, meaning that smaller gas field is needed to supply an MTO unit over the course of its productive life relative to an LNG or GTL plant.

**Table 12.1** Economic comparison of LNG, GTL and GTO processes

| | LNG | GTL | GTP |
|---|---|---|---|
| Gas cost, \$ mm / year | 220 | 134 | 40 |
| Operating cost, \$ mm / year | 93 | 64 | 79 |
| Total cash cost, \$ mm / year | 313 | 198 | 119 |
| | | | |
| Product revenue, \$ mm / year | 965 | 425 | 441 |
| - Transportation costs | 333 | 25 | 24 |
| Net revenue, \$ mm / year | 632 | 400 | 417 |
| | | | |
| Gross profit, \$ mm / year | 319 | 202 | 298 |
| | | | |
| Capex, \$ mm (low) | 2000 | 1000 | 1200 |
| Capex, \$ mm (high) | 2800 | 1500 | |
| | | | |
| Simple ROI (low Capex) | 16% | 20% | 25% |
| Simple ROI (high Capex) | 11% | 13% | |

## 12.3 Methanol to olefins

The goal of the MTO process is to convert methanol to light olefins, in particular ethylene, propylene and butenes. The key by-products of the reaction include the co-product water, $C_{5+}$ hydrocarbons such as aromatics and heavier olefins, coke that remains on the catalyst at process conditions and light paraffins that are the primary sink for hydrogen lost during aromatic formation. Small amounts of $H_2$ and $CO_x$ are also typically observed in the MTO product, although these by-products could arise from feed and product decomposition on the reactor walls and internals at the temperatures that are typically used.

The ability of molecular sieve catalysts to catalyze this transformation was first discovered by workers at Mobil in the early 1970s (1, 2). Over the past thirty years there have been a number of catalyst and process advances that have brought this technology to the verge of commercial realization.

### *12.3.1 MTO catalysts*

Molecular sieve catalysts that have been used for the conversion of methanol to hydrocarbons fall into two general classifications. Most of the initial research was done using ZSM-5 (MFI), a medium-pore size zeolite with a three dimensional pore system consisting of straight (5.6 x 5.3 Å) and sinusoidal channels (5.5 x 5.1 Å). While most of this work was directed at the conversion of methanol to liquid hydrocarbons for addition to gasoline, it was found that the product slate could be shifted toward light olefins by the use of low pressure and short contact times.

In the early 1980s researchers at Union Carbide discovered that small-pore size silicoaluminophosphate (SAPO) molecular sieves were effective for converting methanol to ethylene and propylene. The best performances were obtained with SAPO-34 and SAPO-17 catalysts (6). SAPO-34 has the CHA structure with a three dimensional pore system consisting of large cavities (about 9.4 Å in diameter) separated by small windows (3.8 x 3.8 Å).

The structures of SAPO-34 and ZSM-5, along with small organic molecules that are key to the MTO process, are shown in Figure 12.3. The ionic radii of the oxygen atoms of one of the pore mouths of each structure are highlighted. While linear hydrocarbons, such as ethylene and propylene, are able to pass through the pores of both structures, branched molecules, such as i-butene and benzene, cannot pass through the 8-ring pores of SAPO-34.

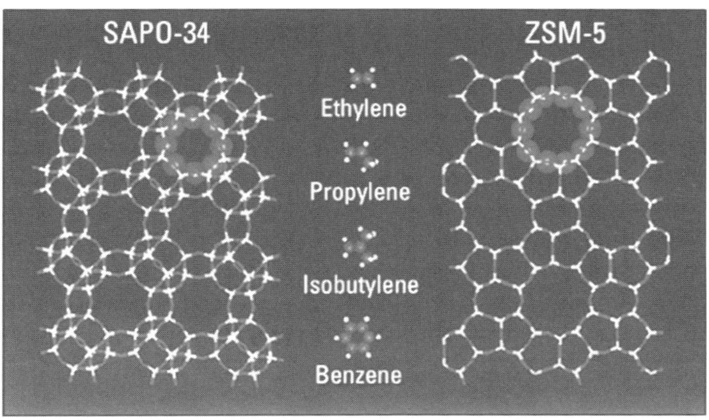


**Fig. 12.3** Framework structures of SAPO-34 (CHA) and ZSM-5 (MFI) molecular sieves

The impact of the structural differences of ZSM-5 and SAPO-34 on representative MTO product compositions is shown in Figure 12.4. The medium pore-size sieve gives propylene as the major light olefin and a significant amount of $C_{5+}$ hydrocarbons, much of which is aromatic and accounts for the high light paraffin production (10). In contrast, the small pore-size SAPO-34 gives predominantly ethylene and propylene with less heavy hydrocarbons and paraffins. Coke formation is higher with SAPO-34, since any aromatics formed within the pore structure are trapped, whereas they are able to diffuse through and desorb from the ZSM-5 structure.

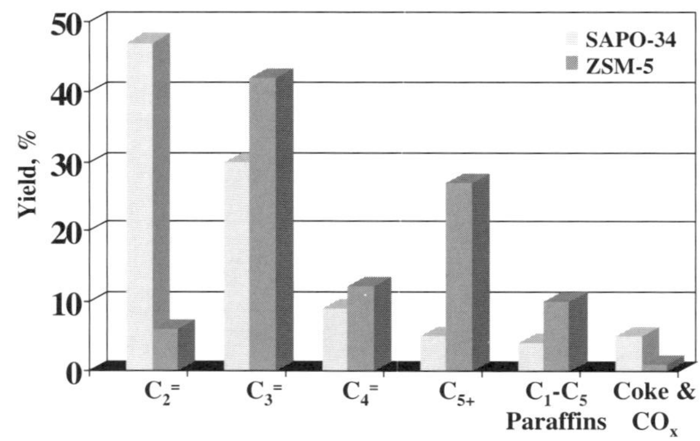


**Fig. 12.4** MTO product compositions for ZSM-5 and SAPO-34

The acid sites of zeolitic molecular sieves arise from the substitution of $Al^{+3}$ for $Si^{+4}$ in the mostly siliceous crystalline lattice. An acidic proton, as a bridging hydroxyl between Si and Al framework atoms, must also be added to the framework in order to balance the charges. In the case of SAPO molecular sieves the basic structure is $AlPO_4$. Substitution of $Si^{+4}$ and $H^+$ for a portion of the framework $P^{+5}$ introduces acid sites to the material. The acid strength of zeolitic materials is typically greater than that of SAPO sieves. The greater electron affinity of aluminum bonded to -O-Si(OR)$_3$ groups in the zeolite, compared to -O-P(OR)$_3$ for the silicoaluminophosphate, more effectively stabilizes the conjugate anion of the bridging hydroxyl. The impact of this acidity difference on MTO performance has been shown by the comparison of the catalytic performances of SAPO-34 and SSZ-13, which is a synthetic silicoaluminate that also has the chabazite structure

(11). Table 12.2 shows that SAPO-34 has significantly better stability due to a lower rate of coke formation than SSZ-13 samples with comparable tetrahedral-atom substitutions and acid site densities. SSZ-13 also shows greater production of light paraffins which is consistent with accelerated hydride transfer for the catalyst with higher acidic strength.

**Table 12.2** Comparison of SAPO-34 and SSZ-13 (Chabazite) catalysts for MTO

| Material | SAPO-34 | SSZ-13 | SSZ-13 | SSZ-13 |
|---|---|---|---|---|
| T atom % | 10% Si | 18% Al | 10% Al | 3.3% Al |
| Selectivities (% at 2 hr) | | | | |
| $C_2$-$C_4$ Olefins | 96 | 69 | 75 | 87 |
| $CH_4$ | 1.4 | | | |
| $C_2H_6$ | 0.3 | | | |
| $C_3H_8$ | 0.9 | | | |
| Stability (hr at >50% conversion) | >40 | 6 | 13 | 7 |
| Coking (% carbon on used catalyst) | 19 after 54 hr | 16.6 after 18 hr | 19.3 after 18 hr | 15.0 after 18 hr |

### *12.3.2 MTO reaction mechanism*

The reaction mechanism for the conversion of methanol to hydrocarbons over molecular sieve catalysts has been extensively investigated over the past 25 years. It is widely accepted that methanol conversion initially proceeds through equilibration with DME. Early work with ZSM-5 showed that light olefins are then the initial hydrocarbon products, followed by heavier olefins, paraffins and aromatics (Figure 12.5) (2).

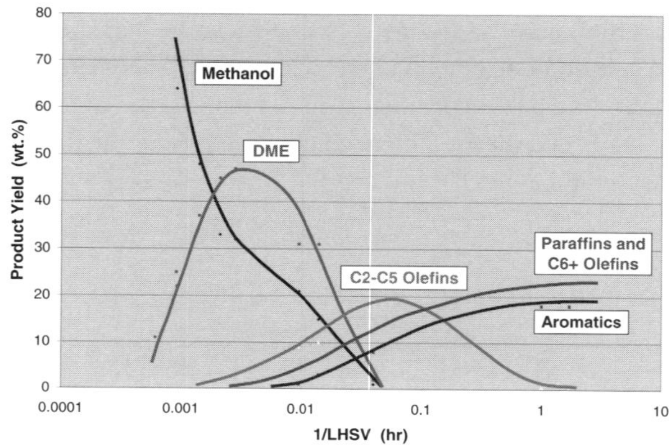


[ MeOH ↔ DME ] → Light Olefins → Paraffins + Aromatics

**Fig. 12.5** MTO Reaction Mechanism: Formation of the first C-C bond

Mechanisms for the formation of light olefins, in particular ethylene, generally fall into two types – they are either an initial product formed by C-C bond forming steps of oxonium ion intermediates or they arise from the cracking of heavier intermediates built up by methylation of adsorbed olefins or aromatics. Low conversion experiments with ZSM-5 suggest that ethylene is a primary product of oxygenate conversion (12). This lead to an extensive effort to determine the mechanistic steps leading to the formation of the first C-C bond from the methanol or DME precursor. While no clear answer has been established to date, the rearrangement of ylid-type species on either an oxonium cation or the catalyst surface (Figure 12.6) are most widely supported (13).

Further work with ZSM-5 at low methanol pressure and varying contact times has shown that the majority of the ethylene actually arises from the secondary equilibration of heavier olefins (14). This work indicates that while the C-C bonding forming reactions may be important for initiating the MTO reaction, the vast majority of the methanol is converted to hydrocarbons by methylation of species already formed within the pore structure of the zeolite. Light olefins are then made by the cracking of these heavier compounds. This type of pathway, where light olefins arise from methylation and cracking of adsorbed intermediates has become known as the 'carbon pool' mechanism (15).

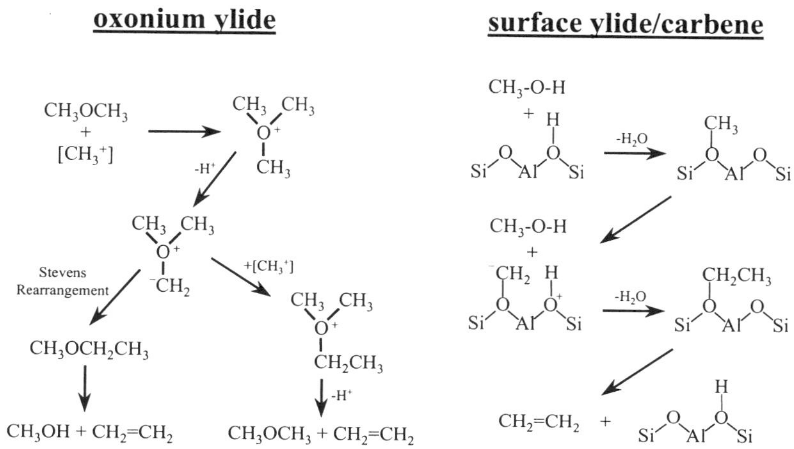


**Fig. 12.6** MTO Reaction Mechanism: Formation of first C-C bond

The original mechanisms proposed for the conversion of methanol to hydrocarbons suggested that the carbon pool consists of higher molecular weight olefins (14). However, work with isotope-labeled alcohol (16) and aromatic (17) co-feeds suggests that the nature of the carbon pool is a more complex mixture of adsorbed compounds that may be more hydrogen deficient. Recent *in situ* MAS NMR spectroscopic investigations of operating catalysts at reaction temperature have shown the presence of adsorbed olefins in the case of ZSM-5 (18) and a mixture of aliphatic and cyclic olefins as well as aromatics on SAPO-34 and SAPO-18 (19). NMR examination of the species remaining in the molecular sieve after it is rapidly cooled from operating temperature to ambient has been used to claim that a dimethylcyclopentyl carbenium ion is the active carbon pool species in the case of ZSM-5 (20), while adsorbed alkylaromatics play this role in SAPO-34 (21). However, it is not clear how dominant these compounds are at actual operating conditions or whether they are only the thermodynamic sink for the non-desorbed molecules remaining in the sieve after the temperature is quenched. Further isotopic labeling experiments have shown that the carbons of methanol are rapidly incorporated into the rings of the aromatics trapped inside SAPO-34, suggesting that aromatic transformations are more complex than just methylation and side-chain cracking (22).

The high selectivity to ethylene at commercially viable conditions is one of the key advantages of small pore size molecular sieves, such as SAPO-34, over other catalysts for the methanol conversion. Co-feeding $^{13}CH_3OH$ with unlabelled propylene over ZSM-5 gives a binomial distribution of the butene isotopomers once conversion is high enough to overcome the effects of the initiation period (23). In contrast the $^{12}C$ contents of the butenes obtained with a SAPO-34 catalyst with the same feed are substantially lower indicating that the propylene is much less involved in the olefin conversion pathway (24). It has been proposed that this is due to slower diffusion of the light olefins though the SAPO-34 pore structure compared with methanol.

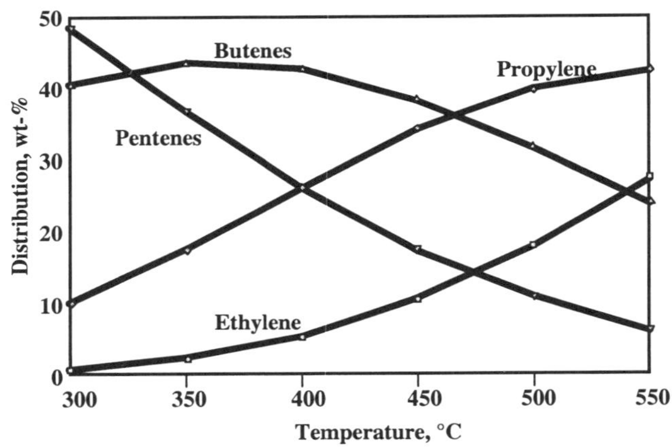


**Fig. 12.7** Equilibrium distribution of $C_2$-$C_5$ olefins at 0 psig

Evidence for the equilibration of light olefins within SAPO-34 prior to diffusion out of the crystalline structure has been obtained by comparing the ethylene/propylene ratio in the MTO product with that calculated from thermodynamic equilibrium. Figure 12.7 shows the thermodynamic ratios of the $C_2$-$C_5$ olefins at 0 psig as a function of temperature. The concentration of ethylene increases at higher temperatures. The influence of equilibrium on the ethylene/propylene product ratio obtained with the MTO-100 catalyst over a range of temperatures and pressures can be seen in Figure 12.8. The relationship is linear with a slope that is significantly greater than 1, indicating that ethylene is in excess of equilibrium. An explanation for this relationship is that ethylene and propylene are in equilibrium within the pore structure of the SAPO-34 molecular sieve. However, because the 4 Å pore diameter of the SAPO-34 is too small to allow propylene to take a straight-line path through the pore mouth, its diffusion is hindered relative to ethylene.

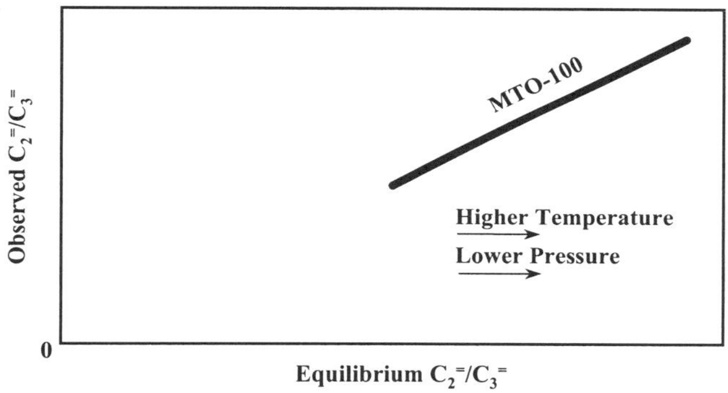


**Fig. 12.8** Relationship of Ethylene/Propylene product ratio over MTO-100 catalyst with thermodynamic equilibrium

## 12.4 UOP/Hydro MTO process

UOP and Norsk Hydro have jointly developed and demonstrated a new MTO process utilizing a SAPO-34 containing catalyst that provides up to 80% yield of ethylene and propylene at near-complete methanol conversion. Some of the key aspects of the work have included the selection of reactor design for the MTO process and determination of the effects of process conditions on product yield. Evaluation of the suitability of the MTO light olefin product as an olefin polymerization feedstock and demonstration of the stability of the MTO-100 catalyst have also been determined during the development of this process.

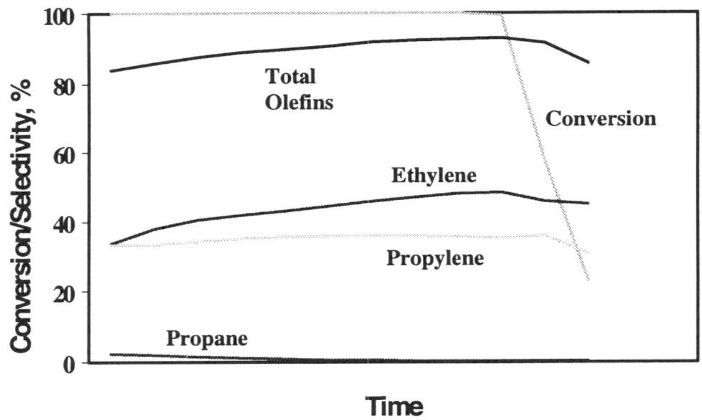


**Fig. 12.9** Fixed bed MTO performance

### *12.4.1 MTO reactor design*

The importance of the design of the MTO reactor can be illustrated with the results of a typical fixed bed pilot plant test of a SAPO-34 catalyst (Figure 12.9). Conversion of methanol and DME is maintained at 100% for a period of time, after which it drops rapidly as the catalyst deactivates due to coke deposition. The composition of the hydrocarbon product changes throughout the run with light olefin selectivities reaching a maximum at the point of methanol/DME breakthrough. A commercial unit utilizing this type of reactor would have to swing frequently between process and regeneration, requiring expensive high-temperature valves. The down-stream separation equipment would also have to be designed to accept the varying product composition.

The use of a fluidized-bed reactor has a number of advantages in the MTO process. The moving bed of catalyst allows the continuous movement of a portion of used catalyst to a separate regeneration vessel for removal of coke deposits by burning with air. Thus, a constant catalyst activity and product composition can be maintained in the MTO reactor. Figure 12.10 demonstrates the stability of a 90 day operation in the fluidized-bed MTO demonstration unit at the Norsk Hydro Research Center in Porsgrunn, Norway. A fluidized-bed reactor also allows for better heat recovery from the exothermic methanol-to-olefins reaction. This type of reactor has been widely used in the FCC area, particularly for catalyst regenerators.

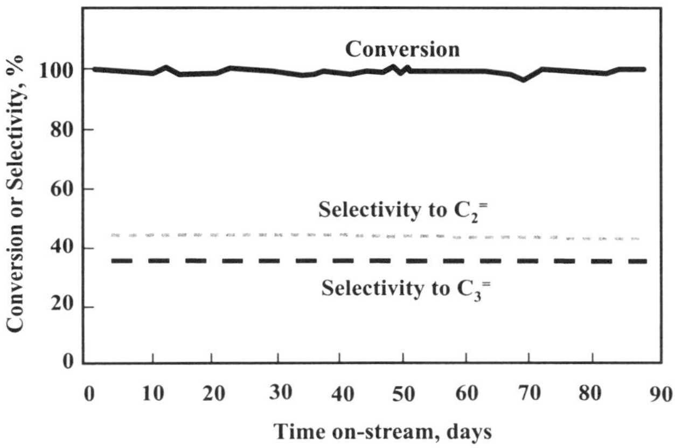


**Fig. 12.10** Fluidized bed MTO performance in MTO demonstration unit

### *12.4.2 MTO product composition*

The UOP/HYDRO MTO process offers a wide range of flexibility for varying the ethylene and propylene product ratio by adjusting the operating severity of the reactor. The MTO process can be designed for an ethylene to propylene ratio between 0.75 and 1.5 (Figure 12.11). Over the range of 0.8 to 1.3, the overall yield of light olefins changes slightly with the highest yields achieved with about equal amounts of ethylene and propylene. This region provides the lowest methanol requirements, but the ratio can be adjusted to reflect the relative market demand and pricing for ethylene and propylene. An example material balance is shown in Table 12.3 for the production of 600,000 MTA of light olefins with equal amounts of ethylene and propylene. Approximately 3 tons of methanol is required per ton of light olefins. This represents a carbon-based yield of almost 80%.

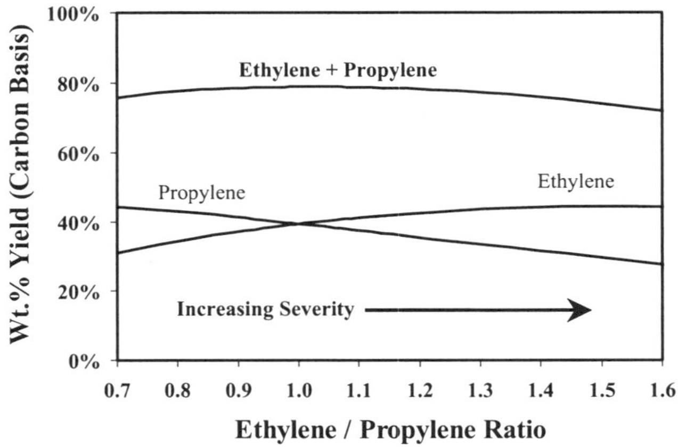


**Fig. 12.11** MTO reactor yields as function of ethylene to propylene ratio and severity

**Table 12.3** MTO material balance for 600,000 MTA ethylene and propylene

| | Feedstocks, MTD | Products, MTD |
|---|---|---|
| Methanol | 5,204 | |
| Ethylene | | 882 |
| Propylene | | 882 |
| mixed butenes | | 272 |
| C5+ hydrocarbons | | 100 |
| fuel gas | | 88 |
| other (water, COx, coke, etc.) | | 2,980 |
| Totals | 5,204 | 5,204 |

Because of the high olefin yields and low light ends make, the MTO process does not require the high cost separation equipment of a conventional ethylene recovery unit. Figure 12.12 shows the overall process design, including product compression, fractionation and treatment that are necessary to obtain polymer-grade ethylene and propylene products.

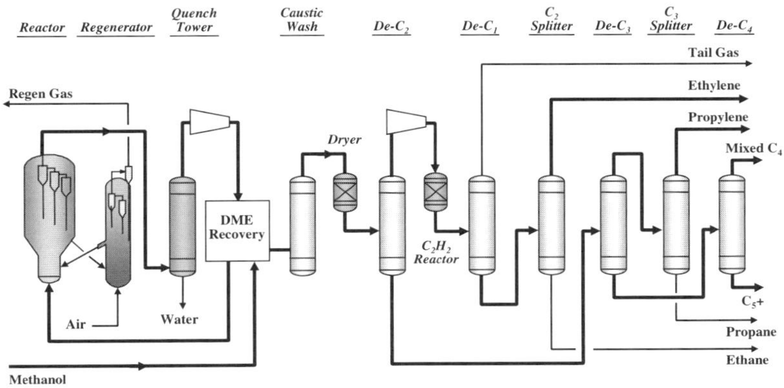


**Fig. 12.12** MTO process flow scheme

Modern olefin polymerization catalysts are sensitive to a wide variety of potential trace impurities in the olefin feedstocks, such as oxygenates, acetylenes/dienes, $O_2$, CO, $CO_2$, COS and arsine. Detailed analyses of the reactor effluent from the MTO demonstration unit has identified the acetylenic, diolefinic and oxygenated compounds present in the MTO product. No unusual by-products have been found. In fact, a comparison of the oxygenates found in the effluents of an MTO unit and an LPG cracker shows that the same oxygenates are present in both streams, although at a higher level in the MTO product (Figure 12.13). The UOP/Hydro MTO Process includes equipment to recover the major oxygenates and recycle them back to the reactor for conversion to additional olefins. The remaining minor oxygenate impurities can be removed by conventional technology, such as the UOP ORU process. Figure 12.14 shows that the ORU treatment is able to remove all oxygenates in the light olefin product to below the 2 ppm detection limit of the GC analysis.

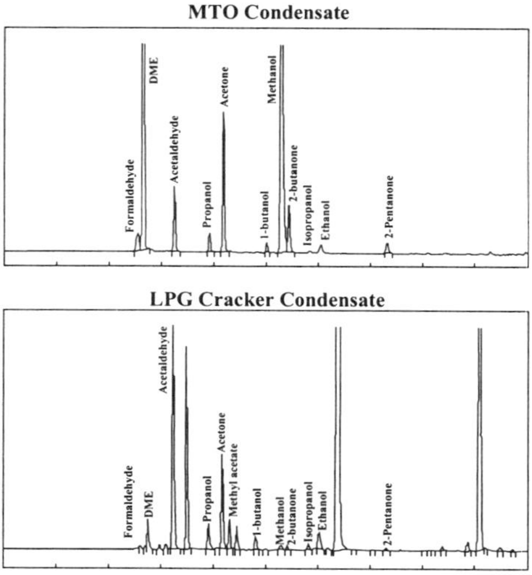


| | Cracker Condensate | MTO Condensate |
|---|---|---|
| Formaldehyde | ✓ | ✓ |
| DME | ✓ | ✓ |
| Acetaldehyde | ✓ | ✓ |
| Propanol | ✓ | ✓ |
| Acetone | ✓ | ✓ |
| Methy acetate | ✓ | ✓ |
| 1-butanol | ✓ | ✓ |
| Methanol | ✓ | ✓ |
| 2-butanone | ✓ | ✓ |
| Isopropanol | ✓ | ✓ |
| Ethanol | ✓ | ✓ |
| 2-Pentanone | ✓ | ✓ |

**Fig. 12.13** By-products in MTO and LPG condensates

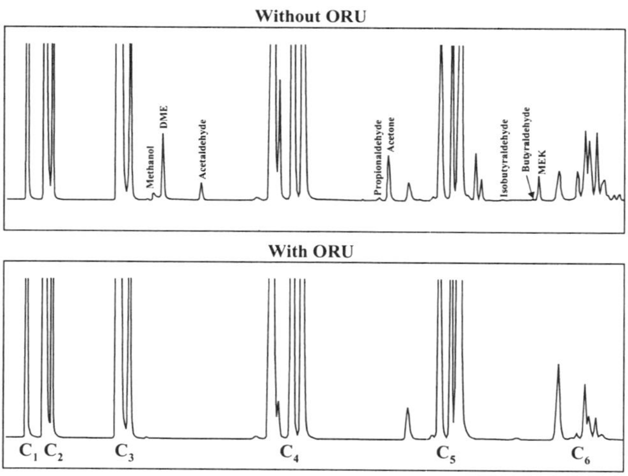


**Fig. 12.14** Oxygenate removal from MTO product by ORU

*12.4.3 Long-term MTO catalyst stability*

The environments in both the MTO reactor and regenerator contain steam at high temperature that has potential for hydrothermal damage to the MTO-100 catalyst. The impacts of any long-term catalyst changes and the catalyst make-up rate required to maintain steady-state activity need to be established in order to demonstrate the economic viability of the MTO process. A sample of the MTO-100 catalyst has been tested through 450 cycles of process-regeneration in order to evaluate its long-term stability. Figure 12.15 shows that catalyst activity slowly declines over this period.

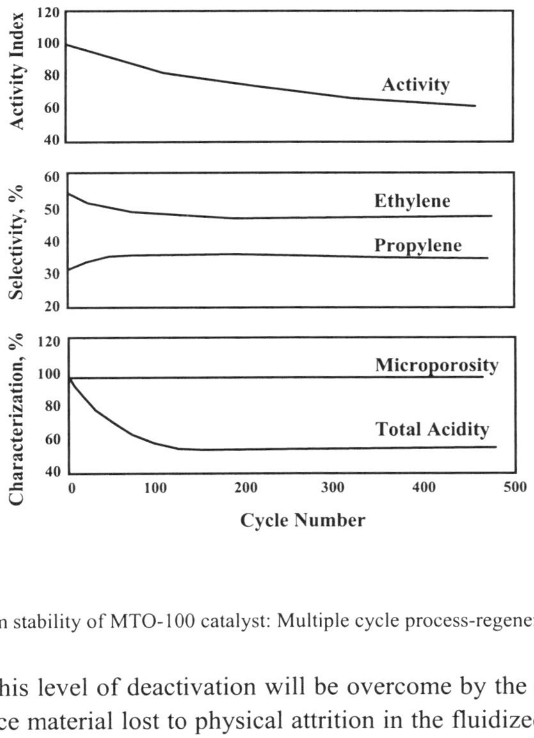


**Fig. 12.15** Long–term stability of MTO-100 catalyst: Multiple cycle process-regeneration

However, this level of deactivation will be overcome by the catalyst make-up required to replace material lost to physical attrition in the fluidized bed reactors. A slight shift in the ethylene and propylene selectivities was also observed during the initial portion of this run. Characterization of the catalyst during the multi-cycle test shows that there is no change in the microporosity of the SAPO-34 molecular sieve, indicating the molecular sieve structure remains intact. X-ray diffraction of the catalyst after 205 cycles also shows no changes in the SAPO-34 crystalline structure. In contrast, a sample of the same catalyst that was treated under severe hydrothermal conditions (650°C, 100% steam) shows evidence of SAPO-34 destruction and the formation of a dense phase $AlPO_4$ with the cristobalite structure (Figure 12.16). There was an initial decline in the total acidity of the MTO catalyst during the multiple cycle test, as measured by $NH_3$ TPD (Figure 12.15), possibly due to some migration of Si out of the sieve framework. However, this acidity loss lines-out after about 100 cycles.

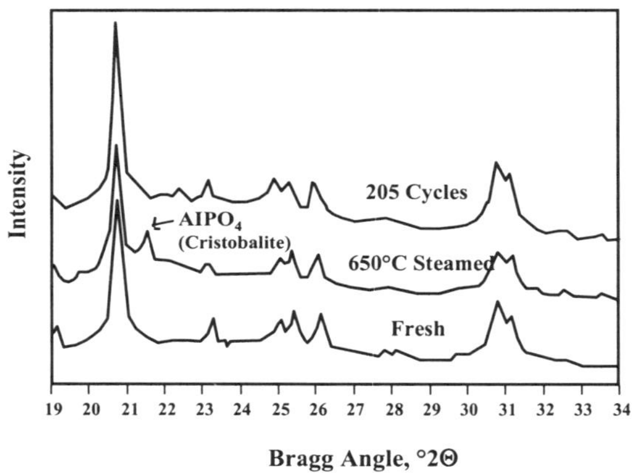


**Fig. 12.16** Long–term stability of MTO-100 catalyst: hydrothermal vs. multiple cycle aging

## 12.5 Other olefin conversion processes

The use of molecular sieve catalysts has also become more widespread in the past decade for the production and inter-conversion of olefins from feedstocks other than oxygenates. The addition of a modified ZSM-5 additive to the Y zeolite-based catalyst can substantially increase the amount of propylene produced in a conventional Fluid Catalytic Cracking (FCC) unit. This has become a very valuable modification, particularly in areas where propylene supplies are tight. More recently, a number of processes have been announced for the direct cracking of $C_{4+}$ olefinic steams to propylene. These processes also use modified ZSM-5 based catalysts. Figure 12.17 illustrates the conversion of $C_5$-$C_7$ olefins to propylene and butenes achieved by Lurgi's Propylur® process (25).

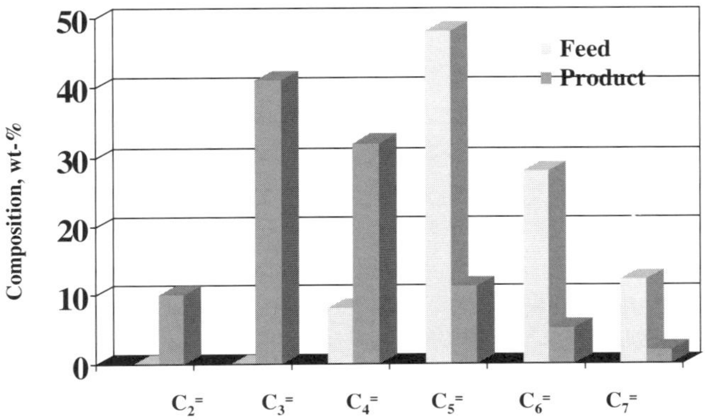


**Fig. 12.17** C5-C7 olefin cracking with the Lurgi Propylur® process

With the introduction of tertiary ethers (MTBE, TAME) for fuel blending over the past ten years to raise octane and oxygenate in the gasoline pool, there has been a need to increase the supply of tertiary olefins as feedstocks for these compounds. Molecular sieve catalysts have been developed by a number of companies for the conversion of straight-chain olefins to tertiary olefins that can be easily etherified, such as n-butene to i-butene. One-dimensional, medium-pore sized zeolites, such as ferrierite (FER) and ZSM-22 (TON), as well as non-zeolitic molecular sieves (NZMSs) have been described in the literature (26-28). Figure 12.18 illustrates the isomerization of feeds containing n-butene and n-pentene using NZMS-based catalysts.

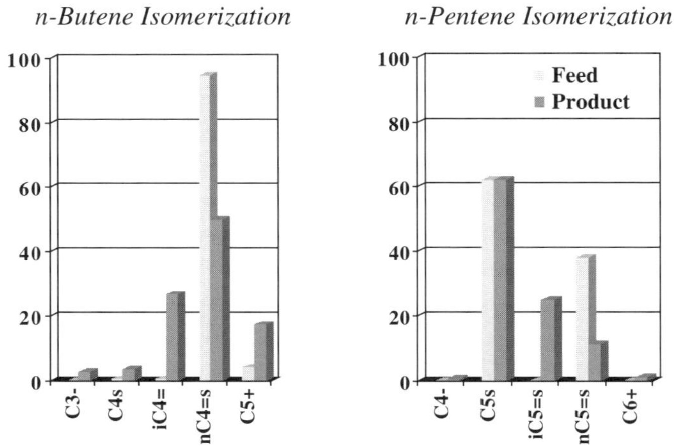


**Fig. 12.18** Olefin isomerization over NZMS catalysts

Olefin skeletal isomerization is also the key step in dewaxing of lube oils to lower their pour point while maintaining viscosity. Chevron's Isodewaxing process utilizes a Pt/SAPO-11 catalyst to give a controlled amount of side-chain branching to long-chain normal paraffins by dehydrogenation to the corresponding olefin and skeletal isomerization (29). The use of the medium-pore size SAPO-11 sieve prevents cracking, which results in yield loss, and over-isomerization.

## 12.6 Summary

The production of high-valued light olefins using molecular sieve catalysts has been an area of active commercial development over the last decade. The development of SAPO molecular sieves with high selectivity to ethylene and propylene has lead to new MTO technology for natural gas utilization that is on the verge of commercialization. The UOP/HYDRO MTO Process provides, for the first time, an economically viable route for the conversion of low-valued natural gas, particularly remote or stranded gas, to high-valued petrochemical feedstocks. Some of the key issues in the development of this process have included demonstration of the fluidized bed reactor design, determination of the suitability of the MTO light olefin products as feeds for olefin polymerization and the long-term stability of the SAPO-34 sieve in the process and regeneration environments.

Other applications for molecular sieve catalysts for the production of olefins include the use of ZSM-5 additives in FCC units to increase propylene yield and the skeletal isomerization of $C_4$-$C_5$ normal olefins for etherification feedstocks. More recently, a number of processes have been announced for the production of light olefins by the selective cracking of heavier olefinic streams that can be recovered from various locations in the refinery. These developments point to an increasing role for molecular sieve catalysts in the petrochemical area.

## References

1. Chang C.D., Lang W.H. and Silvestri A.J., U.S. Patent **4,062,905** (1977).
2. Chang C.D. and Silvestri A.J., *J. Catal.* **47** (1977), 249.
3. Chang C.D., *Catal. Rev.-Sci. Eng.* **26** (1984), 323.
4. Chang C.D., *Hydrocarbons from Methanol* (M. Dekker, New York, 1983).
5. Kaiser S.W., U.S. Patent **4,499,327** (1985).
6. Kaiser S.W., *Arabian J. Sci. Eng.* **10** (1985), 361.
7. Lewis J.M.O., in *Catalysis 1987*, Ed. Ward J.W. (Elsevier, Amsterdam, 1988), 199.
8. Vora B.V. et al., in *Natural Gas Conversion IV*, Ed. De Pontes M. et al. (Elsevier, Amsterdam, 1997), 87.
9. Vora B.V. et al., *Proceedings of the 6th World Congress of Chemical Engineering*, (Melbourne, 2001), in press.
10. Socha R.F. et al., in *Industrial Chemical via $C_1$ Processes*, Ed. Fahey D.R. (ACS, Washington D. C., 1987), 34.
11. Yuen L.-T. et al., *Microporous Mater.* **2** (1994), 105.
12. Wu M.M. and Kaeding W.W., *J. Catal.* **88** (1984), 478.
13. Hutchings G.J., *Catal. Today* **6** (1990), 279.
14. Dessau R.M., *J. Catal.* **78** (1986), 111.
15. Dahl I.M. and Kolboe S., *Catal. Letters* **20** (1993), 329.
16. Mole T., *J. Catal.* **84** (1983), 423.
17. Mole T., Whiteside J.A. and Seddon D., *J. Catal.* **82** (1983), 261.
18. Seiler M, Schenk U. and Hunger M., *Catal. Letters* **62** (1999), 139.
19. Hunger M., Seiler M and Buchholz A., *Catal. Letters* **74** (2001), 61.
20. Goguen P.W. et al., *J. Am. Chem. Soc.* **120** (1998), 2650.
21. Song W., Haw J.F., Nicholas J.B. and Heneghan C.S., *J. Am. Chem. Soc.* **122** (2000), 10726.
22. Arstad B. and Kolboe S., *J. Am. Chem. Soc.* **123** (2001), 8137.
23. Iglesia E., Wang T. and Yu S.Y., in *Natural Gas Conversion V*, Ed. Parmaliana A. et al. (Elsevier, Amsterdam, 1998), 527.

24. Kolboe S. and Dahl I.M., in *Catalysis by Microporous Materials*, Ed. Beyer H.K., Karge H.G., Kiricsi I. And Nagy J.B. (Elsevier, Amsterdam, 1998), 427.
25. Koss U., *Int. J. Hydrocarbon Eng.* **4** (1999), 66.
26. Powers D.H. et al., European Patent **523,838** (1999).
27. Haag W.O., Heck R.H., Santiestaban J.G. and Shihabi D.S., U.S. Patent **5,237,120** (1993).
28. Gajda G.J., Barger P.T. and Abrevaya H., U.S. Patent **5,336,831** (1994).
29. Miller S.J., in *Zeolites and Related Microporous Materials: State of the Art 1994*, Ed. Weitkamp J., Karge H.G. Pfeifer H. and Holderich W. (Elsevier, Amsterdam, (1994), 2319.

CHAPTER 13

# ZEOLITE EFFECTS ON CATALYTIC TRANSFORMATIONS OF FINE CHEMICALS

D.E. DE VOS, P.A. JACOBS
*Centre for Surface Chemistry and Catalysis, K.U.Leuven, Kasteelpark Arenberg 23, 3001 Leuven, Belgium*

## 13.1 Introduction

Reactions for preparation of fine chemicals differ in many aspects from the hydrocarbon reactions that constitute the major application of zeolite catalysts. Fine chemicals reactions often involve complex transformations of molecules with several functional groups. Chemoselectivity is therefore of prime importance. Secondly, reactions must be carried out in mild conditions in order to avoid undesired degradation of the organic molecules. Often this implies that the reactions proceed in the condensed phase instead of in the vapor phase, and the addition of a solvent has major impacts on adsorption of reactants and products on zeolites or other porous catalysts.

The rough match between the dimensions of zeolite micropores or mesopores and those of organic molecules constitutes a challenge and an opportunity. On one hand, obstruction of the porous framework and the ensuing deactivation of the zeolite catalyst is a frequent problem. On the other hand, the various forms of shape selectivity may allow the selectivity of the reaction to be directed away from the thermodynamically most feasible route.

At the basis of the application of zeolites in fine chemicals reactions is the rich variety of catalytic functions with which zeolites can be endowed. Brönsted acidity, Lewis acidity and metallic functions are well known from classical bifunctional chemistry; but for specific reactions, unusual sites, e.g. Lewis acid $Ti^{4+}$ centres, have been introduced into zeolites. Moreover, zeolites can acquire more or less weakly basic properties; metal complexes can be entrapped in zeolite pores or cavities, and enantioselective reactions have been performed by decorating the zeolite surface with chiral modifiers.

There are excellent, almost exhaustive overviews of organic reactions catalyzed by zeolites (1-4). Instead of pursuing completeness, this chapter intends to

illustrate zeolite effects on fine chemicals transformations with a series of well-selected and recent examples. To summarize, some of the main zeolite effects are:
1. **Shape selectivity**: in reactant, transition state, and product shape selectivity, the spatial constraints of the zeolite channels or pore mouths discriminate between molecules with different sizes and forms
2. **Concentration effects**: by selectively adsorbing one or more reactants from a liquid or gaseous feed, high intraporous concentrations are achieved, which often increase rate or selectivity. Examples are Diels-Alder reactions, or selective epoxidation of olefins over TS-1 catalysts (5-6)
3. **Site isolation effects**: catalytically active centres such as metal complexes can be dispersed over the zeolite matrix. In this way, reaction between adjacent metal centres can be impeded for steric reasons (7). In other well-documented cases, intramolecular reactions are preferred over intermolecular reactions due to the spatial isolation of molecules adsorbed in low concentrations (8)

The following illustrative examples have been chosen from many different reaction types. For some specific reactions that are of interest mainly to the fragrance industry, e.g. epoxide rearrangement, reference is made to another chapter in this volume. Moreover, functionalization of aromatic compounds, comprising alkylations, acylations and hydroxyalkylations, and oxidations of phenol or olefins are not dicussed in this chapter, since these topics are covered by other authors.

## 13.2 Zeolite effects in C-N bond formation reactions

### *13.2.1 Synthesis of N-heterocycles by condensation with carbonyl compounds*

The gas phase acid-catalyzed synthesis of pyridines from formaldehyde, ammonia and an alkanal is a complex reaction sequence, comprising at least two aldol condensations, an imine formation, a cyclization and a dehydrogenation (9). With acetaldehyde as the alkanal, a mixture of pyridine and picolines (methylpyridines) is formed. In comparison with amorphous catalysts, zeolites display superior performance, particularly those with MFI or *BEA topology. Because formation of higher alkylpyridines is impeded in the shape-selective environment, the lifetime of zeolites is much improved in comparison with that of amorphous materials. Moreover, the catalytic performance can be enhanced by doping the structure with metals such as Pb, Co or Tl, which assist in the dehydrogenation.

The zeolite catalyst can be tailored depending on the desired product distribution. For instance, pyridine/picoline ratios are higher with H-ZSM-5 catalysts than with other zeolites (10):

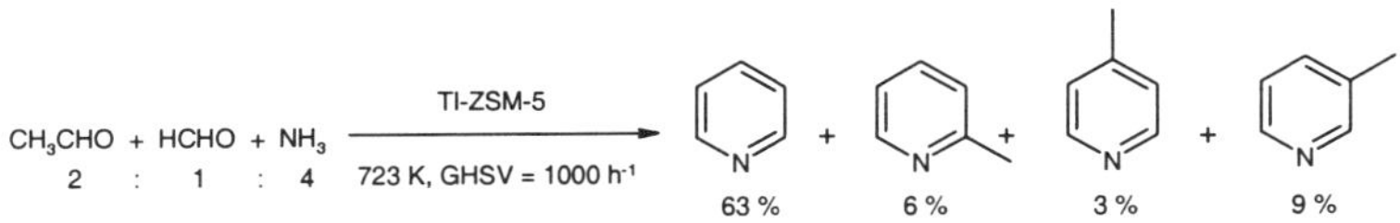


An explanation for this observation was given by McAteer et al., based on results of reactions performed at different formaldehyde/acetaldehyde ratios (9, 11):

$$3\ CH_3CHO + NH_3 \rightarrow \alpha\text{- or }\gamma\text{-picoline} + H_2 + 3\ H_2O$$
$$2\ CH_3CHO + HCHO + NH_3 \rightarrow \text{pyridine} + H_2 + 3\ H_2O$$
$$2\ CH_3CHO + 2\ HCHO + NH_3 \rightarrow \beta\text{-picoline} + 4\ H_2O$$

It is clear that β-picoline formation is a higher order reaction than pyridine formation, since the reactions involve 5 and 4 molecules respectively. Since a lower order reaction is favored in a more shape-selective environment, pyridine production is highest on ZSM-5 zeolites. Alternatively, one might try to maximize the fraction of β-isomers in the picoline products. With a H-Beta zeolite, more than 98% of the picolines consist of β-picoline, which highly simplifies the product purification (12):

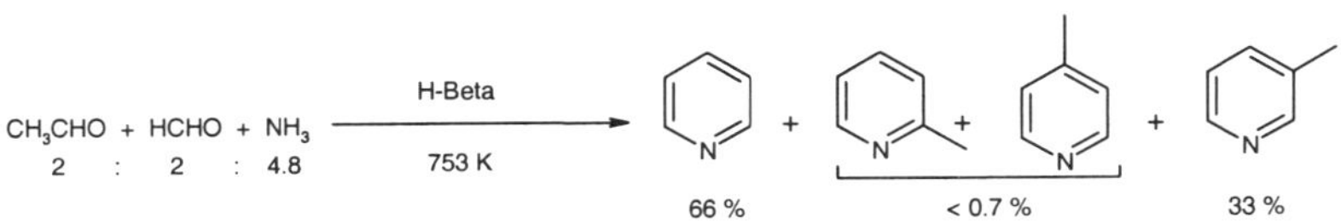


In the same domain of heterocyclic synthesis, much effort has been devoted to understanding the selectivities in the synthesis of indoles starting from phenylhydrazine and ketones, i.e. the Fischer indole synthesis. With phenylhydrazine, 1-phenyl-2-butanone and zeolite H-Beta, the slimmer product 2-phenyl-3-methylindole dominates over the more bulky isomer (13):

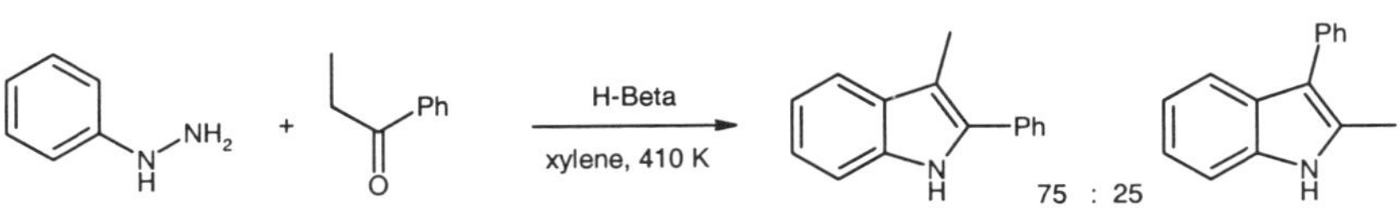


However, there is evidence that the reaction is catalyzed by the external surface, and in homogeneous reactions, the nature of the acid sites plays an important role as well. Hence, the selectivity depends on more factors than zeolite structure alone.

*13.2.2 Benzamine rearrangement*

Picolines (methylpyridines) are useful building blocks for preparation of drugs and agrochemicals. A new zeolite-catalyzed entry to these o-$CH_3$-pyridines is based on the benzamine rearrangement (14). Formally, this reaction is an N-ortho-C exchange process. The first step is likely attack of ammonia on adsorbed aniline, resulting in m-phenylenediamine. Undesired side-products are, e.g. condensed aromatic compounds. Three-dimensional zeolites with medium pores, e.g. H-ZSM-5, give higher selectivities than amorphous materials, because the condensed products cannot be formed inside the pores. Since m-phenylenediamine is the intermediate in the benzamine rearrangement of aniline, it is not surprizing that it is itself a good reactant in the benzamine rearrangement (15):

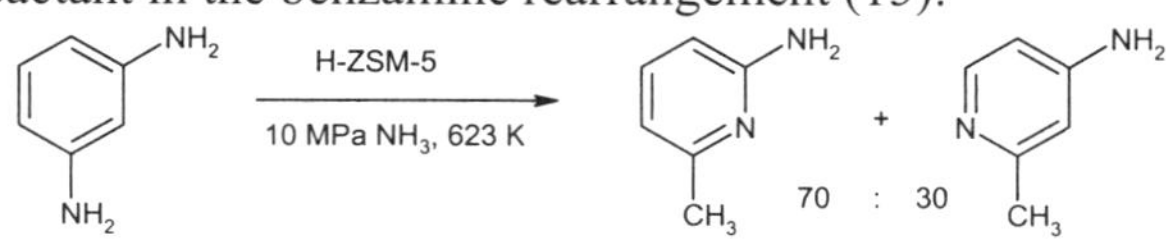


*13.2.3 Beckmann rearrangement*

The rearrangement of cyclohexanone oxime to caprolactam is still an important step in nylon production, and the heterogeneously catalyzed Beckmann rearrangement has been extremely well investigated (4, 16-19). In order to obtain catalysts that couple a high lactam selectivity to long lifespan, careful tuning of the zeolite properties is required. Some important factors are:

- *acid strength*: too strong acid sites catalyze side reactions, e.g. formation of 5-cyanopentene. Therefore, zeolites with mild acidity have been used, such as B-ZSM-5 or SAPO-11;
- *porosity*: poor desorption of the caprolactam product is the main reason for fast coking of the catalyst. Therefore, the pores should be sufficiently narrow to impede free access of cyclohexanone oxime to the pore system. 10-membered ring zeolites, e.g. with FER, MFI or AEL topology are preferred, since the reactions only take place at the outer surface or in the pore mouths.

Strong acid zeolites with 12-MR pores do not seem suitable candidates for the Beckmann rearrangement therefore. However, provided a diluting alcohol such as 1-hexanol is fed together with cyclohexanone oxime, long catalyst lifetimes and high selectivities can be observed, even for H-Beta and H-USY. Clearly, the 1-

hexanol facilitates product desorption by competitive adsorption; moreover, 1-hexanol can adsorb on strong acid sites and thus prevent undesired side reactions. An examplary reaction is given below (19):

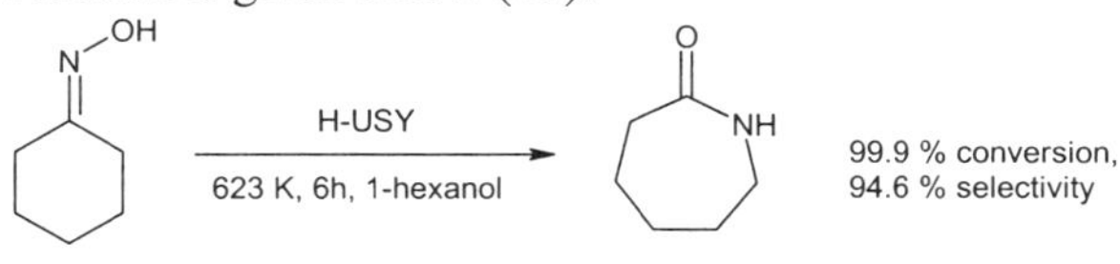


## 13.3 C-O bond formation

### *13.3.1 Etherification*

Zeolite-catalyzed etherification is at the basis of some large scale industrial operations, e.g. in the methanol-to-gasoline process, or in the MTBE or TAME production. On a smaller scale, etherification is important for protection chemistry, e.g. in the formation of 1,3-dioxolanes from carbonyl groups. Remarkable reactant shape selectivity was observed with a catalyst similar to a Beta zeolite (20). Organic phenethyl groups were incorporated in the framework during the synthesis of a siliceous Beta zeolite using tetraethylammonium fluoride as the template, by replacing part of the TEOS Si source by (2-phenylethyl)triethoxysilane. The template can be removed by solvent extraction, and in a next step, the phenyl groups are sulfonated with $SO_3/H_2SO_4$. The resulting material essentially has the microporosity of a Beta zeolite, but contains strong sulfonic acid groups in the inner void volume. In a ketalization with ethyleneglycol, cyclohexanone is easily converted by this material, but the bulky 1-pyrenecarboxaldehyde is not converted:

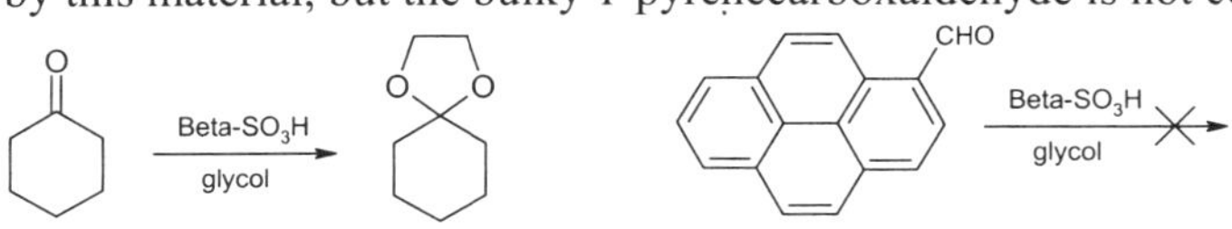


Etherification is also important in the preparation of the new generation of nonionic detergents, the alkyl polyglucosides. Two routes are possible: 1) etherification of glucose with n-butanol, and in a next step, transetherification of the n-butylglucoside with a long alcohol; 2) direct etherification of glucose with a long chain alcohol such as n-dodecanol. Zeolite catalysts may be used in the etherification and in the transetherification (21-22). A particularity of these reactions is that the reactants have widely diverging polarities, and this explains the strong dependence of the rate on the zeolite hydrophobicity, and hence, Si/Al ratio, defects etc. For instance, strong adsorption of glucose on a hydrophilic Y zeolite impedes butanol adsorption; at higher Si/Al ratios, both reactants have better access to the zeolite.

Zeolites have also been successfully applied in the reverse reaction, i.e. the hydrolysis of ethers to alcohols. A relevant example is the splitting of bis(3-hydroxypropyl)ether. This compound is a by-product in 1,3-propanediol synthesis, which can be performed by hydration of acrolein and reduction of 3-hydroxypropanal (23). In the hydrolysis of bis(3-hydroxypropyl)ether, a 20 weight % aqueous solution of the ether is passed over a ZSM-5 catalyst at 240°C:

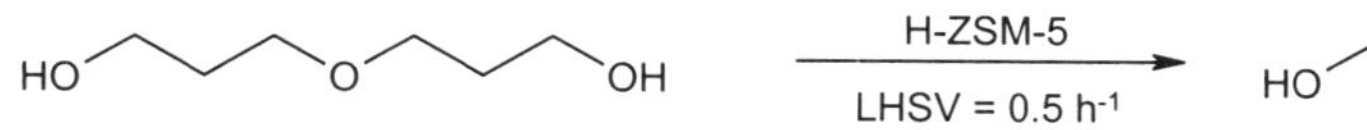


An important advantage of ZSM-5 over other catalysts is its high hydrothermal stability; this enables continuous operation over long periods even in the drastic hydrolysis conditions.

### *13.3.2 Esterification of alcohols or olefins with acids*

Zeolites also confer particular effects to esterification of polyols such as glycerol with long fatty acids. The aim in these reactions is to combine a high selectivity for the desired mono-ester product with a sufficient yield based on the fatty acid. The reaction mixture containing a long fatty acid and glycerol is biphasic. At elevated reaction temperature, the miscibility of the polar and apolar phases increases, and a spontaneous reaction occurs, catalyzed by the weak acid groups of the carboxylic acid. Therefore, if one wants to observe effects of a solid catalyst, the reaction temperature should be sufficiently low (24). In the biphasic reaction mixture, zeolitic catalysts are largely localized in the polar polyol layer. When a zeolite such as H-USY is used as a catalyst, the monoester selectivity is very high, up to 100%:

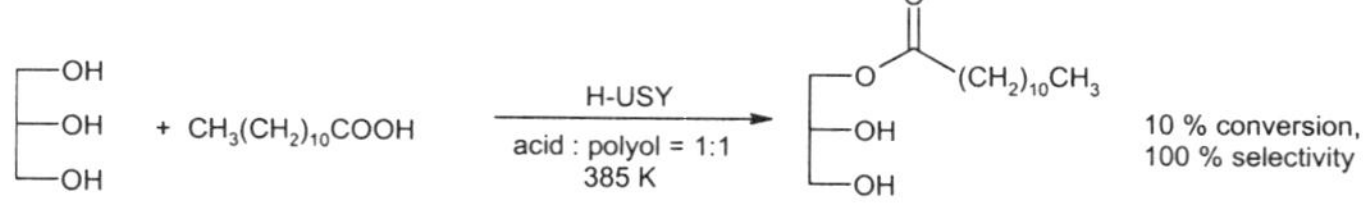


Consecutive reactions to form di- and triglycerides hardly occur because the polyol/monoester ratio is very high at the active sites of the zeolites, e.g. in the micropores. However, since the fatty acid concentration in the micro- or mesopores is low, the reaction only occurs slowly.

Several approaches can be followed to increase the reaction rates and yields while maintaining a high selectivity. First, an increasing Si/Al ratio of the zeolite, e.g. H-USY, strongly increases the activity, because the more hydrophobic zeolite has more affinity for the fatty acid (25). Moreover, desorption of water, which is a product of the reaction, is more facile if the catalytic matrix is more hydrophobic. Only at Si/Al > 35, the activity decreases again, because the number of active sites becomes quite small.

Secondly, strong acid - rather hydrophobic mesoporous catalysts have been designed by the covalent attachment of propylsulfonic acid groups to the pores of a mesoporous MCM-41 material or even simply a chromatographic $SiO_2$ (26). With such materials, conversion is fast; at 363 K and using a glycerol: lauric acid ratio of 3,53% monolaurin yield can be achieved at 80% lauric acid conversion. The monolaurin yield is appreciably higher than with the homogeneous catalyst p-toluenesulphonic acid at the same conversion level.

Another effect of zeolite pore architecture on esterification is found in the lactonization of 15-hydroxypentadecanoic acid. With dissolved acid catalysts or with amorphous $SiO_2$-$Al_2O_3$, dimerization or polymerization are the dominant reactions, but when the reactant is adsorbed within the pores of a dealuminated HY zeolite, only the pentadecanolide is obtained (8):

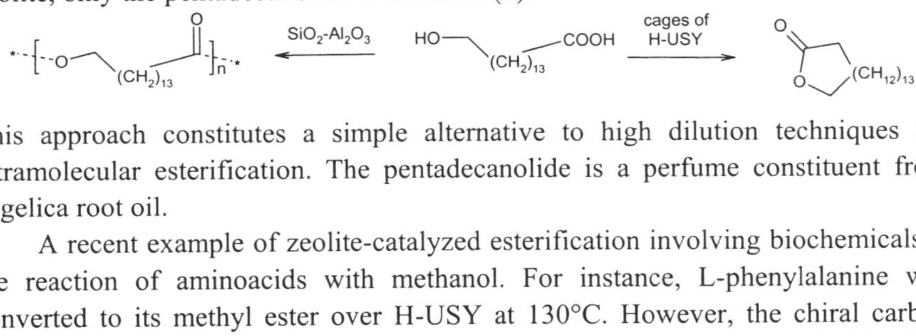


This approach constitutes a simple alternative to high dilution techniques for intramolecular esterification. The pentadecanolide is a perfume constituent from angelica root oil.

A recent example of zeolite-catalyzed esterification involving biochemicals is the reaction of aminoacids with methanol. For instance, L-phenylalanine was converted to its methyl ester over H-USY at 130°C. However, the chiral carbon atom was racemized to a considerable extent, yielding an eventual ee of 52% (27). In the reaction of $\alpha$ ,$\beta$-unsaturated acids with phenols, the esterification over H-Beta is followed by an alkylation of the aromatic ring; for instance resorcinol and acrylic acid react to form 7-hydroxy-3,4-dihydrocoumarin (28).

Finally, esters can also be produced by direct addition of carboxylic acids to double bonds. This reaction has been explored particularly for the gas phase addition of acetic acid to terminal olefins. Evidently, this reaction produces no net water (29).

## 13.4 C-C bond formation

### *13.4.1 Diels-Alder reactions*

Even in the absence of Lewis acid functions, zeolites can accelerate gas phase Diels-Alder reactions. This rate enhancement, for instance in the butadiene cyclodimerization, is attributed to a concentration effect inside the zeolite pores. The effect is however not zeolite-specific; any adsorbent with affinity for dienes, such as a carbon molecular sieve, displays similar effects (5).

Clear-cut examples of effects of zeolite pore architecture on the selectivity of Diels Alder reactions are not easily found. For instance, 4-vinylcyclohexene is formed with high selectivity from butadiene over a $Cu^{I}$-Y zeolite; however, the selectivity is intrinsically due to the properties of $Cu^{I}$, which can be stabilized by the zeolite, and not to the framework as such (30-31). A simple NaY has been used in the cycloaddition of cyclopentadiene and non-activated dienophiles such as stilbene. With such large primary reactants, formation of secondary products can be impeded by transition state shape selectivity. An exemplary reaction is the condensation of cyclopentadiene and *cis*-cyclooctene (32):

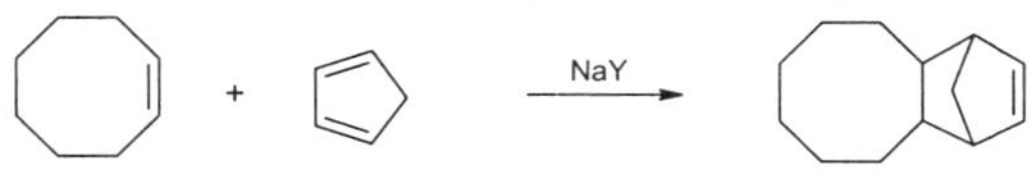


The resulting substituted norbornenes can be interesting as monomers for addition or ring-opening polymerization reactions.

The excellent yields and regioselectivities that are observed in many zeolite-catalyzed Diels-Alder reactions are probably the result of a combination of factors. The nature of the active site plays a prime role, e.g. extraframework $Al^{3+}$ or $Cu^{2+}$ or $Cu^{+}$. The stability of the products towards the catalyst can be important as well. For instance, in the reaction of isoprene with methyl acrylate, a completely selective reaction is observed over zeolite H-Beta (33):

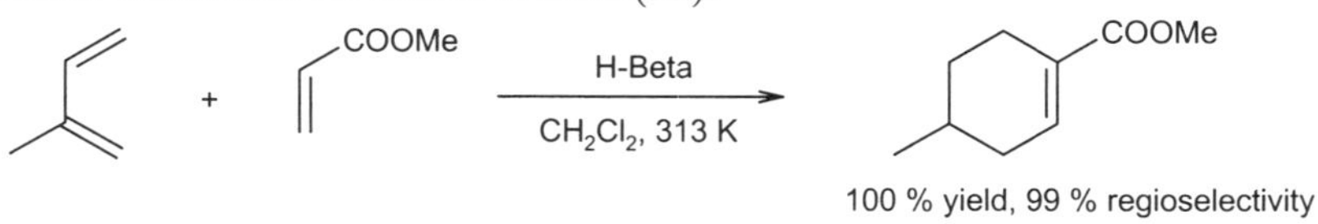


All other zeolites or Lewis acids which were tested yielded inferior results; with $AlCl_3$, the best homogeneous Lewis acid, the maximal yield was 83%, with a regioselectivity of 95%.

### *13.4.2 Base-catalyzed reactions*

While zeolites are mostly used in acid catalysts, there are various procedures to introduce basic sites with variable strength into these materials. Depending on the nature of the active site, one is able to selectively catalyze reactions with different basicity requirements, and this is probably the main virtue of base catalysis with zeolites. For instance, in a classical Knoevenagel condensation, the reaction selectivity can be decreased by a consecutive Michael reaction, since the Knoevenagel product can serve itself as a 'Michael receptor':

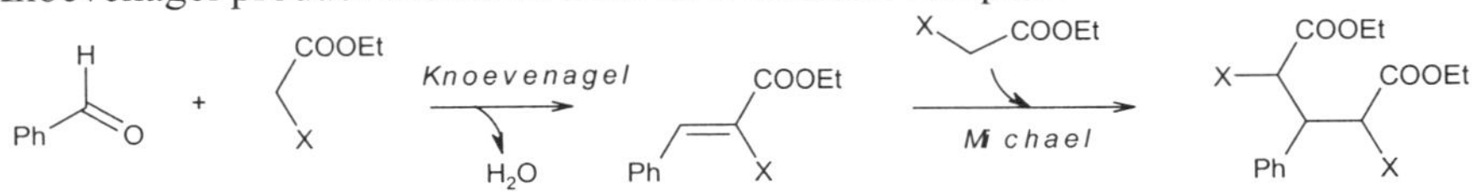


However, the second step requires stronger basic sites than the first step. This demonstrates the importance of being able to tune the base strength. In this sense, zeolites form a valuable complement to other solid base catalysts, e.g. MgO, or mixed oxides derived from layered double hydroxides.

If a zeolite with a low Si/Al ratio is exchanged with a large monovalent cation, e.g. $Cs^+$ in zeolite X, one obtains a weakly basic catalyst (CsX). The basic sites in this case are the oxygen framework atoms of the zeolite itself. Much stronger basic sites are formed when a zeolite or MCM-41 support is overexchanged or impregnated with, e.g. Cs acetate, and subsequently calcined (34). This results in formation of $Cs_2O$ oxidic particles.

CsX is useful for the simple Knoevenagel reaction of benzaldehyde with ethyl cyanoacetate; even a simple NaY is sufficiently basic to form carbamates starting from primary aromatic amines and dialkyl carbonates (35, 36). At contrast $Cs_2O$-MCM-41 can also be used for the addition of $CO_2$ to epoxides, or for Michael addition of one or two molecules of diethyl malonate on neopentylglycol diacrylate (37, 38) :

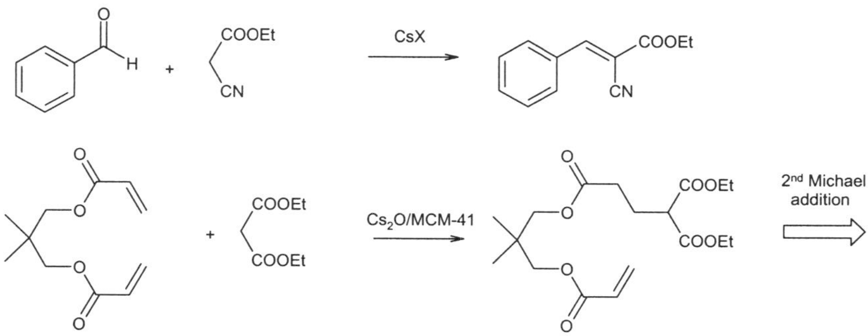


The first reaction is easier not only because it is a Knoevenagel reaction and not a Michael process, but also because the C-H bond in ethyl cyanoacetate is considerably more acidic than in diethyl malonate. It should be noted that there are few, if any, examples of shape-selectivity with such base catalysts.

Zeolite basicity can be further increased by the deposition of metallic clusters with excess electrons in the cages [39]. While there are many procedures to form clusters such as $Na_4^{3+}$ or $Na_6^{5+}$, one of the best accessible procedures seems the thermal decomposition of zeolite-adsorbed $NaN_3$. In an alternative procedure, ammoniacal metal solutions can be used for impregnation of the zeolite (40).

A material such as $Na^0$/NaY catalyzes the aldol condensation of acetone, to form mesityl oxide and eventually isophorone. Another strong base catalyzed reaction is the side chain alkylation of toluene with ethylene. In contrast with acid catalysis, side chain reaction is strongly preferred over ring alkylation. With a $Na^0$/NaX in the gas phase at 473 K, toluene reacts to give n-propylbenzene (66%) and the dialkylated product, 3-phenylpentane (32%) (41).

With such strong base catalysts, carbanions can be formed even from simple olefins. Evidence for carbanionic intermediates is found in butene isomerization. Starting from 1-butene, the rate of *cis*-2-butene formation is twenty times higher than the rate of *trans*-2-butene formation (42):

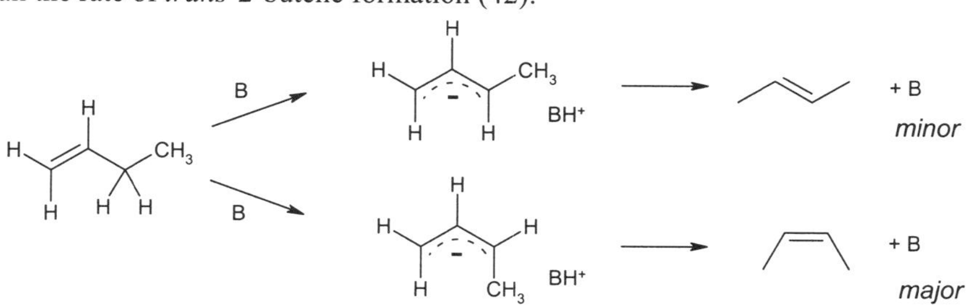

A drawback of these inorganic zeolite base catalysts is that as the basicity of the sites increases, so does the sensitivity of the materials to $CO_2$ and environmental moisture. Only materials such as exchanged CsX can be used in atmospheric conditions. An alternative access to materials with high $pK_{BH^+}$ values but with better moisture resistance is the covalent attachment of organic bases such as guanidines or P-N compounds to inorganic or organic supports. These materials are considerably more hydrophobic; hence their resistance to moisture and leveling by water is much better. For instance, 1,5,7-triazabicyclo[5.5.0]dec-5-ene was anchored to an oxirane-functionalized MCM-41 (43). An example of a reaction catalyzed by a guanidine-type base on MCM-41 is the synthesis of 3-cyanocoumarins. Salicylaldehyde and ethylcyanoacetate first react in a Knoevenagel reaction, which is followed by an intramolecular transesterification:

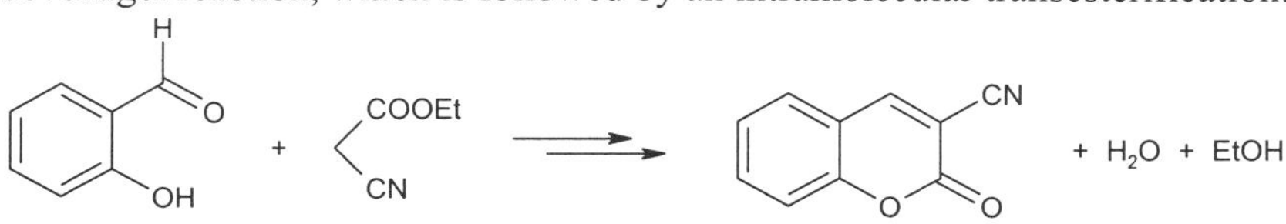


Alternatively, quaternary ammonium residus can be immobilized on solid phases, and anion-exchanged with hydroxyl anions. These simple materials are highly similar to strong basic anion exchangers, but obviously, their basicity can never exceed that of $OH^-$ anions (44). In summary, there is a large scale of zeolite-based base catalysts available, and the catalyst choice can be determined by the base strength desired.

### *13.4.3 Polycyclization*

Some zeolites can mimic the activity of natural polyene cyclases, which effect the cyclization of squalene epoxide to steroid-type structures (45). As a model compound, an epoxypolyene with a TMS-functionalized propargylic group (**1**) was selected. On NaA, the epoxide is rearranged to a ketone, which is the common reaction for basic catalysts. H-Beta and faujasites are subject to coking, due to too strong adsorption of the polyene in the 12-MR structures. However, useful results are obtained with medium pore zeolites, such as mordenite and in particular H-ZSM-5. Compound **1** is converted in about 50% yield to the tricyclic compound **2a** over a H-ZSM-5 with a dealuminated outer surface. While with homogeneous Lewis acids several isomers are formed in roughly comparable amounts, **2a** strongly predominates in the zeolite-catalyzed reaction:

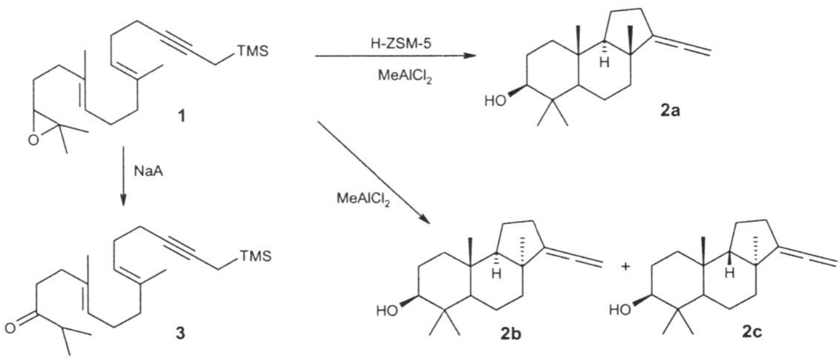


Apparently, **1** is adsorbed only at the pore mouths of the ZSM-5 zeolite, so that desorption and recovery of the cyclized products remain possible. The high selectivity for **2a** points to a highly selective interaction between the zeolite outer surface and the organic reactant, resulting in excellent stereoselectivity.

### *13.4.4 Other carbon skeleton rearrangements*

In some cases, the steric constraints of the zeolite pores can enable or impede secondary reactions. A beautiful example is found in the rearrangement of benzyl allyl ethers over H-Beta zeolites (46). In the normal course of this reaction, the benzyl allyl ether is transformed to a 4-arylbutanal, probably via a five-membered ring intermediate:

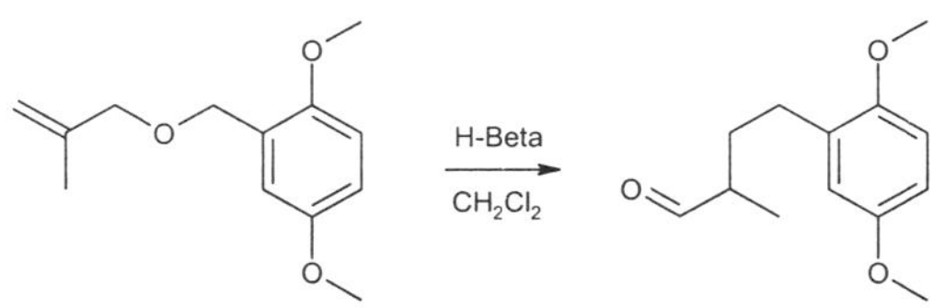


With zeolite Beta and the 2,5-dimethoxybenzyl ether, the reaction stops at the aldehyde. However, with the slimmer 3,4-dimethoxybenzyl ether, the reaction proceeds via a dehydration and a cyclization, to give a dihydronaphthalene:

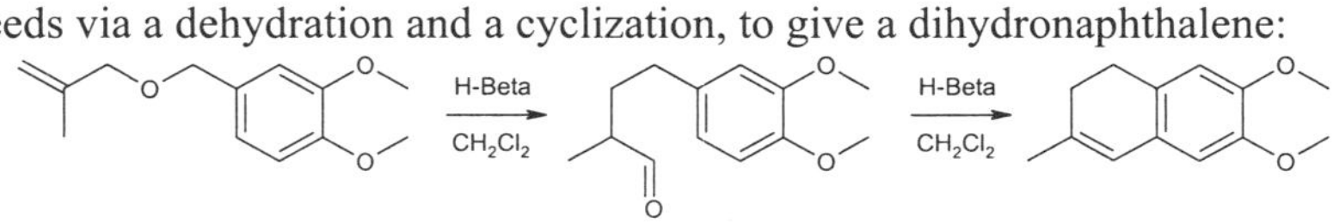


It therefore seems plausible that intraporous reaction occurs.

While zeolites are themselves rarely chiral, adsorption of chiral inductors can enable enantioselective or diastereoselective reactions. For instance, Y zeolites have been modified with chiral dithiane oxides (47); Ramamurthy and his group have focused on ephedrine-modified Y zeolites (48). As a model reaction, they selected the light-induced rearrangement of tropolone ethers:

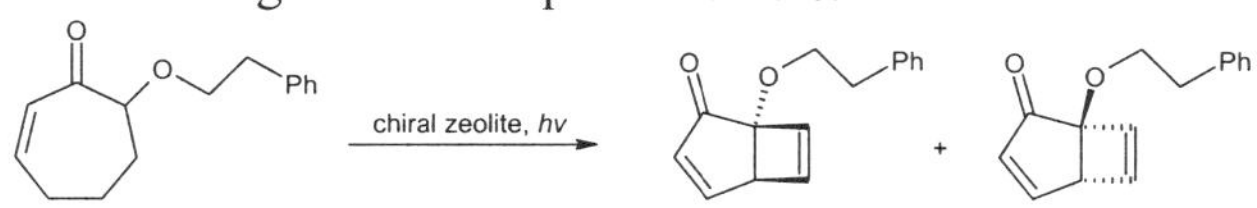


With (+)-ephedrine as a modifier on NaY, an ee of 69 % for one of both enantiomers was obtained. With tropolone ethers already containing a chiral carbon atom, a diastereomeric excess of 90% was observed.

## 13.5 Reductions

There are numerous examples of reactant shape selectivity in the hydrogenation of olefins over noble-metal loaded zeolites (49). This can be important to remove impurities from olefin feedstocks, or as a criterion to assess the location of the noble metal, at the outer or inner surface of the zeolite. However, shape selectivity is also increasingly used in reductive conversion of (poly)functional molecules.

### *13.5.1 Selective hydrogenation of molecules with several unsaturated bonds*

In the chemoselective reduction of a polyfunctional molecule such as cinnamaldehyde, the zeolite lattice can play an important role in improving the chemoselectivity. Pt catalysts often display only moderate selectivity for the allylic alcohol, but the chemoselectivity can be influenced by many factors, such as surface ligands, the presence of a second metal, or the metal cluster size (50). Thus, large Pt clusters (3-6 nm) on graphite allow to obtain 98% chemoselectivity for cinnamyl alcohol at 50% conversion (51). With smaller supported metal clusters, chemoselectivity is usually lower, because the C=C bond can more easily approach a metal surface with high curvature than a large, flat surface. However, when small Pt particles (1 up to 5 nm) are entrapped within the structure of zeolite Y, the cinnamaldehyde can only approach the Pt surface via the terminal aldehyde group, and a high cinnamyl alcohol selectivity is realized (52):

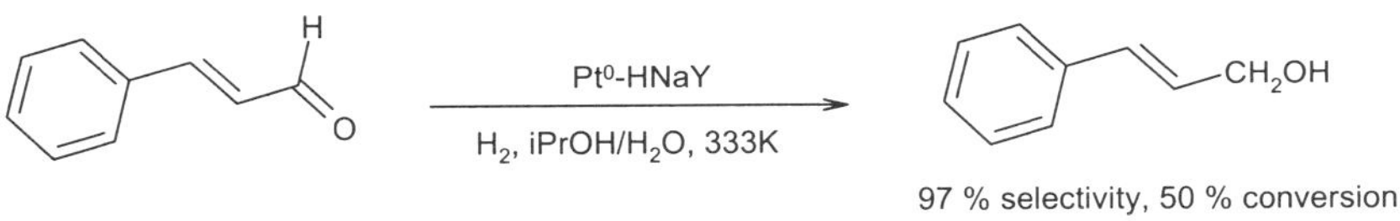

Analogous observations can be made for the cinnamaldehyde hydrogenation over large Pt ensembles in the inner pore volume of Beta zeolites (53).

Similar effects can be observed in the hydrogenation of molecules with two competing C=C double bonds (54). As an example, the hydrogenation of geraniol to citronellol or 3,7-dimethyl-2-octen-1-ol has been performed with supported $Pt^0$. With Pt particles in a non-shape selective environment such as MCM-41, the chemoselectivity for citronellol amounts to 85% at 80% conversion. However, in a shape selective host such as zeolite Pt-HBeta, 100% chemoselectivity is obtained at the same conversion level:

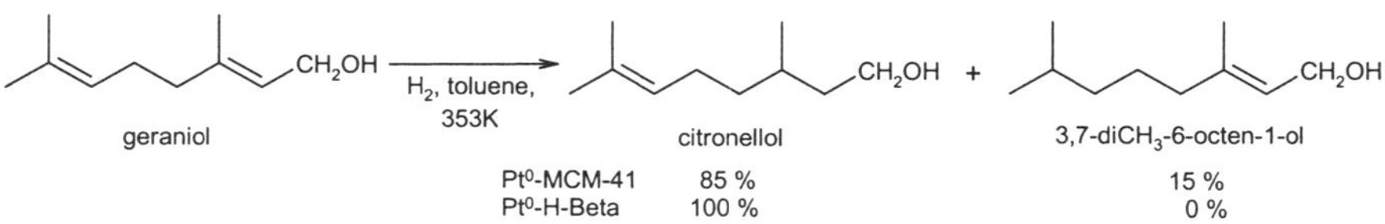


Again, this provides evidence for a zeolite-directed approach of the unsaturated substrate to the metallic surface.

A related example is the selective hydrogenation of an α,ω-dinitrile to an α-amino-ω-nitrile (55). With metals supported on 8-membered ring zeolites, e.g. Ni-SAPO-34, adiponitrile is converted with high selectivity to ε-aminocapronitrile.

A process with potential practical applicability is the hydrogenation of edible oils. Reduction of multiply unsaturated triglycerides with hydrogen over Ni-based catalysts is frequently used to gain autoxidative stability of edible oils. According to the Polanyi-Horiuti mechanism, multiple 1,2 or 1,4 diadsorption of the fatty acid tail with exclusively *cis*-configuration around the double bonds causes *cis-trans* isomerisation, whilst the number of double bonds is being reduced. The *trans*-fatty acid chains have adverse effects on the human metabolism and must be minimized.

When the hydrogenation function is embedded in the crystal voids of an MFI topology, the formation of *trans*-isomers is strongly reduced. After partial reduction of soy bean oil with such catalyst from an iodine value of 140 to 80, virtually no *trans*-isomers are obtained (56). This is the result of pore mouth catalysis combined with zeolite shape selectivity. Due to the bent character of the *cis*-isomer chains in triglycerides, *trans*-configured chains preferentially enter the pore mouths for hydrogenation. In this environment, metal-catalyzed *cis-trans* isomerization is restricted for steric reasons as multiple readsorption is minimal.

### *13.5.2 Enantioselective reactions*

Several zeolites have been used as supports for cinchona modified Pt in the enantioselective hydrogenation of α-ketoesters such as ethyl pyruvate:

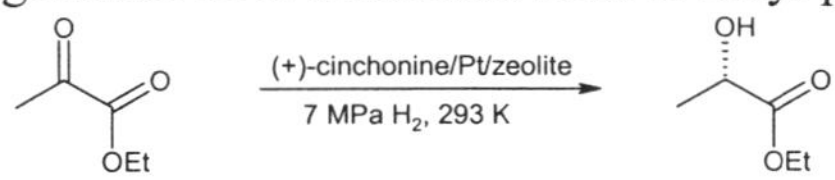


There is no direct relation between the zeolite type and the ee values (75-80%). In this case, the only role of the zeolites is to provide a suitable dispersion of the noble metal (57).

A more specific interaction between a zeolite surface and a chiral catalyst was recently uncovered (58). It was found that the Ru-binap catalyst can be specifically withheld on the outer surface of Beta zeolites. Such a heterogeneous catalyst is relevant for the highly enantioselective hydrogenation of methyl acetoacetate to (R) or (S) 3-hydroxymethylbutyrate, with typical ee values of about 95 %.

### *13.5.3 Hydrogen transfer reductions*

While the use of zeolites as catalysts for C=O reduction with alcohols as a reductant goes back to the eighties, stereoselective versions of this reaction have been developed more recently. In this so-called MPV (Meerwein-Ponndorf-Verley) reaction, the reactant carbonyl compound and the reducing alcohol coordinate simultaneously on a Lewis acid centre. In the reduction of 4-t-butylcyclohexanone, the *trans* product would normally be thermodynamically favored. However, the *cis* alcohol isomer is economically much more interesting. Van Bekkum and co-workers discovered that in the constrained pores of an Al- or Ti-Beta zeolite, more than 95 % of the product has the *cis* configuration (59):

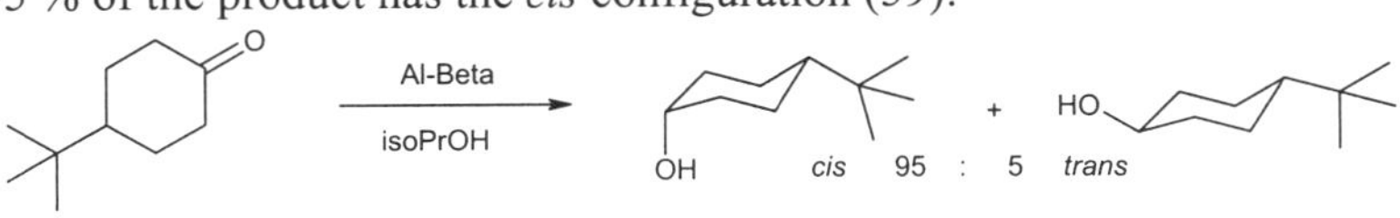


Clearly, the transition state leading to the *trans* product cannot be formed in the Beta pores.

### *13.5.4 Hydrogenation combined with hydrolysis: Carbohydrate reductions*

Upon reduction, hexose sugars like D-glucose, D-mannose and D-fructose yield commercially very attractive hexitols such as D-glucitol (sorbitol) or D-mannitol (see scheme). As high purity hexitols are a prerequisite for food and pharmaceutical applications, in absence of a selective reduction catalyst, a supplementary chromatographic separation or crystallization is required. Typical feedstocks such as starch and (+)-sucrose have to undergo successive hydrolysis and hydrogenation to obtain sorbitol and a theoretical 25/75 mannitol/sorbitol mixture, respectively. This 25/75 ratio originates in the fact that sucrose is hydrolysed to equimolar amounts of glucose and fructose; glucose is hydrogenated to only sorbitol, while fructose normally yields equal amounts of the epimers mannitol and glucitol. Since D-mannitol has particular therapeutic and dietetic properties, it is a very valuable hexitol, and increasing the overall mannitol yield from sucrose is an important research target.

In a bio-chemo combined process, fructose/glucose mixtures resulting from immobilized glucose isomerase treated glucose are hydrogenated with a Pt-on-carbon catalyst into a 50% mannitol in sorbitol mixture (60). Heterogeneous and in particular zeolite catalysis will allow to perform such bifunctional catalytic acts in a single catalyst. Moreover, it has been demonstrated that diastereoselective effects are important in the reduction of fructose (61). For instance, with a $Cu^0$/NaY catalyst, the hydrogenation of fructose yields a 67% selectivity for mannitol and only 33% of sorbitol.

The chemistry of the hydrolysis/hydrogenation process from (+)-sucrose can be schematized as follows:

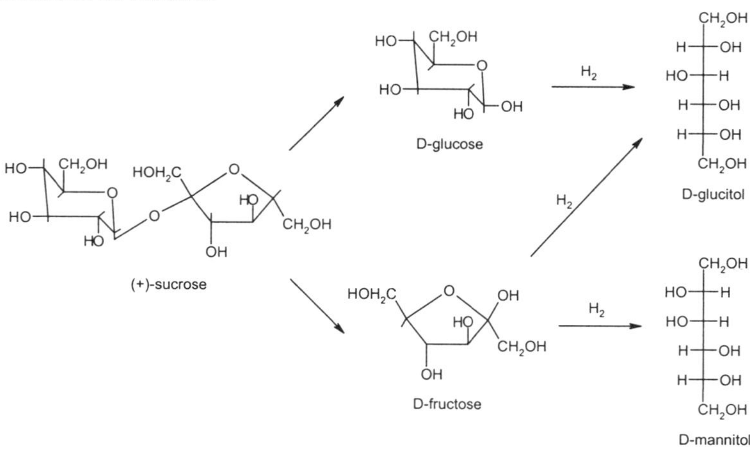


With Ru-H-Beta zeolite as a bifunctional catalyst, the equimolar amounts of the two hexoses formed after hydrolysis of sucrose are converted to between 45 and 37% of D-mannitol at conversions between 20 to 100% with (62). In comparison to any other previously reported system, this represents a clear diastereoselective enhancement of the D-mannitol selectivity exerted by the zeolite topology. The same effect, though less pronounced, is present in zeolite Y. Unfortunately, the understanding at the molecular level of the preference of the zeolite intracrystalline voids for D-fructose over D-glucose is not clear at the moment. Distribution constants between the aqueous and zeolite phase, obtained chromatographically, do not point to the existence of competitive sorption among glucose and fructose.

The typical production of D-glucitol consists of the catalytic hydrogenation of a starch hydrolysate in presence of Brönsted acid. Ru-HUSY zeolite (62) is allowed to do even better as aqueous starch gels can be used as feed. In a single step starch can be transformed into sorbitol of pharmaceutical grade. As the zeolitic protons chop D-glucose entities from the glucose bio-polymer entering the pores, the D-glucose monomers can undergo fast hydrogenation into the thermally more stable D-glucitol, allowing reaction temperatures that exceed the glucose caramelization temperatures. The zeolite effect is only one of enhanced selectivity and reactivity. It is clear that the zeolite topology does not exert any diastereoselective influence as no traces of mannitol are among the products.

## References

1. Hölderich W.F. and van Bekkum H., *Stud. Surf. Sci. Catal.* **137** (2001) 821.
2. Espeel P.H. et al., *Zeolite effects in Organic Catalysis,* in *Catalysis and Zeolites, Fundamentals and Applications,* Ed. Weitkamp J. and Puppe L., (Springer, 2001), 377-436.
3. Hölderich W.F., *Organic Reactions in Zeolites*, in *Comprehensive Supramolecular Chemistry*, Ed. Alberti G. and Bein T. (Pergamon, 1996), 671-692.
4. Sheldon R.A. and van Bekkum H., *Fine Chemicals through Heterogeneous Catalysis* (Wiley-VCH, 2001).
5. Dessau R.M., *J. Chem. Soc., Chem. Commun.* (1986) 1167.
6. Langhendries G., De Vos D.E., Baron G.V. and Jacobs PA., *J. Catal.* **187** (1999) 453.
7. Knops-Gerrits P.P., De Vos D.E., Thibault-Starzyk F. and Jacobs P.A., *Nature* **369** (1994) 543.
8. Tatsumi T., Sakashita H. and Asano K., *Chem. Commun.* (1993) 1264.
9. McAteer C.H. and Scriven E.F.V., in [4], 275-283.
10. Shimizu S. et al., US Patent **4810794** (1989).

11. Sato H., Shimizu S., Abe N., and Hirose K., *Stud. Surf. Sci. Catal.* **84** (1994) 1951.
12. McAteer C.H., Brown D.C., and Davis R.D., US Patent **5780635** (1998).
13. Kunkeler P.J. et al., *Stud. Surf. Sci. Catal.* **105** (1997) 1269.
14. Prins R., in [4], 205-210.
15. Stamm T., Kouwenhoven H.W., Seebach D., and Prins R., *J. Catal.* **155** (1995) 268.
16. Heitmann G.P., Dahlhoff G., Niederer J.P.M. and HölderichW.F., *J. Catal.* **194** (2000) 122.
17. Dahlhoff G., Barsnick U. and Hölderich W.F., *Appl. Catal. A* **210** (2001) 83.
18. Tatsumi T., in [4], 185-204.
19. Dai L..X., Koyama K., Miyamoto M. and Tatsumi T., *Appl. Catal. A* **189** (1999) 237.
20. Jones C. W., Tsuji K., and Davis M.E., *Nature* **393** (1998) 52.
21. Chapat J.F., Finiels A., Joffre J., and Moreau C., *J. Catal.* **185** (1999) 445.
22. Corma A., Iborra S., Miquel S., and Primo J., *J. Catal.* **180** (1998) 218.
23. Haas T. and Arntz D., US Patent **5364987** (1994).
24. Heykants E., Verrelst W.H., Parton R.F., and Jacobs P.A., *Stud. Surf. Sci. Catal.* **105** (1997) 1277.
25. Machado, M. et al., *Appl. Catal. A* **203** (2000) 321.
26. Bossaert, W.D. et al., *J. Catal.* **182** (1999) 156.
27. Wegman M.A. et al., *Green Chemistry* **3** (2001) 61-64.
28. Gunnewegh E.A., Hoefnagel A.J. and Van Bekkum, H., *J. Mol. Catal. A* **100** (1995) 87.
29. Young, L.B., US Patent 4365084 (1982).
30. Maxwell, I.E., Downing, R.S. and van Langen, S.A. J., *J. Catal.* **61** (1980) 485.
31. Maxwell I.E., de Boer J.J. and Downing R.S., *J. Catal.* **61** (1980) 493.
32. Bergström C., Mölsä A., Yli-Kauhaluoma J., WO 9940049 (1999).
33. Eklund L., Axelsson A., Nordahl A. and Carlson R., *Acta Chem. Scand.* **47** (1993) 581.
34. Doskocil E.J., and Davis R.J., *J. Catal.* **188** (1999) 353.
35. Corma, A. et al., *Appl. Catal. A* **59** (1990) 237.
36. Selva M., Tundo P. and Perosa A., *J. Org. Chem.* **66** (2001) 677.
37. Kloetstra R. and van Bekkum H., *Stud. Surf. Sci. Catal.* **105** (1997) 431.
38. Tu M. and Davis R.J. *J. Catal.* **199** (2001) 85.
39. Martens L.R.M., Grobet P.J. and Jacobs P.A., *Nature* **315** (1985) 568.
40. Baba T. et al., *J. Mol.. Catal. A* **98** (1995) 49.
41. Martens L.R.M., Ph.D. Thesis K.U.Leuven (1987).

42. Martens L.R.M., Grobet P.J., Vermeiren W.J.M. and Jacobs P.A., *Proc. 7th IZC,* Ed. Murakami Y., Ijima A. and Ward J.W. (Kodansha-Elsevier, 1986) 935-941.
43. Rao Y.V.S., De Vos D.E. and Jacobs P.A., *Angew. Chem.*, **109** (1997) 2776.
44. Rodriguez I. et al., *Chem. Commun.* (1999) 593.
45. Sen S.E., Zhang Y. Z., Smith S. M. and Huffman J. C., *J. Org. Chem.* **63** (1998) 4459.
46. Wennerberg J., Ek F., Hansson A. and Frejd T., *J. Org. Chem.* **64** (1999) 54.
47. Feast S. et al., *J. Catal.* **167** (1997) 533.
48. Joy A. and Ramamurthy V., *Chem. Eur. J.* **6** (2000) 1287.
49. Creyghton E.J., and Downing R.S., *J. Mol. Catal. A* **134** (1998) 47.
50. Gallezot P., *Handbook of Heterogeneous Catalysis*, Ed. Knözinger H., Ertl G. and Weitkamp J., (Wiley-VCH, 1990), 2209-2221.
51. Giroir-Fendler A., Richard D., and Gallezot P., *Catal. Lett.* **5** (1990) 175.
52. Gallezot P., Giroir-Fendler A. and Richard D., *Catal. Lett.* **5** (1990) 169.
53. Gallezot P., Blanc B., Barthomeuf D. and Pais da Silva M.I., *Stud. Surf. Sci. Catal.* **84** (1994) 1433.
54. Tas D., Parton R.F., Vercruysse K. and Jacobs P.A., *Stud. Surf. Sci. Catal.* **105** (1997) 1261.
55. Vandenbooren F., Bosman H., Van der Spoel J., US Patent **5844114** (1998).
56. Van Looveren L. et al., US Patent **6229032** (2001).
57. Böhmer U. et al., *Catalysis Today* **60** (2000) 167.
58. Van Brussel W. et al., US**5997840** (1999).
59. Creyghton E.J., Ganeshie S.D., Downing R.S. and van Bekkum H., *J. Mol. Catal. A* **115** (1996) 457.
60. Makkee M., Kieboom A.P.G., Van Bekkum H., and Roels J.A., *Chem. Commun.* (1980) 930.
61. Ruddlesden J.F., Stewart A., Thompson D.J., and Whelan R., *Faraday Disc.* **72** (1981) 397.
62. Jacobs P.A. et al., unpublished results.

CHAPTER 14

# FUNCTIONALIZATION OF AROMATICS OVER ZEOLITE CATALYSTS

P. MARION, R. JACQUOT
*CRL Rhodia, 85, rue des Frères Perret, BP 62, 69192 Saint Fons Cedex, France*

S. RATTON
*RHODIA, 26, quai A. Le Gallo, 92512 Boulogne Billancourt Cedex, France*

M. GUISNET
*Catalyse en Chimie Organique, 40, avenue du Recteur Pineau,
86022 Poitiers Cedex, France*

## 14.1 Introduction

There is an urgent need for more environmentally acceptable processes for the synthesis of Fine Chemicals. Indeed, 5 to more than 50 kg waste per kg product are generated, i.e. approximately 10 times more than in the case of Bulk Chemicals (1). Of course, this is because Fine Chemicals synthesis requires various successive steps, but also because many of the steps are carried out in homogeneous phase, stoichiometrically or by using acid catalysts such as $H_2SO_4$ or $AlCl_3$. These acids which are not reusable have to be neutralized, which generates a large amount of valueless salts. The solution to this problem seems obvious: these polluting technologies have to be substituted with cleaner technologies such as those based on heterogeneous catalysis (2, 3). Indeed, solid catalysts offer many advantages: easy recovery of reaction products, easy set up of continuous processes, etc. Among them, zeolites with their shape selective properties and the easy tailoring of their acidity, porosity, etc. are particularly well adapted to selective organic synthesis, with, however, the scope limitation to the size of the synthesized molecules (4-8).

This paper is focused on commercially important syntheses of aromatic ketones and ethers with application in pharmaceuticals, agrochemicals, perfumery, etc. It will be demonstrated using several examples that clean processes using zeolite catalysts can be substituted economically for existing polluting processes. Limitations in the use of zeolite catalysts will also be emphasized, other solid catalysts leading, in certain cases, to much better results in terms of selectivity and stability.

## 14.2 Acetylation of aromatics

### *14.2.1 Overview*

Arylketones are generally prepared by acylation of aromatics (reaction 1) or by the related Fries rearrangement (reaction 2):

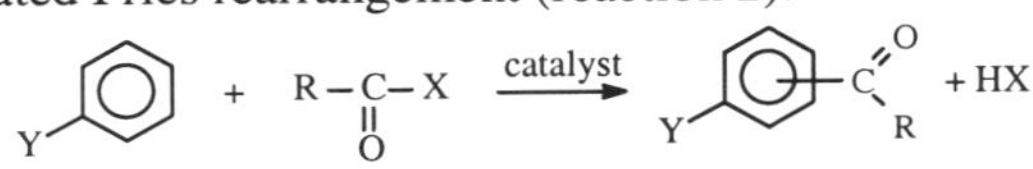

**(1)**

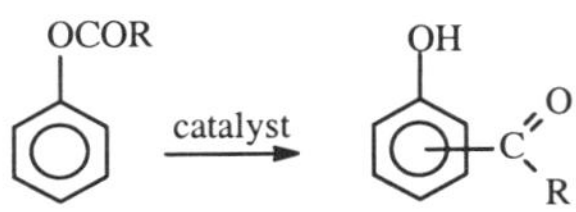

**(2)**

These ketones are important intermediates in the synthesis of fragrances of the musk type, of UV absorbents and of pharmaceuticals such as paracetamol, ibuprofen, S-naproxen. Thus, in the processes of ibuprofen synthesis (9), the first classical step is the acetylation of isobutylbenzene with acetic anhydride in presence of HF (Hoechst process) or of $AlCl_3$ (Boots process):

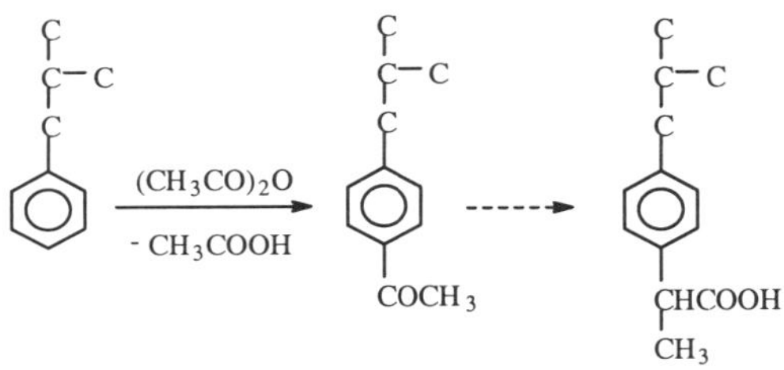

**(3)**

The synthesis of S-naproxen also involves a Friedel Crafts acetylation, but with acetylchloride followed by various reactions, the last one being an asymmetric hydrogenation over a Ru catalyst (10):

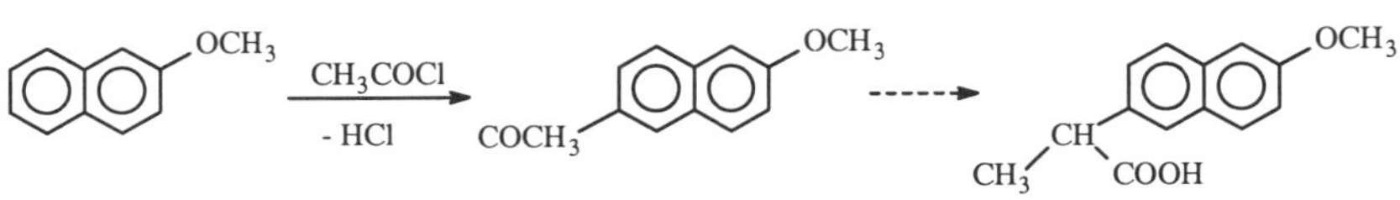

**(4)**

Acylation processes are often carried out in batch reactors by using acid metal chlorides such as $AlCl_3$ as catalysts and acid chlorides as acylating agents. The use of this type of catalysts generates serious environmental problems. Indeed, because the arylketone forms a 1:1 molar adduct with the catalyst, more than stoichiometric amounts are required and the recovery of the ketone requires a hydrolysis step with destruction of the catalyst and production of a large amount of hydrochloric acid (more than 4 mol of $Cl^-$ per mol of ketone produced) with corrosion problems and final production of valueless salts (11, 12). Therefore, the substitution of these polluting and corrosive solutions by solid acid catalysts such as zeolites, which do not present these inconveniences, is very attractive. To make the process still more environmentally and economically friendly, acylchlorides should be substituted for acids or acid anhydrides as acylating agents.

In the past 15 years, a number of research groups have investigated, often in batch reactors, the use of zeolites to catalyse liquid phase aromatic acylation. Good activities and selectivities have been achieved with activated aromatic substrates. Experiments in flow reactors show that deactivation is the main problem to be solved, which can be made by an adequate adjustment of the zeolite characteristics and especially of the operating conditions.

### *14.2.2 Choice of the operating conditions and of the zeolite catalyst*

**Anisole acetylation,** which was one of the main reactions investigated, was first shown to be catalysed by zeolite ten years ago by Bayer (13), which was confirmed by Harvey et al. (14), then by Rhodia (15). Large pore zeolites and especially those with a tridimensional pore structure such as HBEA and HFAU were found to be the most active: at 80°C, in a batch reactor with an anisole/acetic anhydride molar ratio of 5 and after 6 hours reaction, the yield in methoxyacetophenone (MAP) was close to 70% with HBEA and HFAU zeolites, to 30% with HMOR and 12% with HMFI. With all the zeolites and also with clays and heteropolyacids, the selectivity to the para-isomer was greater than 98%, which indicates that this high selectivity is not due to shape selective effects but rather to the reaction mechanism (electrophilic substitution). The lower conversion observed with HMOR can be related to the monodimensional pore system of this zeolite which is very sensitive to blockage by heavy secondary products. Furthermore, limitations in the desorption of methoxyacetophenone from the narrow pores of HMFI are probably responsible for the low activity of this intermediate pore size zeolite.

The reaction over a HBEA zeolite (Si/Al = 10) of an equimolar mixture of anisole and acetic anhydride was investigated at 60°C as a function of time (16). Figure 14.1 shows that whereas the initial rate of p-MAP production is very high, it decreases rapidly with time. This apparent catalyst deactivation was thought to be due to limitation in the adsorption of reactants caused by the strong adsorption of the polar p-MAP product. This was demonstrated by the pronounced inhibiting effect of the adding of p-MAP to the reactant mixture. Moreover, it was concluded from a kinetic study that the adsorption equilibrium constant of p-MAP was much greater than those of the reactants (17). Autoinhibition is a general problem in aromatic acetylation, most of the products being more polar than the reactants.

To limit the inhibiting effect of p-MAP, operating conditions favoring the product desorption have to be chosen: higher temperature, large excess of anisole in the reactant mixture, flow reactor instead of the batch reactor in which the contact time of p-MAP with the zeolite catalyst is very long, low conversion with recycling after p-MAP separation, etc.

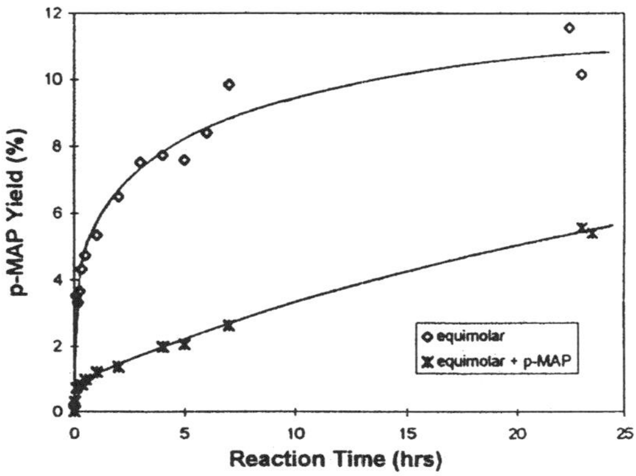


**Fig. 14.1** Reaction of an equimolar mixture of anisole and acetic anhydride (35 mmol of each) on 100 mg of HBEA10 at 60°C in a batch reactor. Influence of the addition of p-methoxyacetophenone (p-MAP) in the reactant mixture (1.5 mmol) on anisole acetylation

The anisole/acetic anhydride ratio has a large effect both on the initial apparent reaction rate and on deactivation. Thus, in a fixed bed reactor with an anisole rich feeding mixture (anisole/acetic anhydride of 5) the maximum conversion of anisole is obtained at short contact time whereas a plateau at only half of the maximum conversion is found at long contact time with an equimolar mixture (Figure 14.2). Moreover, deactivation is much faster in this latter case (16).

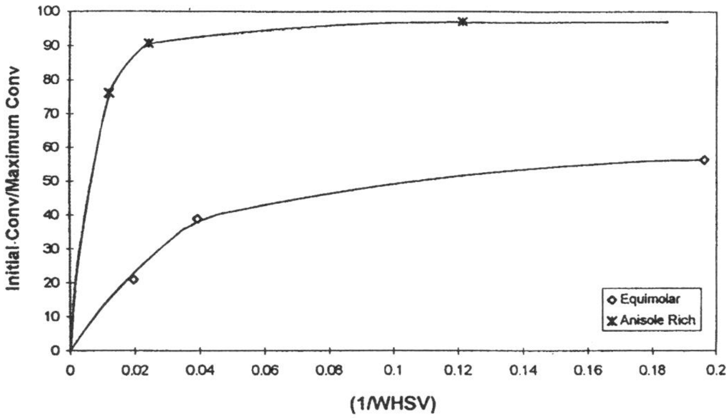


**Fig. 14.2** Acetylation of anisole at 90°C in a fixed bed reactor. Ratio between the initial conversion of anisole and the maximum conversion which can be obtained vs. the inverse of the weight hourly space velocity (WHSV) for anisole

The analysis of the zeolite samples after reaction at different time-on-stream shows that a very large amount of p-MAP is retained in the mesopores (or at the micropore mouth) and that larger molecules (mainly di- and triacetylated anisoles) are trapped in the zeolite micropores (17). As could be expected, this polyacetylation is more significant from the equimolar reactant mixture than from the anisole rich mixture. These large molecules were shown to be responsible for a blockage of the access of nitrogen, hence of the reactant molecules to the zeolite micropores. A complete removal of these products and a complete regeneration of the zeolite catalyst can be obtained by oxidation under air flow, traces of noble metal favoring this oxidation (18). However, partial dealumination of the zeolite by acetic acid with a decrease of the activity of the regenerated catalysts was observed by Derouane et al. (17).

**Acetylation of 2-methoxynaphthalene** (2MN) with acetic anhydride (AA) was also investigated over various molecular sieves: FAU (19, 20), MFI, MOR (20), MTW (19), MCM41 (21) and especially BEA (19, 22-30). With this acetylation, there is an additional problem because of the simultaneous formation of 2-acetyl-6-methoxynaphthalene (II, Figure 14.3), which is the desired product (precursor of naproxen), and of its isomers. Generally, acetylation occurs preferentially at the kinetically controlled 1-position with formation of 1-acetyl-2-methoxynaphthalene (I, Figure 14.3).

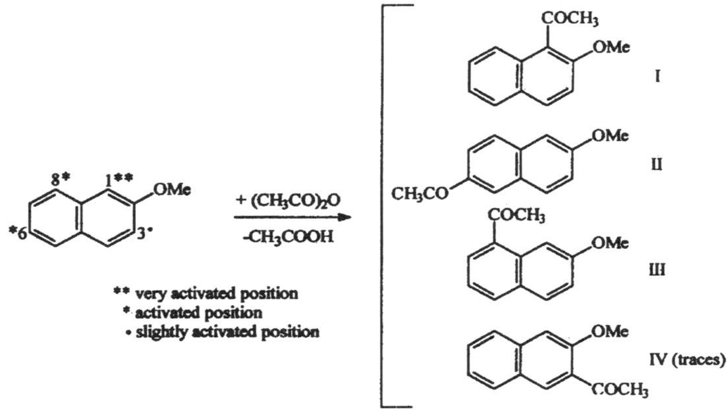


**Fig. 14.3** Acetylation of 2-methoxynaphthalene (2MN). Reaction products.

However, as shown in Figure 14.4, over an HBEA zeolite with a Si/Al ratio of 15 (HBEA15), acetylation is followed by secondary reactions of isomer I: isomerization into II and into 1-acetyl-7-methoxynaphthalene (III) and deacylation. Therefore, the transformation of the 2MN/AA mixture on HBEA15 can be proposed to occur mainly through the following successive scheme (25):

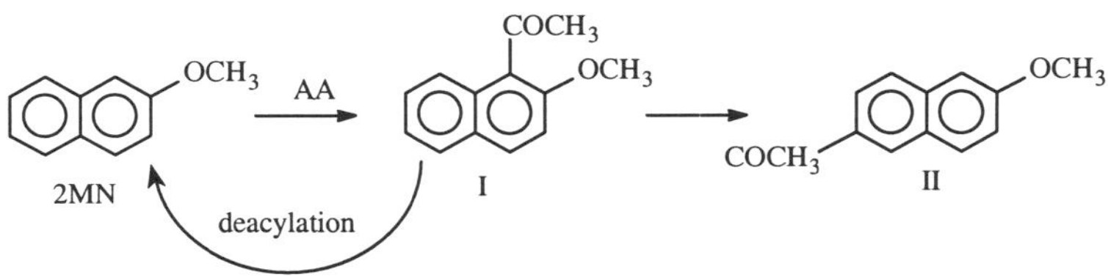


**(5)**

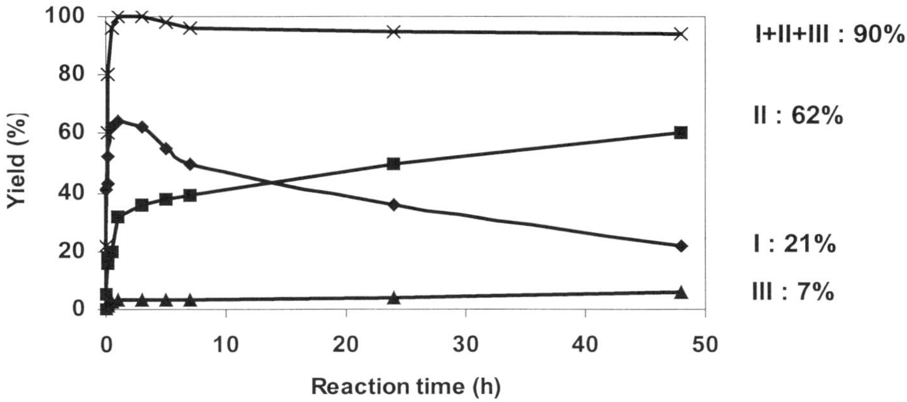


**Fig. 14.4** Acetylation of 2-methoxynaphthalene with acetic anhydride in nitrobenzene solvent. Total yield in acetyl-methoxynaphthalene (I + II + III) (X) and yields in isomers I (◊), II (□), III (Δ).

Isomerization of I into II is much faster in presence than in absence of 2MN, which suggests an intermolecular mechanism for this reaction (transacylation) :

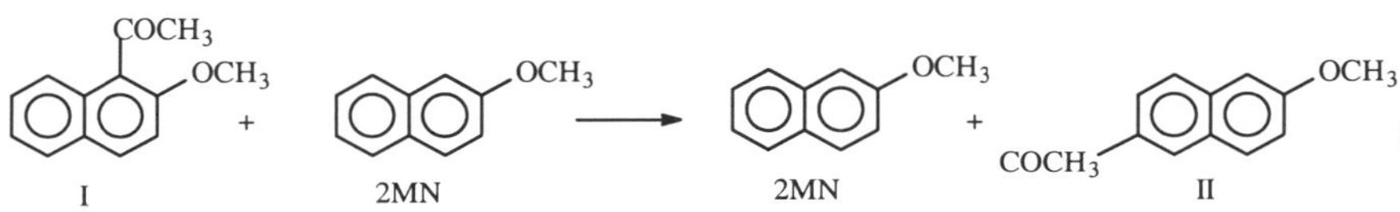


**(6)**

This transacylation mechanism was confirmed by using a mixture of isomer I with a deuterated methoxy group ($OCD_3$) and of light 2MN as a reactant (26).

The solvent polarity has a significant effect on the reaction rates and on the selectivity, solvents with average polarity such as nitrobenzene being the most suitable. Indeed, very polar solvents such as sulfolane compete with the reactant molecules for diffusion into the zeolite micropores and for adsorption on the acid sites, which significantly reduces the reaction rates. On the other hand, low acetylation and isomerization rates and high deacylation rates are found with non-polar solvents such as 1-methylnaphthalene which cannot solvate the acylium ion intermediates. Temperature has a significant positive effect not only on the rate of 2MN acetylation but also on the selectivity to isomer II. Unfortunately, temperature also increases the rate of deacylation. However, a high yield in isomer II (83 %) can be obtained by operating at 170°C and at short reaction times (25).

The location of the HBEA acid sites responsible for the formation of isomers I and II was questioned, certain authors (22-24) proposing that the bulky isomer I can only be formed on the large outer surface of HBEA zeolite (≈ 35% of the total surface area) whereas the slim isomer II would be formed on the outer and on the inner acid sites. However, molecular modelling shows that along the straight channel the interaction energy of isomer I is always negative (barrier of diffusion of approximately 10 $kcal.mol^{-1}$), which suggests that the molecules of I can diffuse inside the micropores and desorb from them. This was confirmed by adsorption experiments carried out with a mixture of I, 2MN and solvent under the reaction conditions. Thus after 4 minutes, the percentage of I was similar (slightly greater) in the mixture present in the micropores and in the reaction mixture. Moreover the percentages of II (and III) in the acetylmethoxynaphthalene mixtures were greater within the micropores than in the reaction mixture. This means that isomerization of I mainly occurs within the micropores, the reaction rate being strongly limited by isomer desorption from these micropores (25).

In a series of zeolite samples resulting from acid dealumination of HBEA15, a maximum in acetylation and isomerization activities was found for a Si/Al ratio between 20 and 35 (30). This maximum in activity can be explained by the opposite effect of the increase with Si/Al of the ease of desorption of the bulky and polar acetylation products and of the decrease in the number of protonic sites. The selective elimination of Lewis acid sites by treatment of HBEA15 with ammonium hexafluorosilicate allows one to confirm that these sites do not directly participate in acylation reactions over zeolites (30).

Therefore the acetylation of aromatics over zeolites proceeds through the two step mechanism formulated by Olah (31):

e.g.:

**(7)**

with displacement to the right of thermodynamic equilibrium by solvation of the acylium ions:

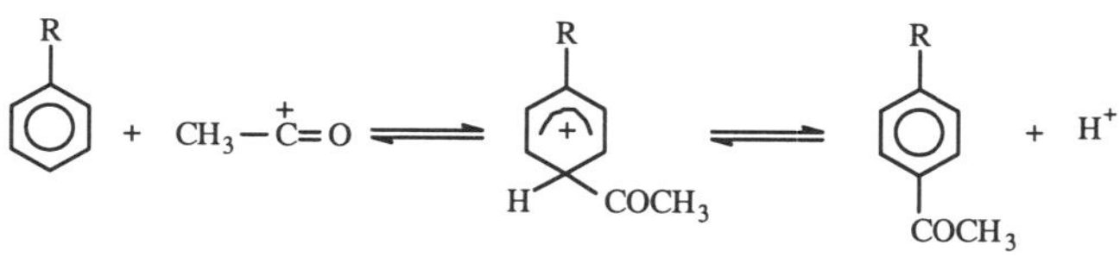


**(8)**

From these two examples of liquid phase acetylation of aromatics over zeolite catalysts the following conclusions can be drawn:

1. Acetylation occurs over the protonic sites of zeolites through a classical mechanism of electrophilic substitution

2. The higher polarity of the product compared to those of the reactants has various negative consequences:
   - Acetylation is autoinhibited, the reaction product limiting the diffusion of reactant molecules into the zeolite micropores and/or their adsorption on the protonic sites
   - Acetylation is often limited by product desorption
   - Due to their long residence time within the zeolite micropores, the reaction products undergo secondary reactions (such as polyacetylation) leading to bulkier and more polar products. These products remain trapped within the micropores, causing a blockage of their access, hence catalyst deactivation
3. Solutions have to be found to limit the negative effect of the product polarity:
   - adequate choice and adjustment of the zeolite crystallite size, of its porosity (including mesopores) and its acidity
   - choice of operating conditions favoring the adsorption of the reactant molecules and the desorption of the products: fixed bed reactor instead of batch reactor, high molar substrate/acylating agent ratio, low conversion rates, high temperature, use of solvents with adequate polarity

### *14.2.3 Industrial processes*

Rhodia is operating an industrial process for acetylation of anisole into paramethoxyacetophenone (pMAP) (18) which is a precursor of Parsol used for sun protection:

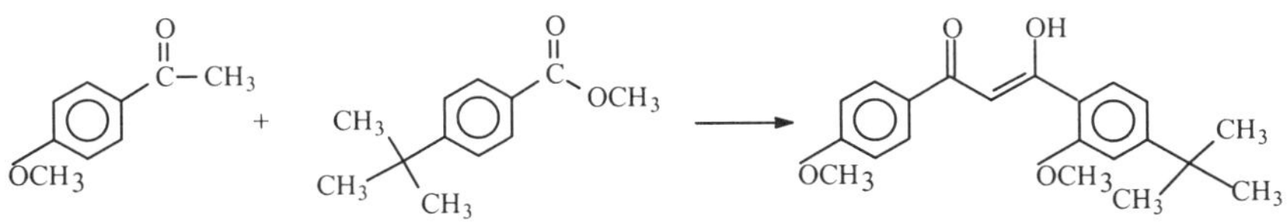


(9)

Tables 14.1 and 14.2 show the dramatic improvement brought by the substitution of the old technology with $AlCl_3$ catalyst and acetylchloride as acylating agent in a batch reactor by the new technology with a HBEA zeolite catalyst, acetic anhydride as acylating agent in a fixed bed reactor (12). The fixed bed reactor process constitutes a major break-through in anisole acetylation.

- This environmentally friendly process is much more simple than the conventional one: two steps instead of 8 (Table 14.2)
- There is a dramatic reduction of water consumption and of aqueous effluents: 35 kg/ton of pMAP instead of 4,500 kg/ton with the old process

The aqueous effluents contain 99 wt % water, 0.8% acetic acid and less than 0.2% of other organics, whereas those of the $AlCl_3$ unit contained more organic compounds (0.7% solvent, 0.8% of acetic acid and 0.8% of other organics) and a large amount of inorganic compounds (5 wt % $Al^{3+}$, 24% Cl)

The consequence is that the new process is not only environmentally friendly but also more economic than the old one.

**Table 14.1** Acetylation of anisole. Characteristics of the old and new technologies

| OLD PROCESS | NEW PROCESS |
|---|---|
| $AlCl_3$ > stoichiometric amount | Zeolite catalyst |
| 1,2-Dichloroethane as solvent | No solvent |
| Batch reactor | Continuous reactor |
| Hydrolysis at the end of reaction | No water |
| Destruction of the catalyst | Periodic catalyst Regeneration |
| Separation of organic/aqueous phases | Distillation of organic phase |
| Treatment and discharge of aqueous phase | |
| Distillation of organic phase | |
| Recycling of solvent | |
| Yield/anisole : 85-95 % | Yield/anisole : 95 % |
| | Higher purity of the final product |

**Table 14.2** Simplified flow chart for the conventional $AlCl_3$ process (**a**) and for the new zeolite catalysed process (**b**)

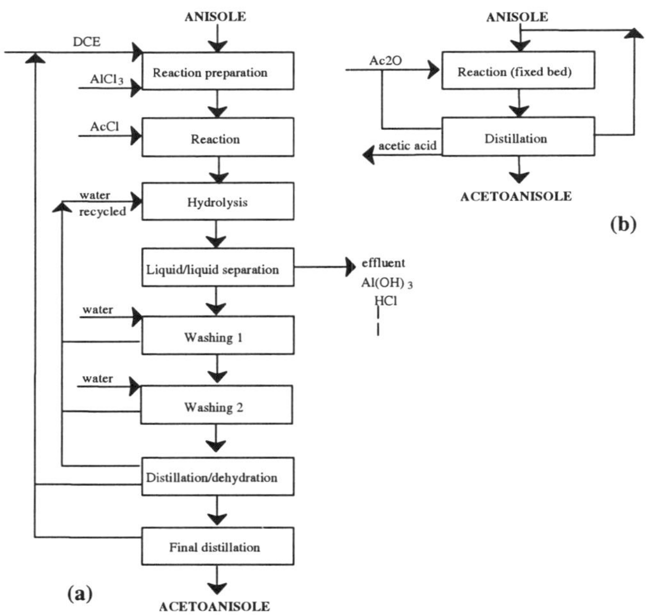

Acetylation of veratrole with acetic anhydride is also operated by Rhodia using a HFAU zeolite as a catalyst. Acetoveratrole which is selectively produced is a precursor of Verbutin, a synergist for insecticides. It should be emphasized that not only the acetylation of veratrole (step 1), but also step 3, which is the synthesis of a benzylic ether, can be catalysed with the same zeolite catalyst (paragraph 14.4.).

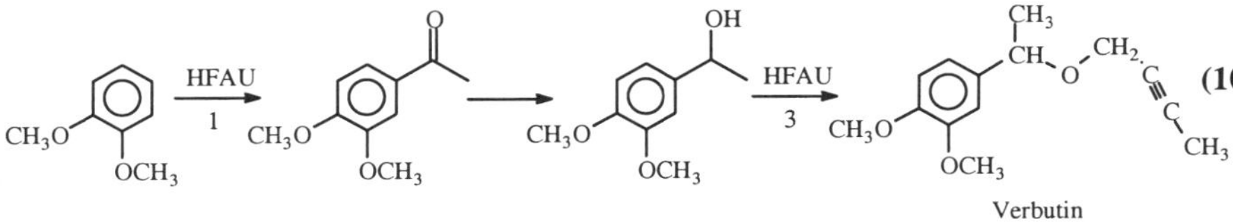


(10)

## 14.3 O-methylation of phenolic compounds

Alkylarylethers have pleasant odors and flavours which made them valuable for the perfume and fragrance industries. They are also important precursors of agrochemicals, pharmaceuticals, antioxidants, etc. (32). Thus, anisole which results from phenol methylation is a precursor of Parsol, a solar protector (paragraph 14.2.). Vanillin results from gaïacol transformation (3, 32):

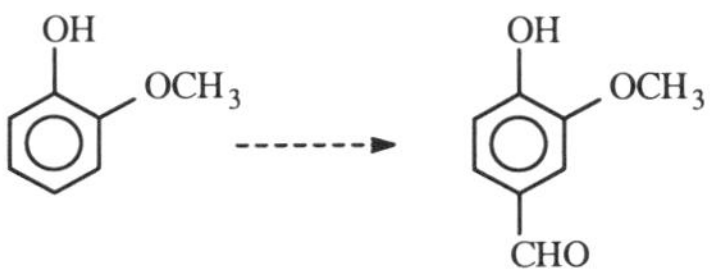


(11)

and papaverine from veratrole transformation (32):

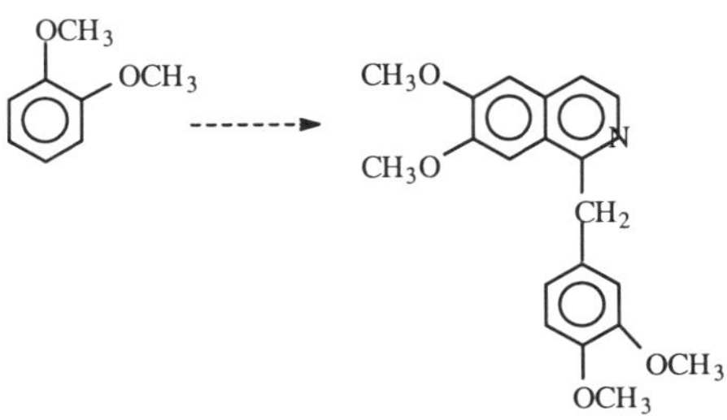


(12)

Phenol methylethers are generally prepared with a high yield (95 %) through the method of Williamson:

(13)

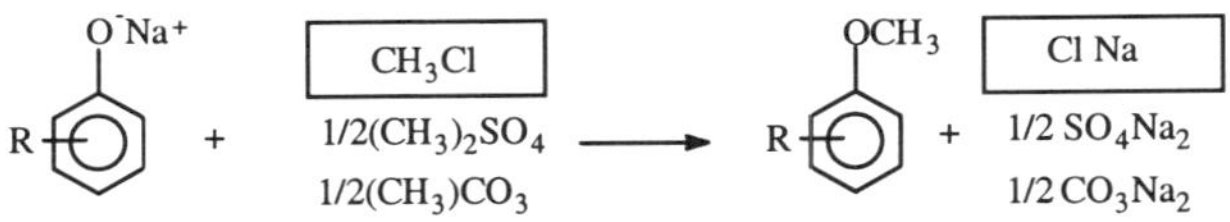


(14)

Whereas a high yield can be obtained (≈ 95%), this stoichiometric process consumes large amounts of sodium hydroxide and methylchoride and produces large amounts of valueless salts. Therefore, it is most attractive to substitute these polluting stoichiometric reactions by an environmentally friendly process.

Methylation of phenolic compounds and especially gas phase phenol methylation with methanol were investigated over various acid and basic solid catalysts (33-46): simple and mixed oxides ($Al_2O_3$, MgO, $SiO_2$-$Al_2O_3$, etc.), phosphates ($La_2(HPO_4)_3$ , $LaPO_4$, etc.), sulfates ($BaSO_4$, etc.), resins, zeolites. With protonic zeolites, anisole, ortho and paracresols were shown to be the only primary products; as it is expected with aromatic electrophilic substitution, the metacresol was not formed initially. High selectivities to anisole were found at low conversion: 70-80% over HFAU with different Si/Al ratios and even 90% with HMFI probably owing to product shape selectivity. However, because of the large significance of secondary reactions, there is a pronounced increase with conversion in the selectivity to cresols as well as the appearance of methylanisoles and xylenols.

The effect of the acid strength on the rate and selectivity of phenol alkylation was examined by Namba et al. (41) using series of HNaFAU and HKFAU samples. As it could be expected, in both series, phenol conversion increases with protonic exchange (e.g. from 55 to 68% with HNaY when the degree of exchange passes from 25 to 80%). Furthermore, a significant decrease in the yield into anisole (from 22 to 4%) is found. From this study, the authors conclude that the weaker acid sites are effective to produce anisole, the moderate acid sites to produce cresols and the strongest ones for the secondary reactions.

A Rideal type mechanism in which phenol from the gas phase reacts with methanol adsorbed on acid sites was proposed for O- and C-methylation of phenol over acid zeolites (46):

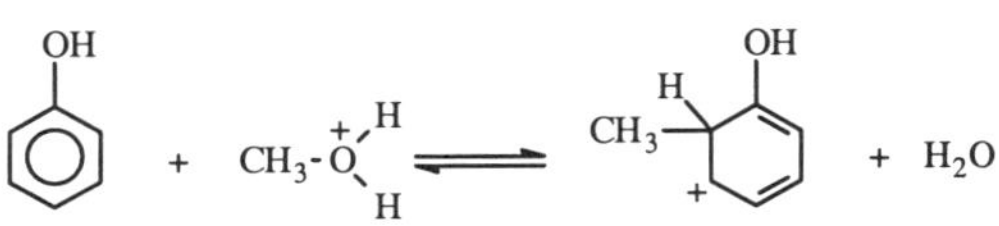


(15)

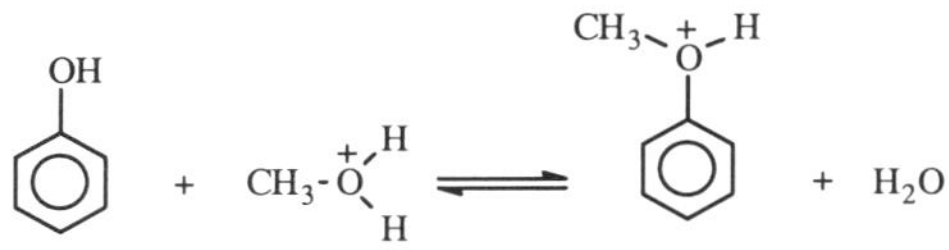

**(16)**

A zero order with respect to phenol was found, which can be related to the strong physical adsorption of phenol into the zeolite micropores (confinement effect). This explains the pronounced increase in phenol conversion with the methanol/phenol ratio. This strong retention of phenols and phenolic products within the zeolite micropores is responsible for a large part of the high apparent ratio of secondary reactions and especially of 'coking', i.e. of formation of heavy products which remain trapped in the zeolite micropores. This fast coking of zeolites is responsible for their rapid deactivation by pore blockage.

Strongly basic catalysts such as MgO and $Ca_3(PO_4)_2$ were found to be selective to C-alkylation, leading predominantly to the ortho-isomer of cresol (37). This high selectivity to ortho-cresol was proposed to be due to a more or less upright posture of phenolate intermediates on the catalyst surface (34).

Vapor phase alkylation of phenols with methanol was carried out at 270-330°C on various phosphate as catalysts (39). The best activities and selectivities in O-alkylation were observed with boron, rare-earth and niobium phosphate. The impregnation of rare-earth phosphates with cesium hydrogenophosphate causes an increase in the selectivity and in the stability (48) without modifying the activity. Thus, the selectivity to gaïacol for a conversion of veratrole of 80% (molar ratio $CH_3OH/C_6H_8O_2$ of 10 in the reactant mixture) increases from 65% on $LaPO_4$ to 94% on $LaPO_4$ with 5% of $Cs_2HPO_4$, i.e. with the quantity necessary to obtain a monolayer.

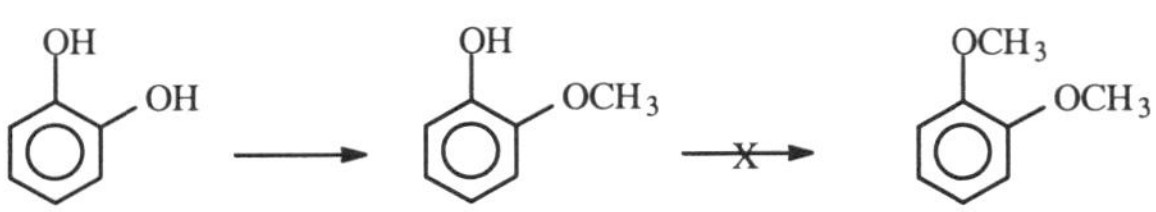

**(17)**

Table 14.3 shows that high selectivities to alkylarylethers can be obtained with $LaPO_4$ impregnated with 8 wt % of $Cs_2PO_4$.

**Table 14.3** O-alkylation of phenols catalysed by $LaPO_4$, $Cs_2HPO_4$

| Reaction | Operating conditions | Results |
|---|---|---|
| 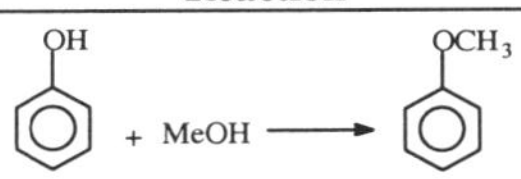 | MeOH/PhOH = 10<br>θ = 360°C<br>tc # 1s. | Conversion = 53 %<br>Selectivity = 90 %<br>o-cresol : 10 % |

| | | |
|---|---|---|
| 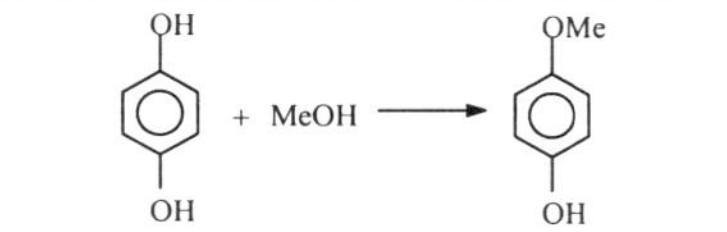 | MeOH/$H_2O$/HQ = 16/7/1<br>$\theta$ = 330°C<br>tc # 1s. | Conversion = 25 %<br>Selectivity = 94 % |
| 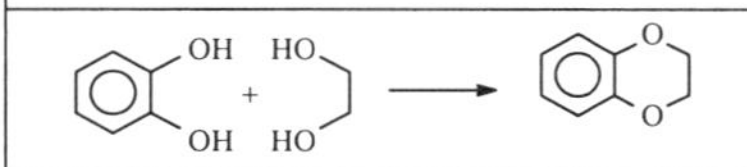 | Ethylene Glycol/PC = 10/1<br>$\theta$ = 330°C<br>tc # 1s. | Conversion = 100 %<br>Selectivity = 98 % |

A bifunctional acid base mechanism (Figure 14.5) was proposed to explain the high selectivity to O-alkylation of rare-earth phosphates. The absence of C-alkylated products by methylation of gaïacol tends to suggest that the guaicolate surface intermediate has a different behaviour on $LaPO_4$ and $LaPO_4$ doped with cesium hydrogenophosphate. In this latter case, the excellent selectivity to O-alkylation could be due to the softness of the cesium ion (39).

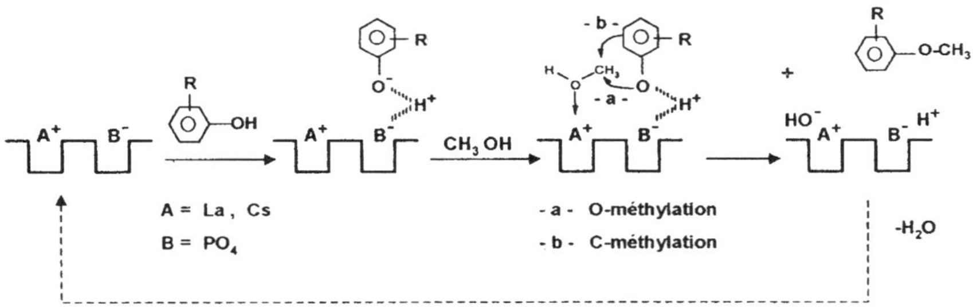


**Fig. 14.5** Acid base mechanism for the alkylation of phenols with methanol over $LaPO_4$ doped with $Cs_2HPO_4$

In summary, O-alkylation of phenols by methanol can be carried out with a relatively high selectivity over zeolite catalysts, but only at low conversions. Indeed, at high conversion, because of the strong retention of phenolic compounds within the zeolites pores, secondary products are formed in large amounts and deactivation occurs very quickly, owing to blockage of the pores by heavy products. Much better results: high selectivity at high conversion, very slow deactivation can be obtained with acidobasic catalysts such as $LaPO_4$ impregnated with cesium hydrogenophosphate (39).

## 14.4 Etherification of benzylic alcohols

Benzylic ethers can be used in the preparation of perfumes, pharmaceuticals, agrochemicals, etc. (32). As was indicated in paragraph 14.3, ethers are generally prepared via the reaction of Williamson with, as a consequence, the formation of a large amount of valueless salts. Moreover, this reaction cannot be used with reactants unstable in basic medium.

Symmetric ethers can also be prepared by using $H_2SO_4$ or $ZnCl_2$ as catalysts (49). The reaction mechanism ($SN_2$ or $SN_1$) depends on the degree of substitution of the alcohol. Thus, with primary alcohols, the reaction occurs as follows ($SN_2$):

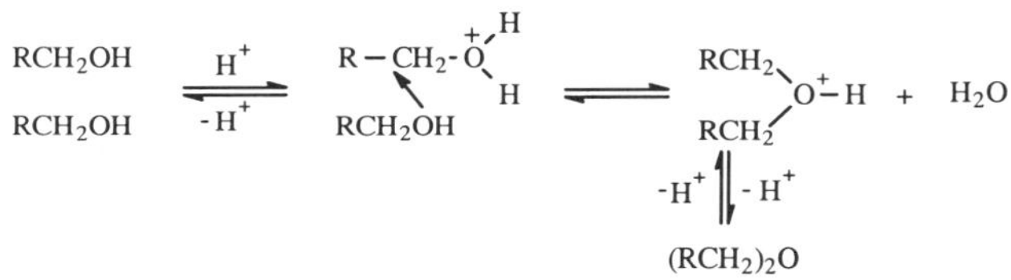


**(18)**

With tertiary alcohols, there is formation of a carbocation which is afterward attacked by the alcohol ($SN_1$).

Unsymmetric ethers cannot be selectively obtained (additional formation of symmetric ethers) except if one of the alcohol is tertiary, hence leads rapidly to a carbocation which reacts with the other alcohol. Furthermore, secondary reactions and especially polymerisation can be observed when the alcohol has very reactive substituents such as dienyl groups.

Obviously, it is very desirable to substitute these modes of benzylic ether preparation by an heterogeneous catalysis process. Clays (50) and resins (51, 52) which were the first solid acid catalysts used have given low or moderate yields. The first experiments with zeolites were carried out by Rhodia (53, 54) on the etherification of vanillic alcohol (A) in a batch reactor over a HBEA zeolite with a Si/Al ratio of 12.5:

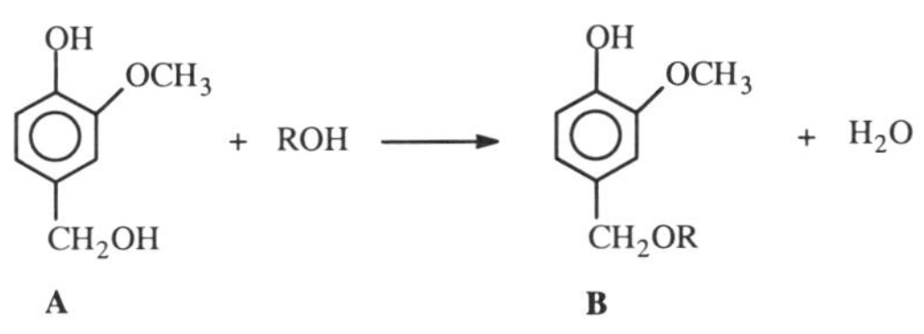


**(19)**

Under the operating conditions: large excess of ROH in order to limit the formation of symmetric benzylethers: 2.5 g for 0.25 g of A, 80°C, 2 hours reaction, a total conversion of A and a 100% yield of B were obtained when ROH is a primary alcohol: methanol, ethanol, 1-propanol or a secondary one such as 2-propanol. With cyclohexanol, the conversion of A is close to 100% but the yield in B equal to 85%.

With tert-butanol, the conversion of A was of 65% and the yield in B of 30% only because of a significant formation of isobutene.

The nature of the benzylic alcohol plays also an important role as shown in Table 14.4 from the results obtained in the reaction of methanol with six different benzylic alcohols

$$CH_3OH + \text{benzylic alcohol} \rightarrow \text{Benzylic ether} + H_2O \quad \textbf{(20)}$$

The conversion of the benzylic alcohol is of 100% and the yield in ether very high when the benzylic alcohol (primary or secondary) is activated in ortho or in para. In absence of activation, no reaction is observed.

The etherification of vanillic alcohol with ethanol was compared over three large pore zeolites: HBEA, HFAU and HMOR and over an average pore size zeolites (HMFI) with different Si/Al ratios. Whatever the large pore zeolite and its Si/Al ratio, a total conversion of the benzylic alcohol and a 100% yield in ether are obtained. With the HMFI samples, conversion and yields were equal to 55% only for an Si/Al ratio of 15 and to approximately 70% for Si/Al = 40. It could be expected that the reaction is limited by diffusion of the bulky reactant and product molecules in the narrow pores of this zeolite.

**Table 14.4** Synthesis of methylethers of various benzylic alcohols over HBEA

| **Benzylic alcohol** | **Conversion (%)** | **Yield** |
|---|---|---|
| 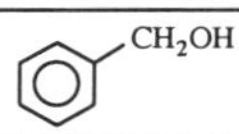  | 0 | 0 |
| 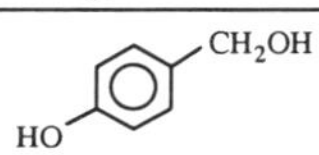  | 100 | 80 |
| 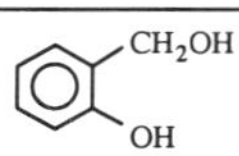  | 100 | 100 |
| 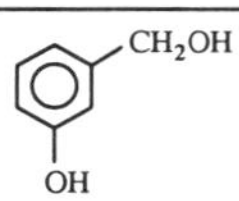  | 0 | 0 |
| 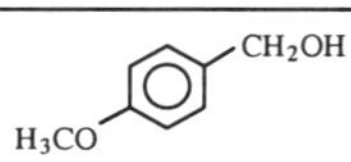  | 100 | 100 |
| 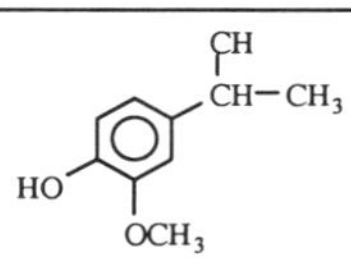  | 100 | 100 |

To estimate the effect of reaction time on activity and selectivity of HBEA (12.5), this zeolite was recovered by filtration after 5 hours reaction and reused several times under the same operating conditions: batch reactor, 80°C, 7.5 g of vanillic alcohol and 35 g of ethanol. Only a small decrease in conversion was observed from 98% to 88%, 86% and 84%, the selectivity remaining close to 100%. Furthermore, the activity is completely recovered after calcination under air flow for 5 hours. The slow deactivation of the HBEA zeolite is most likely due to a partial blockage of the access to the active sites by heavy secondary products ('coke').

This completely selective, stable and clean method was applied to the synthesis of various original benzylic ethers for application in perfumery (Table 14.5). Generally HFAU zeolites which are cheaper than HBEA zeolites were used, another advantage being an easier accessibility of the bulky benzylic alcohol to the acid sites and an easier desorption of the bulky ether. Except for the ether of menthol, the yields estimated by CPV were always higher than 95%. Table 14.5 shows that the yields in pure ether (i.e. obtained after separation from the reaction mixture by distillation, liquid extraction or chromatography) are generally higher than 80%. It should be remarked that the excess of aliphatic alcohol necessary for limiting the formation of symmetrical benzylic ethers can be easily recovered and recycled.

**Table 14.5** Synthesis of various benzylic ethers

| Benzylic alcohol | Aliphatic alcohol | Catalyseur | Yield* |
|---|---|---|---|
| Vanillic | Methanol | HBEA | 98 |
| | Ethanol | HBEA | 98 |
| | Propanol | HBEA | 95 |
| | Isopropanol | HBEA | 90 |
| | Cyclopentanol | HFAU | 90 |
| | Cyclohexanol | HFAU | 85 |
| | ß Citronellol | HFAU | 65 |
| | Menthol | HFAU | 58 |
| Anisidic | Octanol | HFAU | 65 |
| | Heptanol | HFAU | 78 |
| | 3,7-Dimethyloctanol-1 | HFAU | 56 |
| | 3-Pentanol | HFAU | 56 |
| 2-Methoxybenzylic | 2-Propanol | HFAU | 98 |
| 3,4-Dimethoxystyrallic | Ethanol | HMOR | 85 |

*After separation from the reaction mixture.

## 14.5 Conclusion

As was emphasized in the introduction, the main cause of the significant waste production in the Fine Chemicals Industry (5 ≥ 50 kg/kg product (1)) is the large use of homogeneous reactions carried out stoichiometrically or by using stoichiometric amounts of 'catalysts' (e.g. $AlCl_3$ in Friedel Crafts acylation). The examples presented in this chapter show that cleaner, more simple and more economic processes using solid catalysts, especially zeolites, can be substituted for these polluting processes. This was made possible only thanks to a multidisciplinary approach associating specialists in organic chemistry to specialists in mineral synthesis, heterogeneous catalysis, chemical engineering, etc.

More and more the development of safer, more environmentally friendly processes becomes one of the main conditions for the survival of companies involved in Fine Chemicals production. Molecular sieve catalysts with their remarkable properties of shape selectivity and of easy tuning of their pores and active sites, etc. should play in the near future an essential role in the development of these processes.

## References


1. Sheldon R.A., *Chemtech*, March 1994, 38.
2. Heterogeneous Catalysis and Fine Chemicals I, II, III, IV, *Stud. Surf. Sci. Catal.* **41** (1988), **59** (1991), **78** (1993), **108** (1997).
3. Sheldon R.A. and van Bekkum H., *Fine Chemicals through Heterogeneous Catalysis,* (Wiley-VCH. Weinheim, 2001).
4. Holderich W.F. and van Bekkum H., *Stud. Surf. Sci. Catal.* **58** (1991) 631.
5. Venuto P.B., *Microporous Materials* **2** (1994) 297.
6. Perot G. and Guisnet M., *Precision Process Technology,* Ed. Weijnen M.P.C. and Drinkenburg A.A.H., (Kluwer Academic Publishers, 1993) 157.
7. Holderich W.F., Heitnmann G., *Catalysis Today* **38** (1997) 227.
8. Corma A., Garcia H., *Catalysis Today*, **38** (1997) 257.
9. Sheldon R.A., *Chem. Ind.* (London) Dec **7** (1992) 903.
10. Wan K.T. and Davis M.E., *J. Catal.* **152** (1994) 25.
11. van Bekkum H. et al., *Stud. Surf. Sci. Catal.* **83** (1984) 379.
12. Methivier P. in ref. 3 p. 161.
13. Patent to Bayer, US4, **960**, **943** (1990).
14. Harvey G., Vogt A., Kouwenhoven H.W., Prins R., *Proc 9th Int. Zeolite Conf. Vol. II*, Butterworth, Boston (1992) 363.
15. Spagnol M., Gilbert L. and Alby D. in *The Roots of Organic Development,* Ed. Desmurs J.R. and Ratton S., (Elsevier, Amsterdam, 1996) 29.
16. Rohan D., Canaff C., Fromentin E. and Guisnet M., *J. Catal.* **177** (1998) 296.
17. Derouane E.G., Dillon C.J., Bethell D. and Derouane, A bd Hamid S.B., *J. Catal.* **187** (1999) 209.
18. Spagnol M., Gilbert L. Benazzi E. and Marcilly C., Patent to Rhodia, **W09635655** (1996).
19. Harvey G. and Mader G., Collect. Czech. *Chem. Commun.* **57** (1992) 862.
20. Yadav G.D. and Krishnan M.M.S., *Chem. Eng. Sci.* **54** (1999) 4189.
21. Gunnewegh E.A., Gopie S.S. and van Bekkum H., *J. Mol. Catal.* **106** (1996) 151.
22. Harvey G., Binder G. and Prins R., *Stud. Surf. Sci. Catal.* **94** (1998) 397.

23. Heinichem H.K. and Holderich W.F., *J. Catal.* **185** (1999) 408.
24. Kim S.D. et al., *J. Mol. Catal. A : Chem.*, **152** (2000) 33.
25. Fromentin E., Coustard J.M. and Guisnet M., *J. Mol. Catal.* **159** (2000) 377.
26. Fromentin E., Coustard J.M. and Guisnet M., *J. Catal.* **190** (2000) 433.
27. Andy P. et al., *J. Catal.* **192** (2000) 215.
28. Casagrande M., Storaro L., Lenarda M. and Ganzerla R., *Appl. Catal. A: Gen.*, **201** (2000) 263.
29. Botella P., Corma A. and Sastre G., *J. Catal.* **197** (2001) 81.
30. Berreghis A., Ayrault P., Fromentin E. and Guisnet M., *Catal. Letters* **68** (2000) 121.
31. Olah G.A., *Friedel-Crafts and Related Reactions*, (Wiley Interscience, New York, 1963).
32. a) Kirk, *Othmer Encyclopedy*, 4th Edition, (Wiley, New York, 1995) Ch.9
    b) Franck H.G., Stadelhofer J.W., *Industrial Aromatic Chemistry*, (Springer Verlag-Berlin, 1988).
33. Inoue M. and Enomoto S., *Chem. Pharm. Bull.* (Tokyo) **20** (1972) 232.
34. Tanabe K. and Nishizaki T., Proc 6th Intern. Congr. Catal., London (1977) 863.
35. Kotanigawa T., Yamamoto M., Shimokawa K., Yoshida Y., *Bull Chem. Soc.* Jpn **47** (1974) 950.
36. Pierantozzi R. and Nordquist A.F., *Appl. Catal.* **21** (1986) 263.
37. Nozaki F. and Kimura I., *Bull. Chem. Soc. Jpn*, **50** (1977) 614.
38. Kapsi J., Olah G.A., *J. Org. Chem.* **43** (1978) 3142.
39. Gilbert L. et al. in *The Roots of Organic Development,* Ed. Desmurs J.R. and Ratton S, (Elsevier, Amsterdam, 1996) 48.
40. Le Govic A.M. et al. in *The Roots of Organic Development,* Ed. Desmurs J.R. and Ratton S., (Elsevier, Amsterdam, 1996) 62.
41. Namba S., Yashima T., Itaba Y. and Hara N. *Stud. Surf. Sci. Catal.* **5** (1980) 105.
42. Balsama S. et al., *Appl. Catal.* **13** (1984) 161.
43. Marczewki M., Perot G. and Guisnet M., *Stud. Surf. Sci. Catal.* **41** (1988) 273.
44. Parton R.F., Jacobs J.M., Huybrechts D.R. and Jacobs P.A., *Stud. Surf. Sci. Catal.* **46** (1989) 163.
45. Parton R.F., Jacobs J.M.; van Osteghem H. and Jacobs P.A., *Stud. Surf. Sci. Catal.* **46** (1989) 211.
46. Marczewski M. et al., *AICHE Meeting Houston, April 1989*, Session 1.
47. Neves I. et al., *Stud. Surf. Sci. Catal.* **68** (1991) 735.
48. Tirel P.J. et al., *Stud. Surf. Sci. Catal.* **78** (1983) 693.
49. Kim S., Chung K.N. and Yang S., *J. Org. Chem.* **52** (1987) 3917.
50. Salmon M. et al., *J. Mol. Catal. A*, **104** (1995) L127.
51. Inge S., Patent to Merck, **DE 44 34823** (1984).
52. Bong C., *Bull. Korean Chem. Soc.* **13** (1992) 586.
53. Jacquot R. and Spagnol M., Patent to Rhodia **PCT/FR 98 01472** (1998).
54. Jacquot R., Patent to Rhodia **PCT/FR 00 02704** (2000).

CHAPTER 15

# ZEOLITES AND 'NON ZEOLITIC' MOLECULAR SIEVES IN THE SYNTHESIS OF FRAGRANCES AND FLAVORS

W.F. HOELDERICH, M.C. LAUFER

*Chemical Technology and Heterogeneous Catalysis, University of Technology Worringerweg 1, 52074 Aachen, Germany*

## 15.1 Introduction

K. Tanabe and W.F. Hoelderich collected 127 industrial processes which are catalyzed by solid acids (103), solid bases (10) and solid acid-base catalysts (14) (1). In quite a respectable number of these processes, heterogeneous catalysts replaced homogeneous systems. The use of homogeneous Brönsted and Lewis-acids and -bases was one of the main reasons for the production of high amounts of inorganic salts as by-products in industrial syntheses. The ratio of mass unit by-product per mass unit of product is explained as E-factor by Sheldon for several industry segments (2, 3). For example, in the fine chemical industry and for the production of pharmaceuticals sometimes 50-150 kg by-products per kg desired product are formed. Therefore, following the trend, new processes in the production of fine chemicals in order to reduce the high numbers of by-product formation are needed. Production and reactor integrated environmental protection are the key words. Thereby, the use of zeolites will play an important role, as has already happened in the past.

The unique features of zeolites, and the possibility of tuning acidic and basic sites, as well as the creation of multifunctional catalysts, open a wide field of applications in the production of fine chemicals. In this article we present new heterogeneously catalyzed processes for the synthesis of industrially relevant fragrances, flavors and aromas. The emphasis of this review article will remain mainly on solid acids.

Flavors and fragrances have been playing a dominant role stimulating the human senses. Starting as substances used for cultic and religious purposes, they found their way to the individual contribution to the improvement of personal comfort. Incense and myrrh, as well as sandalwood and cedarwood, are counted among the oldest fragrances. Of the more than 5,000 available fragrances nowadays, only 5%

are gained from natural sources (4). In most cases the demand can not be met anymore by natural sources. Therefore, a lot of fragrances, flavors and aromas are based on chemical compounds in the fragrance industry. An important class/category of compounds in this field are in particular aldehydes and ketones as well as alcohols, ethers, thiocompounds, coumarins, indols and esters.

## 15.2 Aldehydes and ketones in the fragrance industry

A very valuable method for the introduction of an aldehyde or ketone group in organic fine chemicals is the epoxidation of olefins and the following rearrangement of the oxiranes in the presence of acidic catalysts. The activation of epoxides for ring opening reactions can be achieved either by Brönsted acidic catalysts via the addition of a proton to the epoxide oxygen or by Lewis acidic catalysts via the coordination of the epoxide oxygen to a multivalent cation.

Homogeneous catalysts such as phosphoric acid, $BF_3$, $FeCl_3$, $ZnBr_2$ and $SnCl_y$, as well as heterogeneous catalysts such as $SiO_2$, $Al_2O_3$, ZnO,$WO_3$, supported metals and various precipitated phosphates have been applied. These homogeneous systems are not as a rule regenerable and they generate voluminous and often corrosive waste streams. The heterogeneous catalysts used in the past had some drawbacks, such as incomplete conversion in most cases, low selectivity because of consecutive aldol condensation to form preferably trimers and low service time. Therefore, the fragrance industry still applies homogeneous catalysts in the manufacturing plants despite its drawbacks with regard to environmental pollution.

Furthermore, in respect to the regioselectivity of the ring opening reaction of oxiranes, electronic as well as steric factors can play a role. These general considerations stimulate the use of zeolites and 'non zeolitic' molecular sieves as heterogeneous catalysts for such rearrangement reactions in the liquid or in the gas phase in a slurry reactor and in a continuous fixed bed reactor, respectively.

### *15.2.1 Isomerization of styrene oxides*

The conversion of styrene oxides **1** (Eq. 15.2.1) yields phenylacetaldehydes **2** which can be used as fragrances with hyacinth or rose odors. Furthermore, phenylacetaldehyde represents a valuable intermediate for the production of other more stable acetals with honey aroma (glycolacetale), with sweet leaf odour (diethylacetale) or with tangy smell (dipentylacetale). In addition, phenylacetaldehydes can also be used for manufacturing pharmaceuticals and insecticides, fungicides, herbicides, in particular when halogenated derivatives are needed.

**Equation 15.2.1** Styrene oxides **1** to phenylacetaldehydes **2**

In the isomerization of styrene oxides in a fixed bed reactor under gas phase conditions, the catalytic performance of various catalysts on the activity, selectivity and service time was screened at 300°C and WHSV = 2 – 3 $h^{-1}$. As shown in Fig. 15.1, zeolites with MFI-structure are superior to other zeolite types and 'non zeolitic' molecular sieves, as well as greatly superior to amorphous metal oxides.

**Fig. 15.1** Catalyst screening at 300 °C (5), 1) 91% or 98% conversion 2) short service time

The high regioselectivity to phenylacetaldehydes may be related to the stabilization of a developing alpha-cation (5). A possible side product is trimeric phenylacetaldehyde (triphenylbenzene), which is formed especially at lower temperatures (10% at 200°C vs. 2% at 300°C).

$B_2O_3$ and $P_2O_5$ modified silica as well as bentonite show similar selectivities as pentasil zeolites, but they deactivate much faster. The reason for that is the formation of trimers by aldolcondensation and aromatisation in the case of the non-shape selective materials. In contrast, the use of MFI-zeolites avoids the trimer formation due to the steric constraints and consecutive reactions resulting in coke. The aim of producing a fine chemical with 100% yield is achieved by the use of a Cs doped boron pentasil zeolite having an extremely weak acidity or basicity.

This knowledge was applied for even more complicated, bifunctional compounds such as phenyl glycidic acid methyl ester **3** to yield alpha-ketone carboxylic acid esters **4** such as 3-phenylpyroracemic acid methylesters (Eq. 15.2.2).

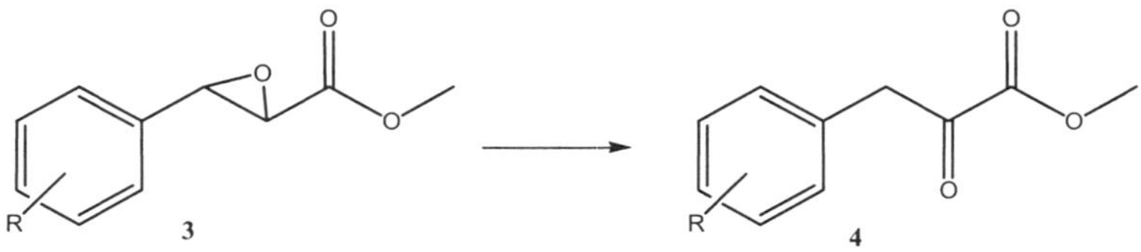


**Equation 15.2.2** Glycidic acid ester **3** to carboxylic acid ester **4**

These esters can be used as fragrances with harmonious fruit odor or as intermediates for the syntheses of herbicides (e.g. triazinones) and of L-amino acids. Also in this reaction, the weakly acidic boron pentasil zeolite and even better the Cs-doped boron pentasil zeolite are the most favored catalysts and are superior to other heterogeneous catalyst systems (Fig. 15.2).

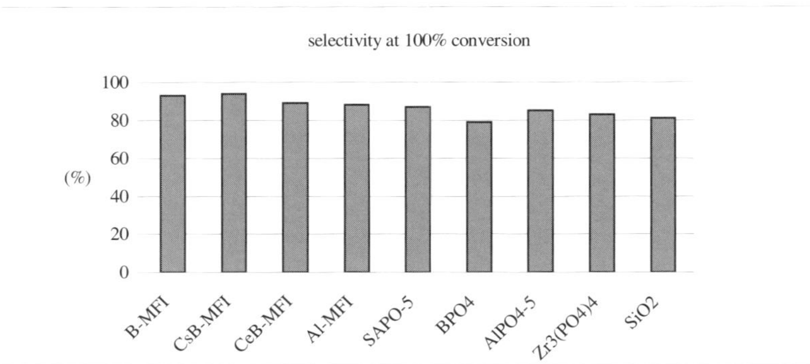


**Fig. 15.2** : Catalyst screening for the conversion of phenyl glycidic acid methyl ester to 3-phenylpyroracemic acid methylester

The advantages of the new process for alpha-ketocarboxylic acid ester under gas phase conditions are high yields, high service time of the catalyst and a one step reaction using easily available starting materials. In contrast, the known processes are very complicated, costly and environmentally hazardo us e.g. starting from oxalic acid ester **5** via Grignard reaction (Eq. 15.2.3).

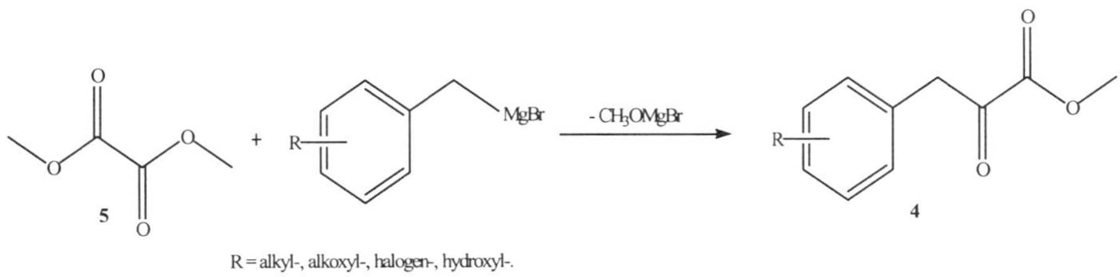


**Equation 15.2.3** Conventional method for alpha-ketone carboxylic acid esters **4**

By increasing the temperature over ≥ 300°C, the phenyl glycidic acid esters are converted to phenyl acid esters. That occurs in a consecutive decarbonylation of α-ketone carboxylic acid esters as intermediate. In the case of p-tert-butyl glycidic acid ester in the presence of a boron-pentasil zeolite at 350 °C the ratio between α-ketone carboxylic acid ester and the phenyl acetic acid ester is 65:35.

The phenyl acetic acid esters are highly valuable intermediates for a lot of applications. Among them, they are applied for the synthesis of fragrances e.g. ethyl ester (honey odour), isobutyl ester (sweet roses odour). The present industrial process for phenyl acetic esters starts from benzylchloride and uses the reactions with KCN to form benzylcyanide **7** (Eq. 15.2.4) and followed by hydrolytic cleavage.

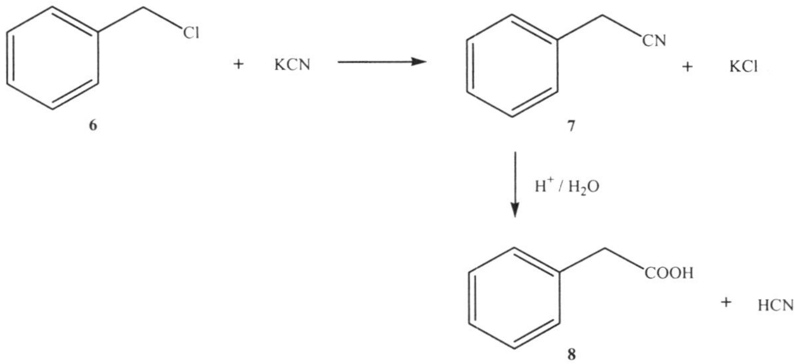


6 7 8

**Equation 15.2.4** Hazardous process for the synthesis of phenyl acetic acid **8** using toxic cyanides

The drawbacks of the conventional route are multistep reaction and use of toxic cyanides. The new route is an environmentally benign alternative.

### *15.2.2 Isomerization of α-pinene oxide*

Alpha-Pinene oxide **9** (Eq. 15.2.5) is known as a reactive molecule which rearranges easily under the influence of an acid catalyst (6, 7). Thereby many products can be formed. For example compounds such as the isomeric campholenic aldehyde **11**, trans-carveol **12**, trans-sobrerol **13**, p-cymene **14** or isopinocamphone **15** are observed as main by-products. At temperatures higher than 200°C more than 200 products can be formed. The industrially most desired compound among these is campholenic aldehyde **10**. It is the key molecule for the synthesis of various highly intense sandalwood-like fragrance chemicals (7, 8).

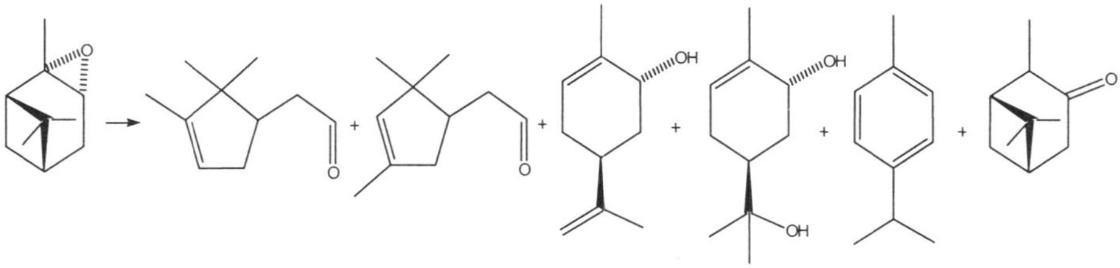


9 10 11 12 13 14 15

**Equation 15.2.5** Product range of isomerization of alpha-pinene oxide **9**

Although $ZnBr_2$ in benzene is known as a well working homogeneous catalyst with selectivities of about 85% to **10** (9), a lot of effort has been made to find a truly heterogeneous catalyst system (10). All the selectivities and yields obtained catalytically never exceeded 60% and have not been competitive with the homogeneous system. The mechanism of the formation of **10** is still being discussed in the literature (11). The main disadvantage is the water pollution with zinc halides causing severe problems in the sludge treatment by destroying the bacteria. To overcome these well known disadvantages of the homogeneous process, a multiple recyclable, heterogeneous catalyst must be found which leads to high yields of **10** at environmentally more benign conditions.

Surprisingly, it was found (7) that some H-US-Y zeolites having a high amount of mesopores are very suitable for this kind of reaction. At the given conditions at a temperature of 0°C, best selectivities of about 75% at a conversion of 30% were obtained with a commercially available, highly dealuminated H-US-Y zeolite (A), kindly provided by The Zeolyst Corporation (Fig. 15.3).

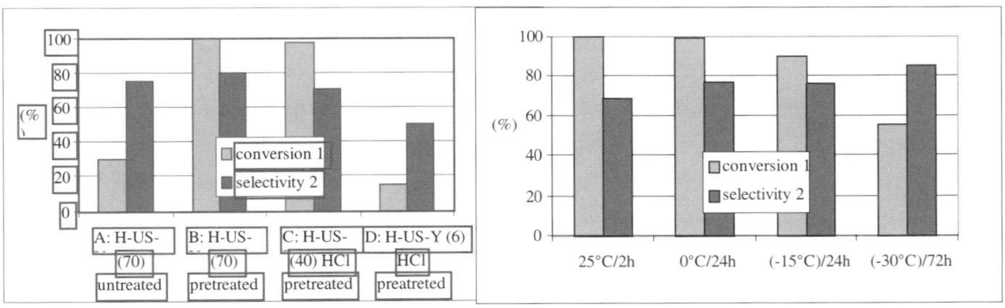


**Fig. 15.3** Left: Influence of acid treatment and $SiO_2/Al_2O_3$ ratio, Right: Influence of temperature with catalyst B, Batch reactor, T = 0°C, t = 24 h, catalyst loading = 7.5 g epoxide / g catalyst

As also shown in Figure 15.3, a pretreatment of the catalyst with diluted acid (0.01 molar HCl at 25°C for 24 h) and subsequent washing and calcination at 550 °C yields a major enhancement in activity without loss in selectivity for the desired aldehyde. The conversion achieved with catalyst B was about 100% and the selectivity of about 80% within 24 h.

The performance of different H-US-Y zeolites strongly depends on the bulk $SiO_2/Al_2O_3$ ratio, as shown in Figure 15.3 for catalysts B to D. The activity as well as the selectivity to campholenic aldehyde **10** increases with decreasing aluminum content. The Brönsted acid sites do not seem to be responsible for the desired reaction, since the number of these sites is equal to the number of aluminum atoms in the framework. From experiments, we cannot rule out that some Brönsted active sites are required, but the performance of these catalysts, especially the good selectivities observed, appears to be caused by Lewis acid sites. They can arise from steaming process and are generally described as extra-framework-aluminium (EFA) species (12). The first intention of washing with HCl was to remove the EFA species to make the inner part of the zeolite framework more accessible and to diminish the influence of diffusion constraints. However, this was not the reason for the enhanced efficiency, as will be shown later on.

With respect to the temperature influence, Figure 15.3 shows increasing selectivity for campholenic aldehyde at lower reaction temperatures with catalyst B. By running the isomerization at lower temperatures down to -30°C, selectivities can be improved to 85%. Surprisingly, our catalyst B shows, even at such unusually low reaction temperatures, an unexpectedly high activity. The conversion at -30°C reaches values up to 55% after 72 h (Fig. 15.3). The enhancement in selectivity with lower temperature can be explained with an improved kinetic reaction control and/or by the avoidance of side reactions of campholenic aldehyde. This is the first example in the literature that a zeolite catalyzed reaction is carried out at such a low temperature, still maintaining high efficiency and not dominated so much by diffusion limitations. This behavior is due to the high mesoporosity.

It was found that the selectivity and conversion are optimal if the pretreatment of the highly dealuminated Y zeolite is pursued at pH = 2. Lower pH values in the pretreatment of the catalyst A cause a decrease of both conversion and selectivity in the catalytic run (7). It is well known that the application of strong acid treatment (pH < 1) leads to removal of framework as well as extra-framework-alumina from dealuminated Y zeolite (13).

To elucidate the reason for the better performance of the acid-treated catalyst (B) we examined samples A and B with various analytical methods such as AAS, FT-IR, BET, pyridine adsorption, $^{29}$Si and $^{27}$Al solid NMR investigations.

AAS of aluminum and FT-IR indicate that there must exist a large amount of EFA species in the zeolite. By measuring the outer T-O-T stretching frequency, it was calculated with the equations disclosed in (14) that only about one Al atom per unit cell is incorporated into the zeolite framework of catalyst B. On the other hand, experimental results derived from Al-AAS show that hypothetically a higher number, i.e. altogether four Al atoms, could be present in the unit cell on the assumption that all Al is incorporated in the framework. This difference tells us that most of the aluminum must exist in the form of EFA. Nitrogen adsorption shows that both zeolites have a large mesopore volume. This allows sufficient diffusion transport of organic molecules even at low temperatures. FT-IR measurements of zeolite samples loaded with pyridine at 150°C were performed (15). The measurements show a strong presence of Lewis sites and a few Brönsted sites. These findings are in agreement with the results from Al AAS.

The $^{27}Al$ and $^{29}Si$ NMR measurements (7) showed that after treatment with 0.01 molar HCl most of the amorphous silica-containing material is removed from the parent catalyst A. This can be understood easily since the maximum solubility of silica (16) is reached at pH = 2. Although the improved performance of the treated catalyst cannot be entirely explained by the removal of less active material, i.e. the increase of the number of Lewis acid sites per mass unit, it is believed that these silica species block most of the catalytically active centers, i.e. the highly dispersed Lewis acidic alumina sites in the micro- and mesopores of the parent US-Y zeolite.

After complete reaction, the catalyst can be reused again without loss of performance. The catalyst can be successfully reactivated by calcination under air atmosphere at 550 °C.

A heterogeneous catalyzed process for the production of campholenic aldehyde from alpha-pinene oxide has been found, which is competitive with the homogeneous $ZnBr_2$ system due to yields up to 85%.

H. van Bekkum et al. (17) reported that the alpha-pinene oxide **9** can be succesfully converted to campholenic aldehyde **10** (Eq. 15.2.5) in the presence of a BEA-zeolite. Ti-BEA proves to be an excellent catalyst for the rearrangement of α-pinene oxide to campholenic aldehyde in both the liquid and vapor phase. This is mainly attributed to the presence of isolated, well-dispersed titanium sites in a Brönsted-acid-free silica matrix. Furthermore, the unique molecularsized pore structure of the zeolite may enhance selectivity by shape-selectivity.

In the liquid phase for example, selectivities of up to 89% in acetonitrile have been achieved. Since alpha-pinene oxide can give undesired bimolecular or polymerization reactions, low intraporous concentrations are preferred for high yields of campholenic aldehyde. Due to the hydrophobic nature of Ti-BEA, apolar solvents are expected to be adsorbed strongly, thus giving a low intraporous alpha-pinene oxide concentration and vice versa. Acetonitrile, the most polar solvent tested, gives the highest initial rate, but the catalyst deactivates rapidly, resulting in a conversion of only 7% after 24 h (89% selectivitiy, TON 139). Dichloroethane proved to be the most suitable solvent with a selectivity of 81% and a TON of 576.

The absence of a solvent when working under vapor phase conditions strongly increases the intraporous alpha-pinene oxide concentration. This leads to a decreased campholenic aldehyde selectivity. A competitive inert co-adsorbate may be added to the reactor feed to control the alpha-pinene oxide concentration. When dichloroethane is chosen as a co-adsorbate, a very high selectivity to campholenic acid (94%) can be obtained, even at complete conversion of alpha-pinene oxide. The use of alkanes as co-adsorbates which have a much lower toxicity also results in very high campholenic aldehyde selectivites of about 85%. Methylcyclohexane and 1,4-dimethylcyclohexane are particulary effective while the bulkier 1,3- and 1,2-dimethylcyclohexane cause reductions in selectivities.

In conclusion, the operation with Ti-BEA at industrially practicable temperatures (90° - 100°C) is of high interest, while for other catalysts based on ultrastable Y-zeolite, low temperatures are required for high selectivity (7).

### *15.2.3 Isomerization of 1,5-dioxaspiro-(2,6) octane and of 4,4,5,8-tetramethyl-1-oxa-spiro-(2,5) octane*

1,5-dioxaspiro-(2,6) octane **16** (Eq. 15.2.6) can be easily converted to 4-formyltetrahydropyran **17** in the presence of zeolites.

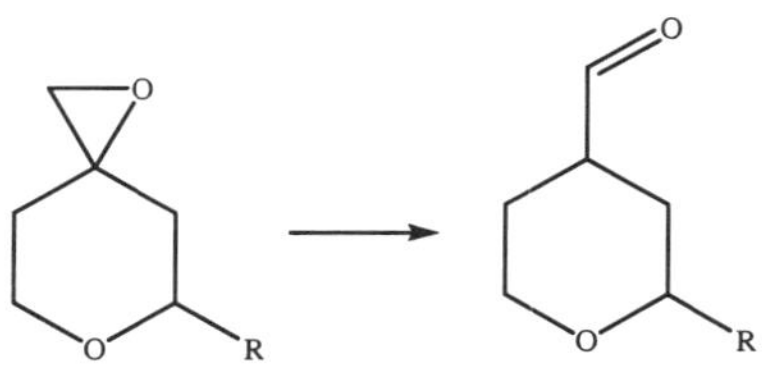


16 17

**Equation 15.2.6** 1,5-Dioxaspiro-[2,6] octanes **16** to 4-formyl-tetrahydropyranes **17** (R = H, alkyl, aryl, heteroaryl, alkoxyaryl)

The product is a valuable intermediate for the production of fragrances. This compound is hardly obtainable by known synthesis routes (Rosemund reduction). Using Al- or B-pentasil zeolite at 300°C the 4-formyltetrahydropyrans can be synthesized in an easy manner and high yields over 97% (18, 19, 20). This reaction was also carried out in the gas phase in a fixed bed reactor (5) with a WHSV = 2 $h^{-1}$ and a ratio of substrate/solvent (THF) = 50/50. Using $SiO_2$ as solid catalyst gave an additional yield of 96.5% (9).

4,4,5,8-tetramethyl-1-oxa-spiro-[2.5]octane **18** (Eq. 15.2.7) is obtained in a multi-step synthesis from easily available (-)-(S)-$\beta$-citronellol or (+)-R-pulegone (21).

It reacts to the industrially desired 2,2,3,6-tetramethyl-cyclohexane-carbaldehyde **19** in the presence of acidic catalysts beside many other products.

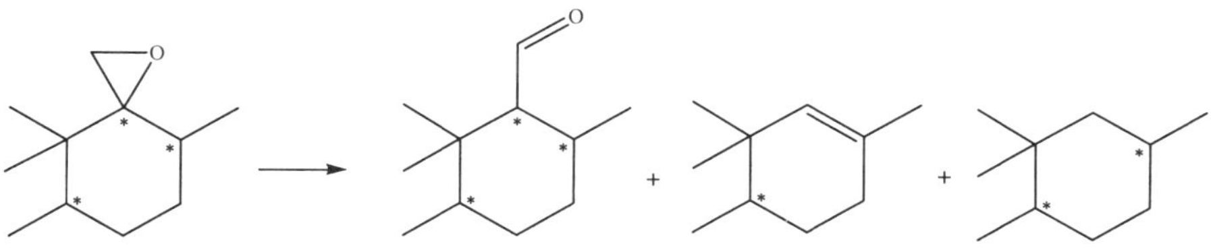

18 19 20 21

**Equation 15.2.7** Main products of the ring opening reaction of **18** under gas-phase conditions

The aldehyde **19** is described as a key molecule for the synthesis of different highly active fragrances like the saturated alcohol 1-(2,2,3,6-tetramethyl-1-cyclohexyl)-3-hexanole) **22** appreciated by perfumers because of its woody-/ambra-like character (22) according to Equation 15.2.8.

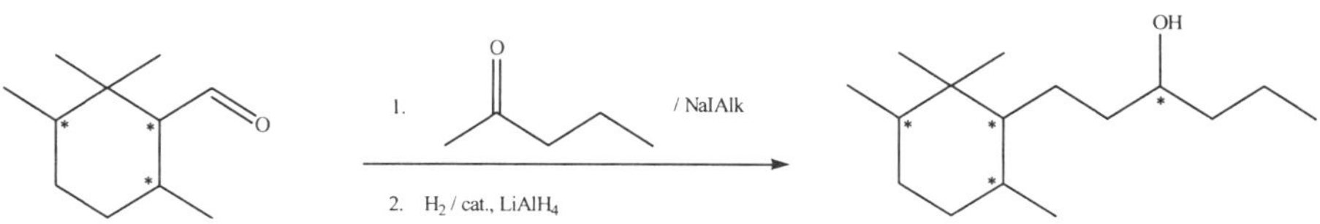


19 22

**Equation 15.2.8** Synthesis of the ambra/woody fragrance **22**

The suitable stereochemistry, not focused herein, is described as playing an important role (22). As reported in patents of Firmenich S.A., the ring opening reaction of **18** (Eq. 15.2.7) was preferably carried out with the Lewis acidic homogeneous catalysts such as $BF_3$-etherate (5), $MgI_2$ (21) or SnCl4 (21) with almost complete yield at room temperature in toluene as solvent. These catalysts, however, imply a process limitation to discontinuous liquid phase conditions. Major drawbacks are given by a relatively complicated separation procedure of products and catalyst in addition to corrosion, loss of catalyst and environmental problems such as high salt formation. Recently, the heterogeneously catalyzed rearrangement of **18** (Eq. 15.2.7) has received attention in literature.

First, following the results of the 1,6-dioxa-spiro[2.5]octane rearrangement (5,19), continuous gas phase conditions were applied in a fixed bed reactor and secondly under liquid phase conditions in a slurry reactor. The catalytic experiments carried out showed that two main reactions took place: rearrangement of **18** to the aldehyde **19** and a oxidative decarbonylation reaction to the olefine 1,3,3,4-tetramethyl-cyclohex-1-ene **20**, which is assumed to be caused by a formaldehyde elimination reaction. Also observed was a deoxygenation reaction to the alkane 1,1,2,5-tetra-methylcyclohexane **21** (Eq. 15.2.7), explained by elimination of CO. There are several other side-products such as 2,2,3,6-tetramethylcyclohex-1-enyl-methanol, ringcontracting compounds and double bond isomers of dimethyl-isopropylene-cyclopentene.

The first tests were carried out to evaluate the behavior of different catalysts under gas phase conditions (Table 15.1). It was observed that conversion of **18** decreased rapidly in the presence of various acidic zeolites H-B-ZSM-5, H-ZSM-5 and H-US-Y. This behavior was even more distinctive with $BPO_4$ and $Nb_2O_5$. The low service times of the catalysts are assumed to be caused by strongly adsorbed compounds as well as coke precursors blocking the acidic sites. Surprisingly, a silica catalyst having gentle acidity showed the best performance. With selectivity to **19** of about 40% there was no drop in conversion after 8 h TOS, even at 230 °C.

**Table 15.1** Conversion of 18 and selectivity to 19, 20, 21 with different catalysts in a fixed bed reactor

| Catalyst (modul) | Conv. **18** / (%) | Sel. **19** / (%) | Sel. **20** / (%) | Sel. **21** / (%) | Drop of conv. (TOS) 2/6h or 2/8h* |
|---|---|---|---|---|---|
| H-B-ZSM-5(34) | 88.9 a | 42.1 | 15.0 | 0.1 | 94/87 |
| H-ZSM-5(60) | 95.6 b | 39.8 | 22.5 | 0.2 | 99/83* |
| H-US-Y(70) | 98.5 b | 29.2 | 25.2 | 0.4 | 100/96* |
| $BPO_4$ | 3.4 b | 29.4 | <0.1 | <0.1 | 52/2* |
| $Nb_2O_5$ | 71.7 a | 25.2 | 17.4 | 1.0 | 81/65 |
| $SiO_2$ D11-10** | 100 a | 39.9 | 32.5 | 1.6 | 100/100 |

Catalyst 2.0 g (1.0-1.6 mm); 20 wt.-% **18** in THF, WHSV = 1 $h^{-1}$, T=180 °C

p=1 bar, a: TOS=4 h, b: TOS=6 h, $N_2$ carrier gas=5 l/h, ** T=180 °C, TOS (Time on stream)

On account of high activity and selectivity of this silica catalyst D11-10, the influence of temperature was studied in the range of 150 °C to 300 °C (Fig. 15.4) (23). It is advantageous to run the reaction at about 250 °C to 300 °C. In this range, complete epoxide conversion is achieved within 6 h TOS in combination with the highest aldehyde selectivities of 45%. The formation of **19** and **21** are increased by increasing the temperature. Compound **20** passes a maximum at around 200 °C.

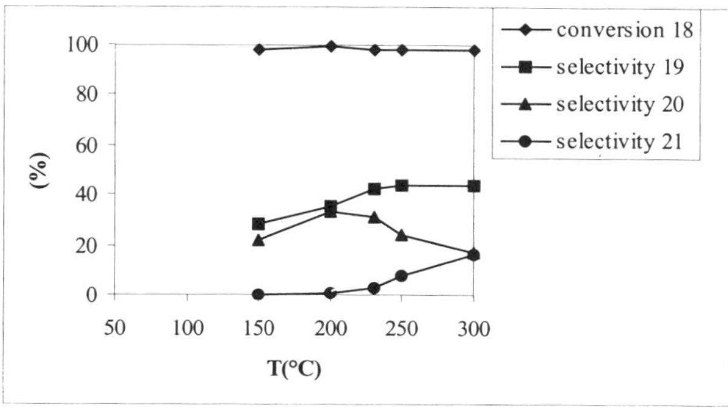


**Fig. 15.4** Influence of temperature on conversion of **18** (20 wt% in THF) and selectivity

Adjusting the residence time respective to the WHSV, a maximum value can be found for WHSV = 2 $h^{-1}$ which results in an increase in the selectivity to 47% at 100% conversion compared to a WHSV of 1 $h^{-1}$ (3 $h^{-1}$) with selectivities of 45.1% (44.6%) respectively. This is a very low sensitivity of the system to this parameter.

A reduced pressure also had an influence on the selectivity (45% at 1 bar and 50% at 100 mbar). A further optimization step includes a change of the solvent THF to the more apolar solvent toluene. Thereby, selectivity to **19** could be enhanced to 53% at complete conversion and no drop of deactivation is observed within 8 h TOS. The positive effect of toluene in comparison to THF might be explained by a different adsorption behavior towards the active sites of the silica catalyst.

It is surprising that silicagel D 11-10 shows better results than zeolites in the gas phase rearrangement reaction. In the case of 1,6 dioxaspiro[2,5]octane, excellent results were obtained in the presence of this silicagel as well as of pentasil zeolites (20). In the present case, the molecular dimensions of the starting material seem to be to bulky, due to the additional methyl groups, and cannot enter the pore volume of the middle sized weakly acid boron pentasil zeolite.

Surprisingly, and in contrast to the gas phase reactions, zeolitic catalysts show a superior catalytic performance in the liquid phase in a slurry reactor. Good conversion and selectivities are obtained over H-US-Y (70) and H-US-Y (70) treated with HCl whose catalytic features are described elsewhere (7). Highest selectivities are obtained at very low temperature of about 0 °C. In order to achieve high conversion, long residence times are necessary. That means the space time yield (STY) in the liquid phase at low temperatures is less than in the gas phase. The HCl treated H-US-Y zeolite shows a better catalytic performance than the untreated one. The reason for that might be the high Lewis acidity and high mesoporosity in this particular material (7).

In summary, we have established a facile heterogeneous catalytic method for the synthesis of the aldehyde intermediate 2,2,3,6-tetramethyl-cyclohexane-carbaldehyde **19** with easy handling, lack of corrosion and ease of separation of the products from catalyst. A silica catalyst is most suitable for the conversion in the gas phase, whereas a H-US-Y (70) treated with HCl shows the best performance in the liquid phase. Furthermore, it was realized that the introduction of an alkyl residue in position 5 of the ring system leads to a clear drop in the 5-alkyl-4-formylpyrane selectivity in comparison to the conversion of the compound without this alkylgroup. The yield obtained in the presence of heterogeneous catalysts is not as high as in the presence of conventionally used catalysts. Improvements of the heterogeneous catalysts are still necessary to become competitive with these homogeneous systems.

### *15.2.4 Isomerization of isophorone oxide*

Terpene epoxides are very reactive compounds. Some products formed by isomerization of such epoxides are valuable raw materials for perfumes, synthetic flavourings and pharmaceuticals, and also provide useful intermediates in organic syntheses. The isomerization of isophorone oxide **23** (Eq. 15.2.9) was originally investigated by H.O. House and R.L. Wasson using boron trifluoride etherate as a homogeneous catalyst (24).

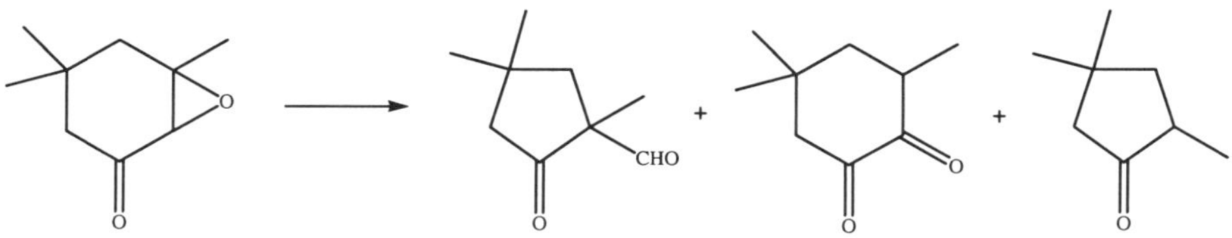


23 24 25 26

**Equation 15.2.9** Products in the isomerization of isoporone oxide **23**

They found that at room temperature using benzene as the solvent only 3% consisted of the diketone **25** and mainly the ring-contracted products were obtained: 33% of keto aldehyde **24** and 28% of ketone **26**. From an industrial point of view the desired compound is the keto aldehyde **24**, which is an interesting intermediate for the synthesis of other cyclopentanone derivatives with floral and fruity smells. The acid catalyzed reaction mechanism leading to the synthesis of keto aldehyde **24** had been discussed earlier (25). Therefore, it is of interest whether the product distribution changes in the presence of a heterogeneous catalyst system and also whether the decarbonylation of the compound **24** to compound **26** can be suppressed.

In the presence of zeolites, the keto aldehyde **24**, accompanied by the ketoenolic form of the alpha-diketone,(26) is mainly formed as reported by Sheldon et al. and Hölderich et al.(27-30).

The reports on catalytic isomerization using various zeolitic catalysts, in comparison to the conventional catalysts previously used, gives results of reactions carried out discontinuously in a batch reactor in liquid phase, as well as for those carried out continuously in a fixed bed reactor in the vapor phase. The results in the liquid phase over heterogeneous catalysts are summarized in Table 15.2.

**Table 15.2** Conversion and selectivities obtained with used zeolites in the liquid phase isomerization of isophorone oxide 23

| Catalyst | Conversion (%) **23** | Selectivities (%) **24** | **25** | **26** |
|---|---|---|---|---|
| H-ZSM-5 (28) | 88 | 68 | 14 | 1 |
| H-ZSM-5 (60) | 95 | 68 | 15 | 0 |
| H-BEA (25) | 100 | 68 | 13 | 14 |
| H-US-Y (70) | 96 | 73 | 12 | 0 |
| H-US-Y (96) | 100 | 74 | 11 | 1 |
| H-US-Y (96)-HCl | 100 | 71 | 17 | 4 |
| H-FER (18) | 100 | 81 | 12 | 0 |

t = 6 h; T = 110 °C; solvent: toluene; loading: 10 g isophorone oxide/g catalyst; 1 atm. pressure; the numbers in parentheses correspond to the $SiO_2/Al_2O_3$ or $SiO_2/B_2O_3$ ratios

The blank test in liquid phase yields less than 1% conversion. Unlike non-zeolitic catalysts, except for H-Nafion, most zeolites yield complete conversion. A high yield of the keto aldehyde **24** up to 81%, was surprisingly attained by using H-FER as a heterogeneous catalyst. For the rearrangement of isophorone oxide, the presence of acidic sites is necessary for the catalytic activity. The reactivity of H-FER can be explained by the acidic outer surface of the catalyst. Molecular modeling showed that the isophorone epoxide is too bulky for the small pore size of ferrierite.

The ratio between the two main products **24** and **25** depends on the type and acidity of zeolite used. For example, H-BEA with a higher acid outer surface demonstrated a high activity but a low selectivity of **24** in comparison to H-FER or H-US-Y catalysts.

High amounts of **24** were also obtained by the use of H-US-Y (96) followed by H-US-Y (96)-HCl. The H-US-Y (96)-HCl used, a modified highly dealuminated ultrastable Y zeolite, was pretreated with diluted acid according to the method described before (28, 31). This zeolitic catalyst, unlike many others, remains active at lower temperatures and also at high loading, as was previously demonstrated in the isomerization of alpha-pinene oxide using this heterogeneous catalyst (28, 31).

The data were correlated to that obtained by Sheldon et al.(30) in the isophorone oxide rearrangement. There, a dealuminated mordenite was used and a selectivity of 85% to **24** in benzene as a solvent was found. The results are comparable to our findings over H-FER or US-Y-zeolites.

To our knowledge, no vapor phase isomerization utilizing heterogeneous catalysts has been previously described in the literature for this reaction. Continuous vapor phase reactions should be preferred over discontinuous liquid phase reactions, as this yields a higher efficiency (space time yield) as well as an easier separation of the catalyst from the reaction mixture.

A general view of the used zeolites in the vapor phase isomerization and their conversions and selectivities is shown in Table 15.3. The blank test yields less than 2% conversion.

**Table 15.3** Conversion and selectivities obtained with used zeolites in the liquid phase isomerization of isophorone oxide 23

| Catalyst | Conversion (%) **23** | Selectivity (%) **24** | **25** | **26** |
|---|---|---|---|---|
| H-ZSM-5 (28) | 99 | 69 | 18 | 2 |
| H-ZSM-5 (60) | 100 | 72 | 18 | 1 |
| H-[B]-ZSM-5 (38) | 90 | 82 | 10 | 1 |
| H-US-Y (70) | 100 | 66 | 19 | 6 |
| H-US-Y (96) | 99 | 69 | 21 | 5 |
| H-US-Y (96)-HCl | 100 | 68 | 23 | 3 |
| H-FER (18) | 100 | 66 | 24 | 3 |

TOS = 4 h; T = 250 °C; solvent: toluene; toluene/isophorone oxide 2:1 wt/wt; nitrogen flow 8 l $h^{-1}$; WHSV = $1h^{-1}$;1 atm. pressure; the numbers in parentheses correspond to the $SiO_2/Al_2O_3$ or $SiO_2/B_2O_3$ ratios

Using zeolites with the same structure but in different ratios of $SiO_2/Al_2O_3$, we found an increased tendency towards selectivity to **24** with higher $SiO_2/Al_2O_3$ ratios, as was previously observed in liquid phase isomerization. All tested zeolites yielded high conversions, except for Na-ZSM-5 and silicalite. A shorter contact time seems to have a positive effect on the aldehyde formation, possibly due to reduced decarbonylation. The influence of 'solvents' in the gas phase is similar to that of the liquid phase. Best yields were obtained with toluene, anisole and o-xylene.

In conclusion, the use of zeolites as catalysts in the isomerization of isophorone oxide **23**, yields up to 86% keto aldehyde **24**. The formation of **26** by decarbonylation of **24** could be reduced by increasing the catalyst loading in a liquid phase batch reactor or by conducting the reaction under short contact time in the gas phase. The heterogeneously catalyzed isomerization of terpene epoxides over zeolites is a suitable, non-polluting method of preparing relevant and useful aldehydes for the synthesis of perfumes and synthetic flavors.

## 15.3. Alcohols, ether, thiocompounds

### *15.3.1 Synthesis of 2-phenylethanol*

2-Phenylethanol is one of the most widely used fragrance chemicals, with worldwide production in excess of 7,000 tons per annum. 2-Phenylethanol is often used in high quantities as fragrance with sweet and rose odour. It is particularly well suited as a fragrance for the production of soap because of its stability towards alkalines.

If the epoxide rearrangement (see chapter **15.2.1**) of styrene oxide is carried out in the presence of hydrogen and by use of a bifunctional boron-pentasil zeolite catalyst having a hydrogenation component such as Cu, then 2-phenylethanol is obtained in one step. This hydro-isomerization renders high yields (> 85%) at 250 °C under the gas phase conditions. It is an example for multifunctional catalysis in a one pot-reaction, that means simultaneous rearrangement and hydrogenation.

This route might be an environmentally benign alternative to the conventional method for the production of the very important fragrance 2-phenylethanol. This compound is still manufactured out of ethylene oxide and benzene in the presence of a high excess of $AlCl_3$ (2,5 molar), yielding unwanted toxic by-products such as dioxanes and several inorganic salts accompanied by corrosion problems.

### *15.3.2 Synthesis of 4-tert-butyl cyclohexanol*

The important fragrance intermediate 4-tert-butylcyclohexanol **27** (Eq. 15.3.1) is available commercially as an isomer mixture containing about 70% of the preferred *cis*-form **27**. H. van Bekkum et al.(32) reported on the first stereoselective Meerwein-Ponndorf-Verley reduction (MPV) of 4-tert-butylcyclohexanone to the desired *cis*-4-tert-butylcyclohexanol catalyzed by zeolite BEA.

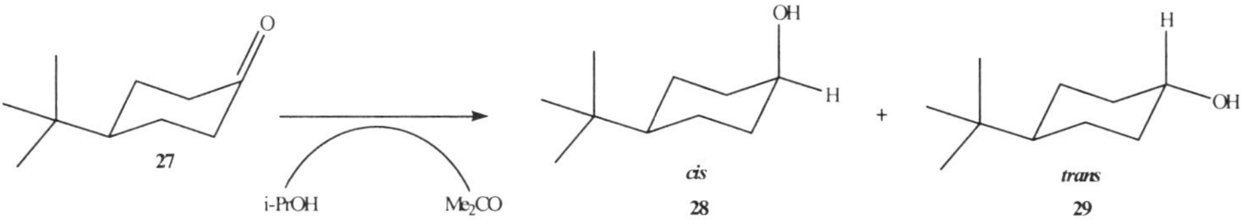


**Equation 15.3.1** Meerwein-Pondorf-Verley reduction of 4-tert-butylcyclohexanone **27**

By conducting the MPV reduction of **27** in refluxing isopropanol at 80°C in the presence of a number of different catalysts, H. van Bekkum et al(32) found that in addition to BEA (in both H- and Na-forms) gamma-alumina, silica alumina and the mesoporous MCM-41 also proved active. These last catalysts, however, give the thermodynamically favoured *trans*-compound **29** in 80-90% isomeric selectivity, high *cis*-selectivity only being found with BEA (Table 15.4).

**Table 15.4** MPV reduction of 4-tert-butylcyclohexanone with isopropanol to 4-tert- butylcyclohexanol over various heterogeneous catalysts[a]

| Catalyst[b] | Si/Al | Conv.(%) | Sel. (%) | ***Cis : trans*** |
|---|---|---|---|---|
| Na-BEA(1) | 12.5 | 32 | >95 | 96 : 4 |
| $NH_4$-BEA(1) | 12.5 | 30 | >95 | 95 : 5 |
| All-Si BEA | >5000 | 0 | - | - |
| Na-Y | 2.5 | 0 | - | - |
| $NH_4$-Y | 2.5 | 0 | - | - |
| USY | 2.5(bulk) | 24.4 | Ca. 85 | 9 : 91 |
| $NH_4$-MOR[c] | 6.5 | 0 | - | - |
| Deal-MOR[c] | 32 | 0 | - | - |
| H-MCM-41[c] | 15 | 10 | Ca. 80 | 10 : 90 |
| Na-MCM-41[c] | 15 | 10 | >95 | 10 : 90 |
| H-MCM-22 | 15 | 33 | Ca. 25 | 24 : 76 |
| HA-HPV | 2 | 19 | >95 | 9 : 91 |
| Gamma-$Al_2O_3$ | - | | >95 | 9 : 91 |

[a]reaction time 6h ; [b]activation temp. 500°C ; [c]activation temp. 430°C

Zeolites Y, MOR and all silica-BEA showed no activity, although activity is shown by US-Y (a dealuminated, ultra stable form of Y), which contains some extra framework Al and the less crystalline MCM-41, which react with the alcohol to form an activated alkoxide species.

Based on the catalytic results, in combination with the information obtained by $^{27}Al$ Mas-NMR and FT-IR, the following reaction mechanism (32) has been proposed. The initial step is chemisorption of the sec-alcohol on a Lewis-acid site, consisting of coordinatively unsaturated Al. This results in the formation of surface alkoxide. The coordinative interaction of the carbonyl of the ketone with the same aluminium center allows the formation of a six-membered transition state in which hydride transfer can occur (Fig. 15.5).

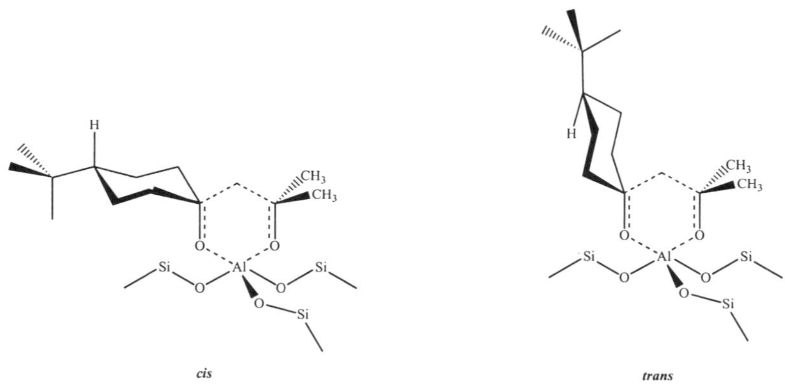


**Fig. 15.5** Transition states for the formation of cis- and trans-4-tert-butylcyclohexanol in the proposed reaction mechanism over zeolite BEA

The transition states leading to the *cis*- and *trans*-alcohols differ substantially in size. That for the *cis*-isomer is more or less linear in form and aligned with the BEA-channel, while the formation of the *trans*-alcohol requires an axially oriented (bulkier) transition state.

Further tuning of the selectivity could be realized by changing the cation in a series of alkali-metal exchanged BEA catalysts which is another indication that the reaction proceeds in the micropores. In the presence of the larger alkaline earth cations almost absolute stereoselectivity is obtained. Figure 15.6 shows the increase in *cis*-selectivity from 95 to 99% on ascending the series from Li to Cs.

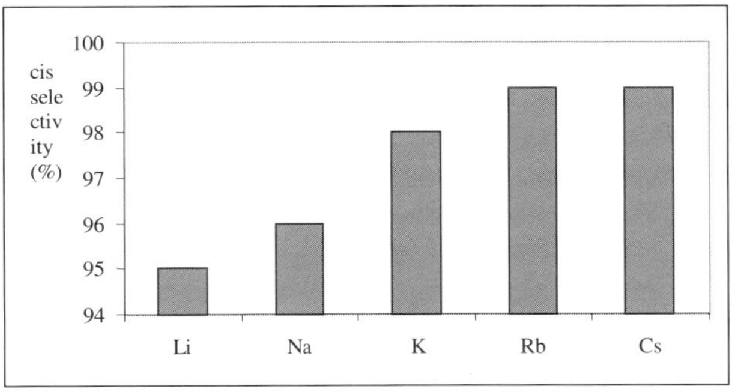


**Fig. 15.6** Meerwein-Pondorf-Verley reduction of 4-tert-bytylcyclohexanone 27 catalyzed by alkali-metal exchanged BEA; effect of cation on *cis*-4-tert-bytylcyclohexanol selectivity

The MPV reduction of 4-tert-butylcyclohexanone provides an excellent example of stereoselectivity by zeolite BEA, and also an example of unequivocal distinction between external-surface and internal micropore activity.

*15.3.3 Synthesis of α-terpineol and of borneol*

Treatment of α-pinene **30** (Eq. 15.3.2) with aqueous mineral acids affords a complex mixture of isomerization and hydration products (33).

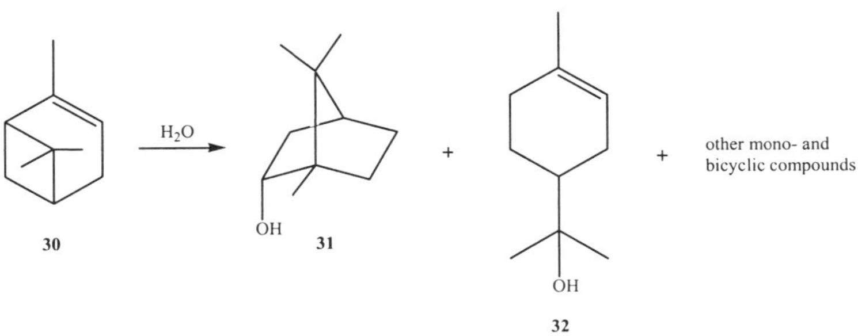


**Equation 15.3.2** Isomerization/hydration products from alpha-pinene 30

The main products are monolitic terpenes such as alpha-terpineol **32**. Minor amounts of bicyclic compounds (34), e.g. borneol **31,** can be observed. Some of these products are of commercial importance as fragrances with lilac and nutmeg odor (monocyclic alcohols like alpha-terpineol or 1-terpinene-4-ol) or with campher-like and extremely delicate pepper odor (bicyclic compounds). In current industrial practice, the bicyclic alcohol borneol **31** is synthesized by a multi-step procedure. Much research has been done to develop clean processes which have high selectivity towards one of the products starting directly from alpha-pinene **30**.

The hydration/isomerization of alpha-pinene catalyzed by zeolite H-BEA is fast and leads mainly to monocyclic terpenes and alcohols with alpha-terpineol **32** as the principal product (up to 48%) (35). The selectivity to bicyclic products is about 26% which, while still too low, is significantly better than the 5% observed for $H_2SO_4$.

### *15.3.4 Synthesis of 8-alkoxy-1-p-menthene*

1-methyl-4-[alpha-alkoxy-isopropyl]-1-cyclohexenes **34** (Eq. 15.3.3) are used as flavors and fragrances for perfume and cosmetic products, as additives for pharmaceuticals and agricultural chemicals, as well as in the food industry (36-39). They are also very important for the synthesis of specialty fine chemicals and natural-like products (40, 41).

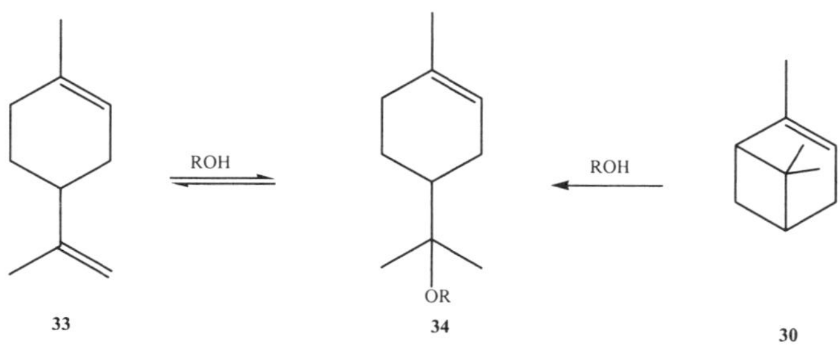


**Equation 15.3.3** Reaction of alpha-pinene 30 and of limonene 33 with alcohols

Among the hydroformylated 1-methyl-4-[alpha-alkoxy-isopropyl]-1-cyclohexenes, i.e. 8-methoxy-p-menthan-2-carboxyaldehyde, the methyl ether **36** (Eq. 15.3.4) smells mangofruit-like and might be used as a perfumery material. The ethyl derivative has a green citrous fragrance, the propyl one has a fresh grass flavour and the butyl one smells woody.

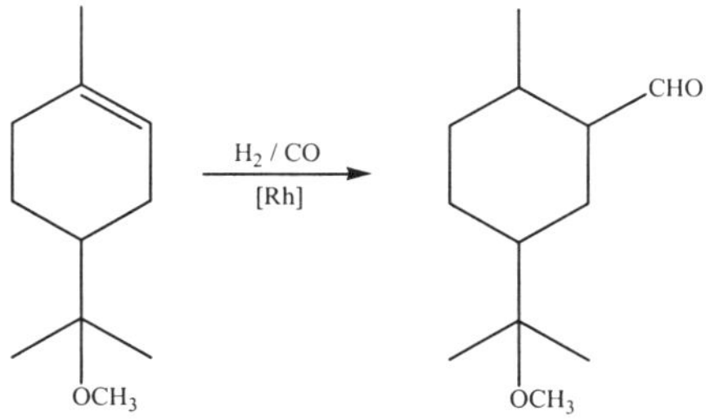


**35** **36**

**Equation 15.3.4** Synthesis of 8-methoxy-p-menthane-2-carboxaldehyde 36 having mangofruit flavour

Several ways to produce such alpha-terpinyl alkyl ethers are known, but limonene and alpha-pinene are the most commonly used feedstock molecules. Alkoxylation of limonene and alpha-pinene over homogeneous or heterogeneous catalysts, such as e.g. strong acids HCl, $H_2SO_4$ or p-toluenesulfonic acid (42-47), aluminium trichloride and boron trifluoride etherate (47) and acidic cation exchange resin (48) has already been reported. The reaction of alpha-terpinyl chloride (which can be obtained by the addition of dry hydrogen chloride to limonene) with protic solvents ($C_1$-$C_5$ alcohol) over zinc or silver compounds as catalysts leads to the corresponding ethers with good yields (up to 95%) (42-52). Due to corrosion and environmental problems, almost all the catalysts for the production of 1-methyl-4-[alpha-alkoxy-isopropyl]-1-cyclohexenes are unacceptable for the industry.

The zeolite-catalyzed alkoxylation of limonene (53, 54) and alpha-pinene (55, 56) over acid-treated mordenite, clinoptilolite and ferrierite as catalysts has already been reported in the literature. The best results were obtained for methoxylation of limonene in the presence of a clinoptilolite-type zeolite (60% yield). The alkoxylation of alpha-pinene with methanol in the presence of mordenite also achieved the highest yields of 66% for 1-methyl-4-[alpha-methoxy-isopropyl]-1-cyclohexene. Syntheses of 1-methyl-4-[alpha-alkoxy-isopropyl]-1-cyclohexenes via zeolite-catalyzed alkoxylation of other terpenes were reported in a review paper (57).

The aim is still: a) improvement and optimization of the already existing zeolite-catalyzed processes with regard to conversion and selectivity; b) regeneration of the zeolite and c) run of the reaction in a continuous process to make an industrial application feasible.

### *15.3.5 Alkoxylation of limonene*

Methanol reacts with limonene over acidic catalysts in a batch reactor to 1-methyl-4-[alpha-methoxy-isopropyl]-1-cyclohexene (alpha-terpinyl methyl ether) as the main reaction product (see Eq. 15.3.3: R- = $CH_3$-). Besides the desired methoxylation, isomerisation reactions leading to terpinolene and traces of alpha- and gamma-terpinene can be observed. Furthermore, the addition of methanol to the methylterpinylether leads to the undesired *cis*- or *trans*-1,8-dimethoxy-p-menthane. The amount of unidentified products does not exceed 1%. At high temperatures and long reaction times the reverse reaction of the alpha-terpinyl methyl ether and the other addition products to limonene and its isomers can be observed. The reaction scheme of the alkoxylation of limonene is illustrated in Equation 15.3.5.

The reaction pathway is identical to that of alpha-pinene alkoxylation except no bicyclic compounds are formed. As mentioned in the literature (53, 54), alkoxylation of limonene to 1-methyl-4-[alpha-alkoxy-isopropyl]-1-cyclohexene can be carried out only in the presence of acidic catalysts. After a catalyst screening using various zeolitic and non-zeolitic acid heterogeneous catalysts, we found that beta zeolite is the best candidate.

Addition of methanol to limonene in the presence of a beta zeolite produces the highest selectivity to 1-methyl-4-[alpha-methoxy-isopropyl]-1-cyclohexene of about 93% at 91% conversion. Surprisingly, the highest yield of about 85% has been obtained at room temperature. The other zeolites and solid acids applied for the alkoxylation of limonene reveal considerably lower conversion and selectivity.

It is interesting to note that the destruction of the structure of beta zeolite by treatment with strong acids or high temperature leads to a complete deactivation of the catalyst for limonene alkoxylation. By using a higher reaction temperature only isomerisation and polymerisation products have been obtained. 1-methyl-4-[alpha-methoxy-isopropyl]-1-cyclohexene or other addition products cannot be found.

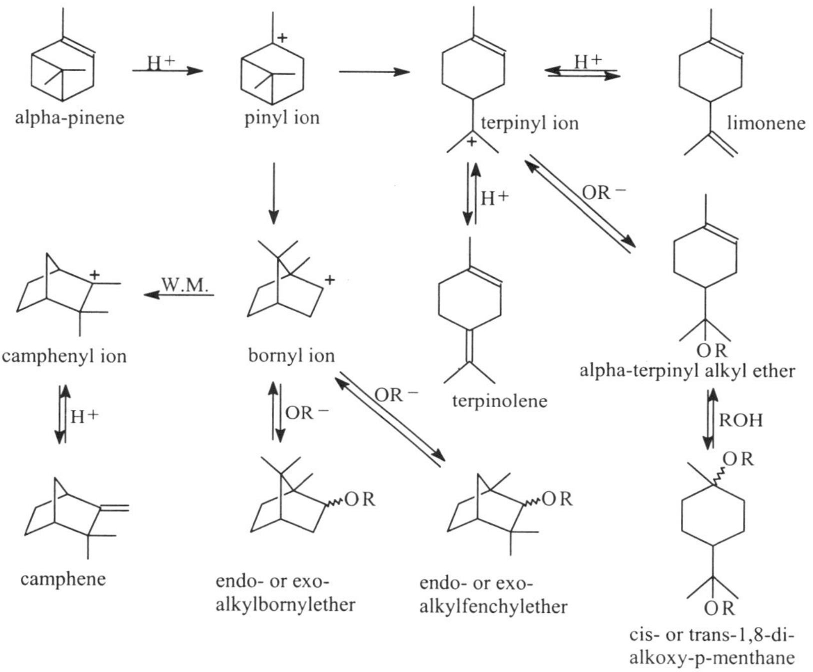


**Equation 15.3.5** By-products from alpha-pinene

The most likely reason for the high activity of zeolite BEA is the relatively high BET surface area of the catalyst (750 $m^2/g$). Furthermore there are hints by temperature-programmed desorption (TPD) of ammonia that a large amount of acid sites are present. We assume that the alkoxylation of limonene takes place inside the pore structure of the beta zeolite. The high selectivity of zeolite BEA might originate from suitable acid sites in pores of its defined size and shape.

Several beta-zeolite samples have been used as catalysts (A,B,C). All of them show the XRD-diffraction pattern of beta-zeolite. However, it is an astonishing result that all of them showed excellent selectivities (> 92%) while the conversion varied from 40 to 90% (Fig. 15.7).

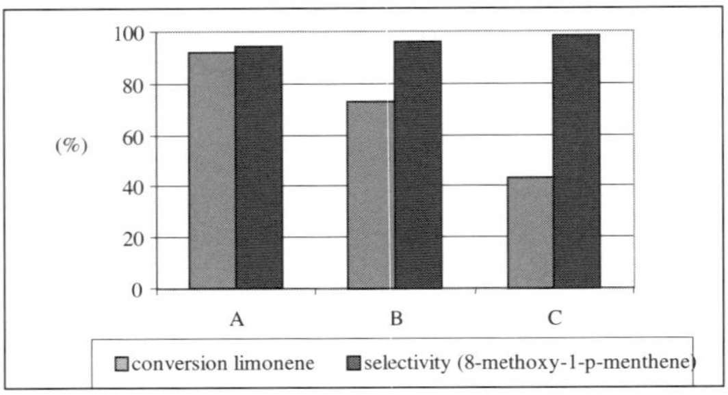


**Fig. 15.7** Conversion and selectivity with different BEA zeolites

The characterization of these beta-zeolite samples via $N_2$-Sorption (BET) revealed that catalyst C only had micropores and a BET surface area of 640 $m^2/g$, whereas the two catalysts with a higher conversion had meso- and macroporous volumina resulting in a faster mass transport. The difference in the performance of samples A and B can be attributed to the highest BET surface area of A (750 $m^2/g$) in contrast to B (680 $m^2/g$). Additionally, TPD measurements revealed that the total amount of acidic sites on sample A was still higher than on sample B (58). A shape selective effect in the microporous system might be also involved since the dialkoxylation of limonene occurs only to a minor extent. In order to achieve high selectivity it is also necessary that the reaction products can leave the zeolitic material very fast via the meso- and macropores. On account of high activity and selectivity of beta zeolite and good product yields for limonene alkoxylation in the presence of this catalyst, the influence of various reaction conditions to optimize the reaction procedure are studied by using this beta zeolite (59). The dependence of limonene conversion and selectivity to 1-methyl-4-[alpha-methoxy-isopropyl]-1-cyclohexene over beta zeolite in a batch reactor on temperature shows that the optimum has been found at 40°C for the reaction time of 5 h. At this temperature, the highest selectivity of about 92% at 93% conversion has been obtained. Above 40°C the selectivity is considerably reduced due to either consecutive reactions and decomposition of 1-methyl-4-[alpha-methoxy-isopropyl]-1-cyclohexene or other side reactions occurring even at a shorter reaction time. At a reaction temperature of 60°C and a reaction time of 2 hours a maximal conversion of 94% is reached. A higher temperature of 80°C increasingly promotes the reverse reaction of the alpha-terpinyl methyl ether to limonene and its isomers under elimination of the methanol, so that the conversion remains almost constant while the selectivity to the desired product decreases. Lower temperature results in a drop of the limonene conversion.

However, the latter can be compensated by a longer reaction time. The selectivities obtained at lower reaction temperatures are up to 98%.

As in the case of limonene methoxylation, the addition of higher alcohols also delivers the corresponding alpha-terpinyl alkyl ethers as the main reaction products. However, the conversions of limonene by using ethanol, 1-propanol, 1-butanol or 1-pentanol are suprisingly lower than that with methanol (59). It would have been expected that in a electrophilic addition, the longer-chain alcohols would be more reactive. The higher conversion when methanol is used is possibly due to less diffusion constraints. The maximum selectivity obtained for ethoxylation of limonene is 94% at 42% conversion. By increasing the chain length of the linear alcohols, a decrease in the selectivity was observed. This might be caused by steric hindrance in the zeolite pore structure in such a way that longer alcohols do not easily form the transition state. Good product yields are achieved only in the case of the linear alcohols, whereas for iso alcohols such as 2-propanol, 2-butanol or 2-pentanol, a remarkable drop in limonene conversion and selectivity to corresponding alpha-terpinyl alkyl ethers is observed; with 2-methyl-2-propanol, no addition products have been found. The reason for those effects might again be sterical hinderance and diffusion restrictions in the elliptical pores of beta zeolite resulting in a lower reactivity.

For methoxylation of limonene in the presence of beta zeolite in a continuous flow-type apparatus with an integral fixed bed reactor, the same tendencies as for the batch reactor are observed in general. The great advantage of the reaction in an integral continuous fixed bed reactor is the easy separation of the products from zeolite and a higher productivity than in a batch process. Surprisingly, at the highest (80 °C) and the lowest (25 °C) reaction temperature the conversion is almost constant. The highest limonene conversion obtained is about 95% at 80 °C. The selectivity to the alpha-terpinyl methyl ether increases within 9 h TOS. However, the selectivity is lowered to 75% at 80 °C while at a lower temperature it increases to 95% during 9 h TOS.

At the limonene conversion of about 60% and a reaction temperature decreased to 40 °C, selectivities higher than 95% are achieved. A temperature lower than 40 °C causes a noticeable decrease of the limonene conversion. In spite of this, selectivity to alpha-terpinyl methyl ether remains on the same level of about 95%. The product distribution is nearly the same as in the batch reactor.

The chosen mass ratio between methanol and limonene of 2:1 is advantageous, because the feedstock and the products are well dissolved at this concentration. However, the excess of methanol can react with 1-methyl-4-[alpha-methoxy-isopropyl]-1-cyclohexene to *cis*- or *trans*-1,8-dimethoxy-p-menthane in a consecutive reaction. This reaction is strongly accelerated when the limonene conversion exceeds 90%. That consecutive reaction together with the decomposition of the desired product and other side reactions decrease the selectivity at a high limonene conversion.

The selectivity to the alpha-terpinyl methyl ether increases with the reduction of the residence time. The experiments have revealed that not only higher reaction temperature and the limonene conversion above 90%, but also a longer contact time caused by a lower WHSV favor the consecutive reaction of the excess methanol with 1-methyl-4-[alpha-methoxy-isopropyl]-1-cyclohexene to *cis*- or *trans*-1,8-dimethoxy-p-menthane.

The application of various linear, iso- and branched alcohols for the addition to limonene in the integral fixed bed reactor gives the same results as those in the batch reactor (59).

The catalyst service time can be increased by using a higher temperature as deactivating compounds desorb more easily at elevated temperatures. However, after an initially small increase, the limonene conversion decreases in one to two days. It was found that the catalyst could be regenerated (59).

### *15.3.6 Alkoxylation of alpha-pinene*

The addition of methanol to alpha-pinene in the presence of the above mentioned beta zeolite as catalyst in the batch reactor results in the cleavage of the cyclobutane ring and yields 1-methyl-4-[alpha-methoxy-isopropyl]-1-cyclohexene (alpha-terpinyl methyl ether) as the main reaction product. The most common by-products to be found are isomerization compounds like camphene, limonene and terpinolene, and several bicyclic and double addition products, e.g. *endo*- or *exo*-methylbornylether, *endo*- or *exo*-methylfenchylether and *cis*- or *trans*-1,8-dimethoxy-p-menthane.

Depending on the reaction conditions, over 50% selectivity to the desired product can be achieved. The other methoxy compounds formed by the secondary addition of methanol to the product or to various isomerisation products are obtained with 28% yield. By summarizing all addition products, the selectivity of the methoxy compounds increases to about 79%. The reaction mixture contains only 20% of terpenes including the unreacted alpha-pinene. The amount of unidentified products does not exceed 1% and can therefore be neglected.

The influence of temperature has been investigated within the range from 25 °C to 80 °C. The optimum temperature for alpha-pinene methoxylation has been found at 40 °C at a reaction time of 5 h. At this temperature, the highest selectivity obtained over beta zeolite is about 54% at 92% conversion. A temperature above 40 °C gives significant reduction in selectivity due to occurrence of consecutive reactions or decomposition of 1-methyl-4-[alpha-methoxy-isopropyl]-1-cyclohexene and other side reactions at even shorter reaction times. A lower temperature causes a drop of the alpha-pinene conversion. However, this can be compensated by a longer reaction time (59). The dependencies of temperature and reaction time are consistent with the above observations for limonene methoxylation. However selectivities up to 98% are obtained for limonene methoxylation, whereas the highest selectivity achieved for methoxylation of alpha-pinene amounts to 54%. On account of the above mentioned additional formation of various bicyclic addition products, the selectivities for all methoxy compounds increase to about 80%. This indicates that for the production of 1-methyl-4-[alpha-methoxy-isopropyl]-1-cyclohexene beta zeolite is more selective for the addition of methanol to limonene than to alpha-pinene. The activity of the catalyst, however, is nearly the same for both reactions.

Methoxylation of alpha-pinene in the continuous flow-type apparatus with an integral fixed bed reactor over beta zeolite as catalyst shows very similar results as in the batch reactor. The dependence of alpha-pinene conversion and selectivity to the alpha-terpinyl methyl ether on temperature and time-on-stream are in agreement with observations for limonene methoxylation.

The heterogeneously catalyzed alkoxylation of alpha-pinene and limonene over beta zeolite provides excellent results in both a discontinuous batch reactor and a continuous flow-type apparatus with a fixed bed reactor. In both reactors, the use of methanol as addition compound and limonene as feedstock gives 1-methyl-4-[alpha-methoxy-isopropyl]-1-cyclohexene with the yield of 85% (conversion: 93%, selectivity: 92%). By means of variation of the reaction parameters, the limonene conversion can be adjusted within the range 40 – 90%. The selectivity to 1-methyl-4-[alpha-methoxy-isopropyl]-1-cyclohexene always remains at about 95%.

For mixtures of methanol and alpha-pinene, the highest yield of corresponding alpha-terpinyl methyl ether is only of about 50% (conversion: 99%, selectivity: 51%) in both reactor types, as several bicyclic and double addition products are formed in parallel. The selectivity of all additional products can reach values of about 85%.

The most likely reasons for the high activity of beta zeolite are the relatively high BET surface area of the catalyst (750 $m^2/g$) and the large amount of acid sites measured by temperature-programmed desorption (TPD) of ammonia(58).

We assume that the alkoxylation of limonene takes place inside the pore structure of the beta zeolite. We also assume that the high selectivity of beta zeolite originates from suitable acid sites in pores of its defined size and shape.

### *15.3.7 Sulfur-containing fragrances and flavors*

1-p-menthene-8-thiol which is a part of the grapefruit aroma and 1-p-menthene-8-thioethers which are used as ingredients for perfumes and cosmetics can be synthesized from terpenes (e.g. alpha-pinene and limonene) with hydrogen sulfide using $AlCl_3$ and $AlBr_3$ as catalysts (60, 61). The disadvantages of this type of catalyst are known and frequently named herein.

Acid zeolite catalysts offer a very good alternative for the clean synthesis of these sulfur containing substances. A suitable feed stock is the 4-isopropenyl-1-methyl-1-cyclohexene. In the presence of a commercial beta-zeolite (25) hydrogen sulfide is added to the autoclave at a reaction temperature of 50°C at a pressure of 17 bar. The conversion is 65.1% and the selectivity to 1-p-menthene-8-thiol is 43.9%. These are very promising results and they can be improved by using a commercial H-US-Y zeolite which rendered a conversion of 76.8% and a selectivity of 64.3% (62).

## 15.4 Coumarins, Indoles

### *15.4.1 Coumarins*

Coumarins are an important family of fragrance chemicals with herbaceous, hay-like odors. The conventional methods for coumarin synthesis require drastic conditions. For example, 4-methyl-7-hydroxycoumarin **39** (Eq. 15.4.1) is prepared via the Pechmann reaction by stirring a mixture of resorcinol **37** and ethyl acetoacetate **38** in $H_2SO_4$ for 12-14 h (63).

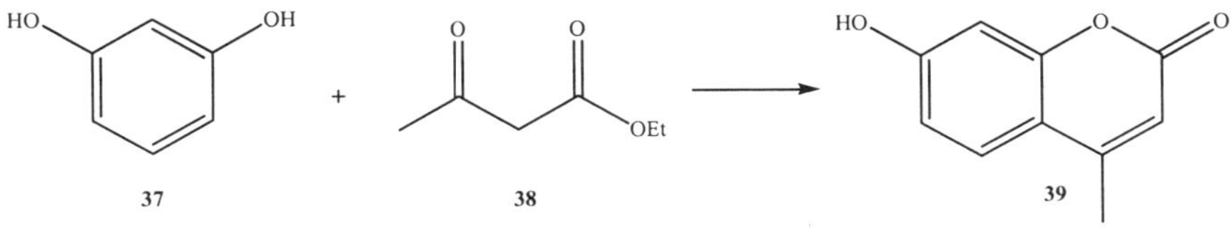

**Equation 15.4.1** Pechmann condensation of ethylacetoacetate 38 with resorcinol 37

The Pechmann reaction catalyzed by solid acid catalysts (e.g. zeolite H-BEA) in refluxing toluene has been reported by H. van Bekkum et al. (64). The reaction proceeds via transesterification and intramolecular hydroxyalkylation, followed by dehydration (Eq. 15.4.2).

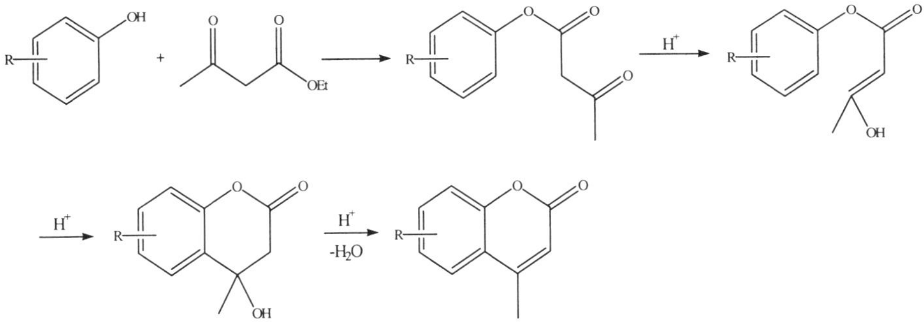

**Equation 15.4.2** Reaction steps of the Pechmann reaction

7-Hydroxy-4-methylcoumarin **39**, which is also useful as a starting material for the preparation of insecticides (hymecromone), is obtained in yields of about 80% by applying zeolite H-BEA or Amberlyst 15. In comparison with Amberlyst 15, zeolite H-BEA is the preferred catalyst because of its excellent regenerability and high turnover numbers.

Recently, we found that a new class of solid acid catalyst, a Nafion resin/silica nanocomposite effectively catalyzes the Pechmann reaction (65). Nafion resin is a tetrafluoroethylene and perfluoro-2-(fluorosulfonylethoxy)propyl vinyl ether copolymer and can be made more suitable for organic catalysis. In the novel nanocomposites of Nafion resin, the nanometer-sized Nafion particles are entrapped within a porous silica matrix. These materials combine the excellent solid-acid catalyst properties of Nafion resin with the desirable porous support characteristic of silica, resulting in increased accessibility of the Nafion resin-based acid sites. A number of very promising applications of these materials have been investigated, ranging from olefin isomerizations (66), dimerizations (67), Fries-rearrangements (68), esterifications (69), acylations (70) and alkylations (71).

The compound 7-hydroxy-4-methylcoumarin **39** was obtained in very high yields up to 81% over the 40% Nafion on silica composite (SAC 40) and almost quantitatively (> 96% yield) over the 80% composite (SAC 80) (Fig. 15.8) in refluxing toluene. We reduced the amount of used catalyst to half of the amount of H-BEA used by H. van Bekkum et al(64).

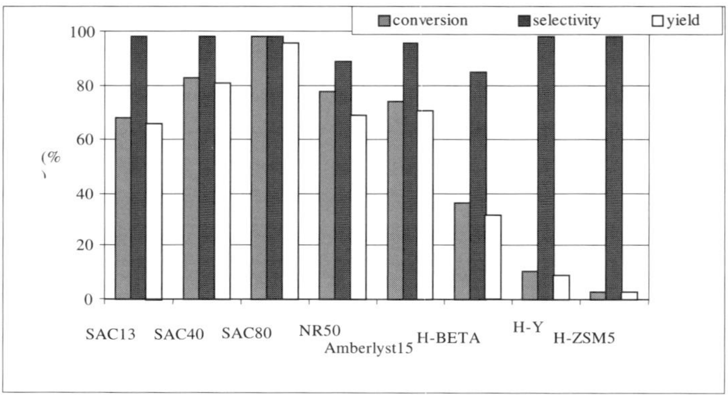


**Fig. 15.8** Results of the Pechmann reaction with different catalysts

In comparison to the results with different amounts of Nafion, the conversion of resorcinol raises Nafion containing material (SAC 13) from 68% for 13 wt% to 83% for the 40 wt% containing composite (SAC 40). Catalysis with the pure Nafion (NR 50) showsn only 78% conversion of resorcinol.

Due to the high conversion obtained with SAC 40 and especially with SAC 80, only a small amount of catalyst contamination was observed and the catalyst could be reused several times or regenerated by a washing procedure with acetone and diluted nitric acid solutions. Compared with the conversion of resorcinol using lower amounts of other solid acid catalysts such as Amberlyst 15 (74%), H-BEA (36%), H-US-Y (10%) and H-ZSM-5 (3%), the SAC 40 and SAC 80 Nafion/silica nanocomposites are the favoured catalysts.

The Pechmann reaction over Nafion/silica composites is an example that in some cases, zeolites show no advantage over other solid acids. Further progress, not only in the area of zeolites but also in the field of other solid acids like acidic ion-exchange resins, is unavoidable.

### *15.4.2 Indoles*

The Fischer indole synthesis is a versatile method for preparing 2,3-substituted indoles. Indoles itself, in small concentrations, has a floral odour while many of 2,3-substituted indoles are biologically active and also find applications as pharmaceuticals and plant growth regulators.

In the Fischer synthesis, indoles are prepared from a ketone **40** and a phenylhydrazine derivative **41** (Eq. 15.4.3). The two-step-reaction, starting with a condensation towards a phenylhydrazone **42** and followed by an acid-catalyzed cyclisation with elimination of ammonia, is usually combined in a one-pot procedure (72).

As shown in Equation 15.4.3, the use of non-symetrical ketones in the Fischer synthesis results in two isomeric indoles, **45** and **46**. For some applications, it is advantageous to obtain a single isomer in high selectivity.

Prochazka et al. (73) recognized that the spatially restricted reaction environment of zeolite pores can be the source of shape selectivity influencing the ratio of indole isomers formed and favouring the less bulky isomer.

H. van Bekkum et al. (72) studied a number of catalysts in the Fischer synthesis starting from 1-phenyl-2-butanone **40** (with $R_1$ = Ph, $R_2$ = $CH_3$) and phenylhydrazine. The isomeric products are the bulky 2-ethyl-3-phenylindole **45** (with $R_1$ = Ph, $R_2$ = $CH_3$) and the linear 2-benzyl-3-methylindole **46** (with $R_1$ = Ph, $R_2$ = $CH_3$). Catalysis of the inolization of **40** by soluble as well as solid (e.g. Amberlyst 15) catalysts typically yielded a mixture of the two isomers in a bulky/linear ratio of about 75/25. Zeolite BEA reverses this bulky/linear ratio giving 75% of the linear isomer **46**, a result interpreted in terms of restricted transition-state selectivity. Although in zeolite BEA the intraporous formation of **45** is largely suppressed, it is in fact probably not completely inhibited.

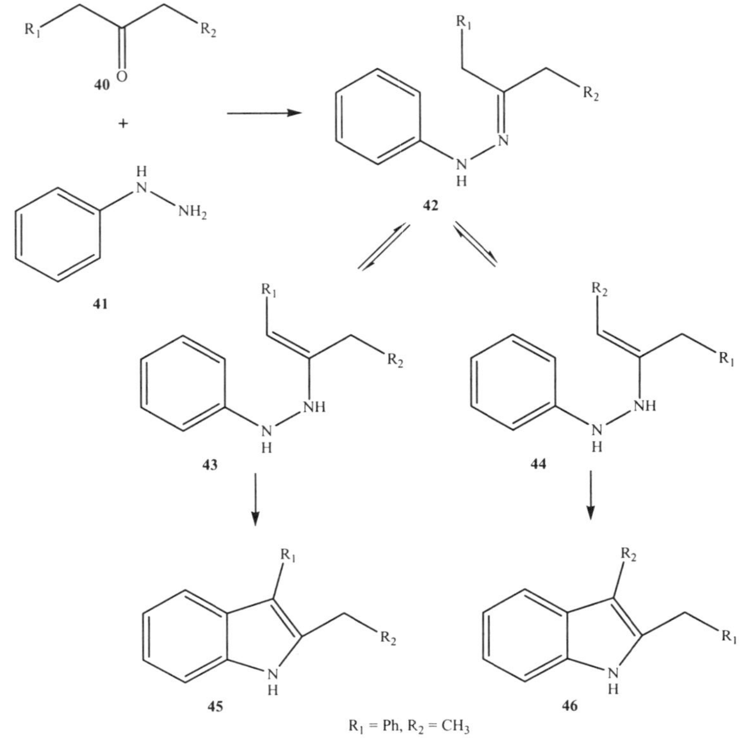


**Equation 15.4.3** Reaction pathways for the Fischer indole synthesis of two isomeric indoles 45 and 46

The selectivity of the catalytic reaction should only depend on the relative rates of formation of the enehydrazines **43** and **44** (Eq. 15.4.3). The sorption data and molecular geometries indicate that, although the formation of both enehydrazines **43** und **44** inside the channels of zeolite BEA should be possible, enehydrazine **43** is severely hindered in adopting the conformation reguired for enolization.

The selective Fischer synthesis of 2-benzyl-3-methylindole from 1-phenyl-2-butanone and phenylhydrazine and catalyzed by zeolite BEA can be considered as an interesting example of restricted transition state selectivity.

## 15.5 More bulky esters

### *15.5.1 Esterification of dicyclopentadiene with carboxylic acids*

The esterification of cyclic olefins such as dicyclopentadiene or naturally occurring terpenes with carboxylic acids yields compounds of industrial value (74). For instance, the esterification of dicyclopentadiene **47** (Eq. 15.5.1) with saturated carboxylic acids, such as acetic acid, leads to a starting material **48** for the flavour and fragrance industry, as well as the resulting alcohols in consecutive hydrolysis (75).

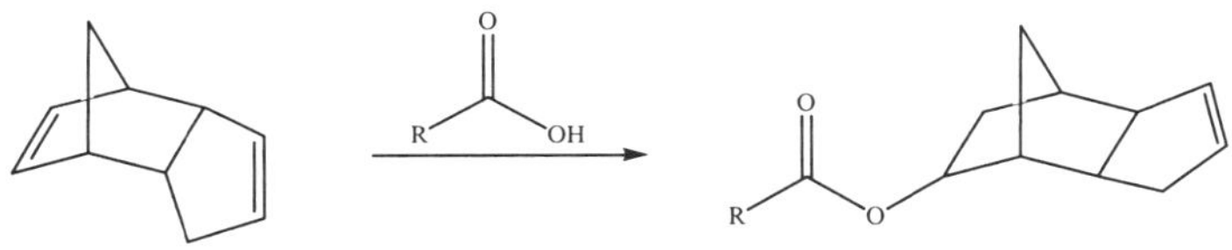


47 48

**Equation 15.5.1** Esterification of dicyclopentadiene 47 with different carboxylic acids (reaction scheme 1)

The esterification is known to be catalyzed quite well by homogeneous catalysts (76). Several traditional catalysts such as $BF_3$ render the desired compounds in high yield (77). However, toxicity, corrosion and difficult separation procedures of the catalysts are well known disadvantages of the homogeneously catalyzed reaction in industrial applications. Up to now several heterogeneous catalysts have been found, which show a good performance in this reaction (78). However, only very reactive olefins are effectively catalyzed heterogeneously (79). In the case of less active olefins in order to realize high yields, very high carboxylic acid olefin ratios or very high catalyst loadings are needed (80). Unfortunately, raising the reaction temperature to enhance the reaction rate causes a decrease in selectivity.

The use of Nafion/silica composites and zeolite catalysts in this kind of reaction has been studied. These composites containing different amounts of Nafion catalyst are compared with the original material and with Amberlyst 15, a well known macroreticular ion-exchange resin, which achieves remarkable results in several acid catalyzed reactions (81).

During the acid catalyzed esterification of dicyclopentadiene **47**, a rapid Wagner-Meerwein type rearrangement occurs simultaneously, according to the reaction scheme 2 (Eq. 15.5.2). The ester compound contains exclusively the dicyclopentenyl component in the *exo*-form.

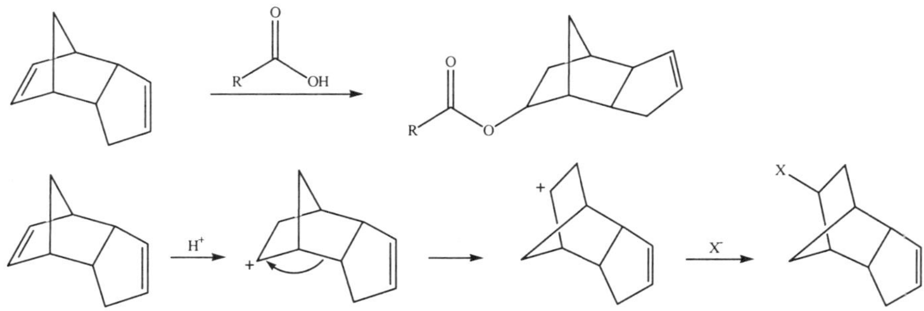


**Equation 15.5.2** Esterification of dicyclopentadiene 47 with different carboxylic acids (reaction scheme 2)

Figure 15.9 shows the esterification of different saturated carboxylic acids with dicyclopentadiene. In all depicted experiments the carboxylic acid/dicyclo--pentadiene molar ratio is about 4 and the amount of catalyst is 10% by weight of dicyclopentadiene at a reaction temperature of 80°C. In this reaction the Nafion/silica composite catalyst is more active compared to the Amberlyst resin, in particular with respect to the amount of the acid groups on the resin.

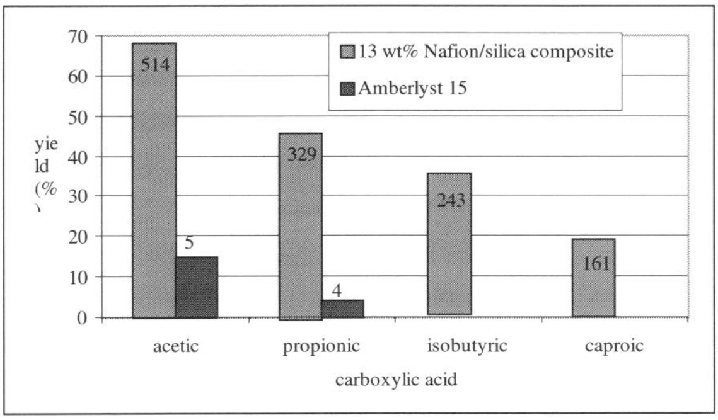


**Fig. 15.9** Esterification of 47 with different carboxylic acids over 13 wt% Nafion/silica

The ion-exchange capacity of pure Nafion is about 0.89 mmol/g, thus the composite capacity is only about 0.12 mmol/g, whereas the Amberlyst resin provides 4.8 mmol/g. As a result, in the case of acetic acid as nucleophile, the turnover number (TON) of the sulfonic acid groups in the Nafion based material are two magnitudes higher than the ones of the Amberlyst material. By using carboxylic acid with longer chain-length, the difference in activity between the Nafion/silica composite and the Amberlyst seems to be even higher. With propionic acid the Amberlyst catalyst renders only just about 5% yield, and with both isobutyric and caproic acid only traces of the products are obtained. The Nafion based material leads to more than 45% yield by using propionic acid and almost 20% for caproic acid.

Carrying out experiments with equal ratios of starting material **47** to sulfonic acid group of the specific catalysts reveals the same pattern as obtained in the experiments described above. Figure 5.2 shows the results obtained over some Nafion/silica composite materials with different amounts of Nafion compared to the Amberlyst ion-exchange resin.

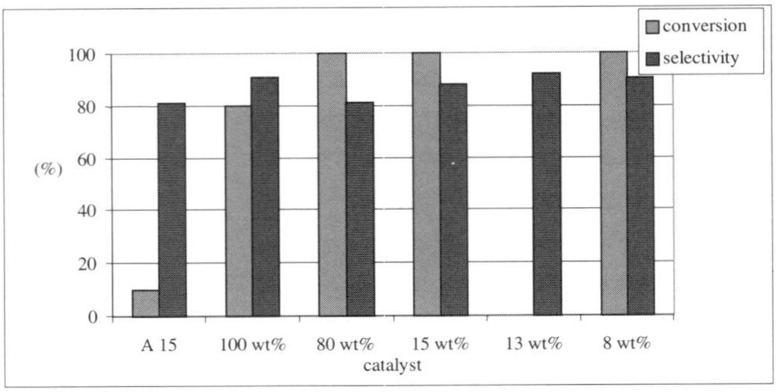


**Fig. 15.10** Esterification of 47 with different carboxylic acids over Nafion/silica composites with different amount of Nafion and over Amberlyst 15

All Nafion/silica composite catalysts show complete conversion independently of the amount of Nafion resin. In contrast, a conversion of only about 80% is obtained over the pure polymer. Again, the Amberlyst material achieves only poor conversion of about 12%.

It seems, due to the very high activity of the Nafion/silica based catalysts, that these reactions conditions are not suitable for observing any differences in activity between the various Nafion/silica materials. All composites with a ratio of starting material to sulfonic acid group of about 1,000 reveal complete conversion.

## 15.6 Conclusion

With several examples we could demonstrate that heterogeneous catalysts are becoming very valuable for the clean production of fine chemicals on the example of the fragrance and flavour industry.

The isomerization of styrene oxide to phenylacetaldehyde yields 100% using modified ZSM-5 zeolites, thereby the highest target achieved by catalysis has been fulfilled. A new process has also been found for the heterogeneously catalyzed production of campholenic aldehyde from alpha-pinene oxide. By using low reaction temperatures of 0 °C and below in combination with HCl treated H-US-Y zeolites, up to 85% yield is achieved. Ti-BEA also shows high activity and selectivity at industrially practicable temperatures of 90°-100°C in the formation of campholenic aldehyde from alpha-pinene oxide. Both processes are competitive with the homogeneous $ZnBr_2$ system.

In several cases, such as the MPV reduction, the specific pore structures of zeolites lead to selectivities not realizable with other catalysts. Nevertheless, in some cases, a zeolite shows no advantage in activity over other solid acids, except for the superior regenerability provided by their thermally stable inorganic matrix.

For instance, nafion composites give promising results. In the Pechmann reaction as well as in the esterification of dicyclopentadiene, the Nafion/silica composites are superior to pure Nafion, zeolites and ion exchange resins. We find high conversions at low temperatures, which can be attributed to accessible acidic sites with high acidic strength.

Despite successes, many goals remain as yet unrealized, such as a widely applicable chiral solid catalyst for performing asymmetric synthesis of enantiomerically pure flavors, pharmaceuticals, etc. Two examples to elucidate the importance of the separate synthesis of enantiomers are limonene and asparagine: while the (S)-enantiomer of limonene has citrus odour the (R)-enantiomer smells like orange; while the (S)-enantiomer of asparagine has a bitter taste the (R)-enantiomer is sweet.

Heterogeneous catalysts, in particular zeolites, with their various properties, contribute extensively to the environmental protection in the synthesis of fine chemicals. For that a broad and very impressive range of acidic and basic catalysts is already available, having all levels of properties between super acidity and super basicity. Also the possibility preparing bifunctional catalysts will gain in importance.

Catalyst research is better equipped than ever to meet the challenges of the new century.

## References


1. Tanabe K. and Hoelderich W.F., *Appl. Cat. A: Gen.* **181** (1999) 399-434.
2. Hoelderich W.F. in *New Frontiers in Catalysis: Proceedings of the 10th Int. Congress on Catalysis* Ed. Guczi K.L. et al., (Budapest, 1992) and *Stud. Surf. Sci. Catal.*, **75** (1993) 127.
3. Sheldon R.A., *Chem. Ind.*, **1** (1997) 12; Tanaka Y., Yanashima H., Minobe M. and Suzukamo G., *Appl. Surf. Sci.*, **121/122** (1997) 461-467.
4. Aebi H., Baumgartner E., Fiedler H.P. and Ohloff G., in *Kosmetika, Riechstoffe und Lebensmittelzusatzstoffe*, (G. Thieme Verlag, Stuttgart, Germany 1978) 32.
5. Hoelderich W.F., Goetz N., *Proceedings of the 9th International Zeolite Conference,* Ed. van Ballmoos et al. (Montreal 1992), (Butterworth-Heinemann, USA, 1993) 309.
6. Kaminska J., Schwegler M.A., Hoefnagel A. and van Bekkum H., *Rec. Trav. Chim. Pays-Bas*, **111** (1992) 432.
7. Liebens A., Mahaim C. and Hoelderich W.F., *Proceedings of the 4th Int. Symposium on Heterogeneous Catalysis and Fine Chemicals, Sept. 8-12,* Basel, Switzerland and *Stud. Surf. Sci. Catal.*, **108** (1997) 587.
8. EP **155591**, (1988) Firmenich S.A.
9. Arbusow B., *Chem. Ber.*, **68** (1935) 1430.
10. Arata K. and Tanabe K., *Chem. Lett.*, (1979) 1017.
11. Carr G., Dosanjh G., Millar A.P., Whittaker D., *J. Chem. Soc. Per. Trans.*, **2** (1994) 1419.
12. Freude D., Fröhlich T., Hunger M., and Scheler G., *Chem. Phys. Lett.*, **98** (1983) 263.
13. Szostak R., *Stud. Surf. Sci. Catal.*, **58** (1991) 153.
14. Sohn J.R., DeCanio S.J., Lunsford J.H. and Donnell D.J.O., *Zeol.*, **6** (1986) 225.
15. Parker L.M., Bibby D.M. and Burns G.R., *J. Chem. Soc. Faraday Trans.*, **87** (1991) 3319.
16. Holleman A., Wiberg N., *Lehrb. Anorg. Chem.*, (New York, de Gruyter, 1985).
17. Kunkeler P.J. et al., *Catal. Lett.*, **53** (1998) 135-138.
18. Hoelderich W.F. in *Acid-Base Catalysis Kodansha – VCH*, Ed. Tanabe K. et al. (Tokyo, 1989) 1-20.
19. Spiegler W. et al., EP **259 814** and US **4 824 973**, (1986) BASF AG.
20. Goetz N., Hoelderich W.F., Hupfer L., EP **332 981**, US **4 968 831**, (1988) BASF AG.

21. Chapuis C., Margot C., Schulte-Elte K.H., Pamingle H., EP **374509**, (1990) Firmenich S.A.
22. Schulte-Elte K.H., Chapuis C., Simmons D., Reichlin D., EP **457022**, (1991) Firmenich S.A.
23. Liebens A., Laufer W. and Hoelderich W.F., *Catal. Lett.*, **60** (1999) 71-75.
24. House H.O. and Wasson R.L., *J. Am. Chem. Soc.*, **79** (1957) 1488.
25. Bach R.D. and Klix R.C., *Tetrahedron Letters*, **26** (1985) 985.
26. Langin-Lanteri M.T. and Huet J., *Synthesis* **8** (1976) 541.
27. Sheldon R.A. et al., *4th Int. Symposium on Heterogeneous Calaysis and Fine Chemicals, Abstr. O27*, (Basel 1996).
28. Liebens A., Mahaim C. and Hoelderich W.F, *Stud. Surf. Sci. Catal.*, **108** (1997) 587-594.
29. Kaminska J., Schwegler M.A., Hoefnagel A.J. and van Bekkum H., *Recl. Trav. Chim Pays-Bas*, **111** (1992) 432.
30. Elings J.A., Lempers H.E.B. and Sheldon R.A., *Stud. Surf. Sci. Catal.*, **105** (1997) 1165.
31. Hoelderich W.F., Röseler J., Heitmann G. and Liebens A.T., *Catal. Today*, **37** (1997) 353.
32. Creyghton E.J., Ganeshie S.D., Downing R.S. and van Bekkum H., *J. Mol. Catal.*, **115** (1997) 457.
33. Alberts R.M., Traynor S.G. and R.L. Webb in *Naval Stores, PULP Chemical Association*, Ed. Zinkel D.F. and Russel J. (New York, 1989) 479.
34. Valkanas G. and Inconomou N., *Helv. Chim. Acta*, **46** (1963) 1069; Williams C.M. and Whittaker D., *J. Chem. Soc. (B)*, (1971) 672.
35. van der Waal J.C., van Bekkum H. and Vital J.M., *J. of Mol. Catal. A: Chemical*, **105** (1996) 185-192.
36. Mussinan C.J. et al., US **4 163 068**, (1979), **4 173 543**, (1979), **4 246 287**, (1981), **4 252 828**, (1981), **4 255 460**, (1981), Int. Flavors and Fragrances Inc. (New York).
37. Kare M.R. and Agric J., *Food Chem.*, **17** (1969) 677.
38. Shaw P.E. and Caleman R., *J. Aric. Chem.*, **19** (1971) 1276.
39. Chang S.S., *J. Food Sci.*, **42** (1977) 298.
40. Ohloff G. and Klein E., *Tetrahedron*, **18** (1962) 37.
41. Delay F. and Ohloff G., *Helv. Chim. Acta.*, **62** (1979) 369.
42. Royals E.E., *J. Am. Chem. Soc.*, **71** (1949) 2568.
43. Johny K.P. and Verghese J., *J. Curr. Sci.*, **39** (1970) 63.
44. Burczyk B., *Przem. Chem.*, **49** (1970) 225.
45. Burczyk B., *Chem. Stosow. Ser. A*, **14** (1970) 91.

46. Burczyk B., *Rocz. Chem.*, **44** (1971) 1381.
47. Lyons J.E. and Hosler P., DE 2 706 879, (1977).
48. Bordenca C., US **3 401 051**, (1968) SCM Corp.
49. Yoshiharu M. and Masahiro M., JP **75 131 948**, (1976) Jpn. Kokai.
50. Bunton C.A., Cori O., Hachey D. and Leresche J.P., *J. Org. Chem.*, **44** (1977) 3238.
51. Anandaraman S., Gurudutt K.N. and Ravindranath B., *Tetrahedron Letters*, **21** (1980) 2189
52. Gurudutt K.N., Ravindranath B., and Srinivas P., *Tetrahedron*, **38** (1982) 1843.
53. Xiao S., Jiang W. and Zhou P., *Linchan Huaxue Yu Gongye*, **9** (2) (1989) 9.
54. Nomura M. et al., *Yukagaku*, **43** (1994) 1089.
55. Zhou P. et al., *Chin. Sci. Bull.*, **34** (2) (1989) 125.
56. Cao W., Li F., Li X. and Yiang L., *Huaxue Yu Nianhe*, **1** (1995) 14.
57. Tan Z. and Xiao S., *Youji Huaxue*, **13** (1993) 1.
58. Hensen K., Mahaim C. and Hoelderich W.F., *Proceedings of the 11th Int. Zeolite Conf.,* (Aug. 12-17, 1996, Seoul, South Korea) and *Stud. Surf. Sci. Catal.*, **105** (1997) 1133.
59. Hensen K., Mahaim C. and Hoelderich W.F., *Appl. Catal. A: General*, **149** (1997) 311.
60. Janes J. et al., *Flavour Fragrance J.*, **8** (1993) 289.
61. Masuda H., Kikuiri H. and Mihara S., **JP 63 201 162**, (1988) Jpn. Kokai.
62. Hoelderich W.F. and Hensen K., **CH 0326-97**, (1997) Firmenich S.A.
63. Horning E.C., *Org. Synthese, Coll. Vol. III,* (John Wiley & Sons, New York 1955) 281.
64. Hoefnagel A.J., Gunnewegh E.A., Downing R.S. and van Bekkum H., *J. Chem. Soc. Chem. Commun*, (1995) 225.
65. Laufer M.C. and Hoelderich W.F., unpublished results.
66. Sun Q., Harmer M.A. and Farneth W.E., *Chem. Commun.*, (1996) 1201.
67. Heidekum A., Harmer M.A. and Hoelderich W.F., *Catal. Lett.*, **47** (1997) 243.
68. Heidekum A., Harmer M.A. and Hoelderich W.F., *ACS Polym. Prepr.*, (1997) 763; Heidekum A., Harmer M.A. and Hoelderich W.F., *J. Catal.*, **176** (1998) 260.
69. Heidekum A., Harmer M.A. and Hoelderich W.F., *J. Catal.*, **181** (1999) 217.
70. Heidekum A., Harmer M.A. and Hoelderich W.F., *J. Catal.*, **188** (1999) 230.
71. Harmer M.A., Farneth W.E. and Sun Q., *J. Am. Chem. Soc.*, **118** (1996) 7708.
72. Rigutto M.S., et al. in *Heterogeneous Catalysis and Fine Chemicals III, Stud. Surf. Sci. Cat.,* Vol. 78, Ed. Guisnet M. et al. (Elsevier Service, Amsterdam, 1993) 661.

73. Prochazka M.P., Eklund L. and Carlson R., *Acta. Chem. Scand.*, **44** (1990) 610.
74. Riondel A., **EP 759423**, (1997) Elf Atochem.
75. Gude F., and Bellut H., **DE 3723891**, (1987) Huels AG.
76. Bruson H.A. and Riener T.W., *J. Am. Chem. Soc.*, **67** (1945) 723.
77. Schuster L., Dockner T. and Ambach E., **DE 4136660**, (1993) BASF AG.
78. Kleine W.H., **DE 3619797**, (1987) Huels AG.
79. Gscheidmeier M., **US 5596127**, (1997) Hoechst AG.
80. Ndong Mebah J.-M., Mielosynski J.-L. and Paquer D., *New J. Chem.*, **17** (1993) 835.
81. Chakrabarti A. and Sharma M.M., *React. Polym.*, **20** (1993) 1.

CHAPTER 16

# POLLUTION ABATEMENT USING ZEOLITES: STATE OF THE ART AND FURTHER NEEDS

G. DELAHAY AND B. COQ

*Laboratoire de Matériaux Catalytiques et Catalyse en Chimie Organique, UMR 5618 ENSCM-CNRS, 8, Rue de l'Ecole Normale, 34296 Montpellier, France*

Environmentally damaging issues from anthropogenic activities have been growing since the beginning of the industrial era. The consequences are pollution of lands, water resources and air through the fabrication of harmful products and the use of energy inefficient and/or dangerous processes. There is now a great concern to implement drastic measures to limit environmental damage. This chapter will deal mainly with the aspect of pollution abatment by the treatment of contaminated lands, water and gaseous emissions, through the use of zeolites as catalysts and adsorbents. Focus will be on NOx and $N_2O$ abatments.

## 16.1 Introduction

It is a truism to assert that man-made activities have an impact on the environment. But the negative effect of this impact has grown exponentially since the beginning of the industrial era. That concerns the development of harmful products, of dangerous and/or energy-inefficient processes, and unsafe waste streams. Many of these environmentally damaging associated issues are of chemical origin, one should therefore state "what a chemist knows how to make, he has to know how to unmake". To that concern, catalysis is of vital importance to promote greener and/or energy-saving processes, cleaner fuels, and to reduce pollutant emissions in gaseous and liquid streams. Focus will be on this latter aspect.

The availability of pure water will certainly be one of the major environmental issues of the 21$^{st}$ Century. Water contamination can originate from domestic, agricultural, agroindustrial or industrial activities, and accidental damages. The major pollutants are heavy metals, radionuclides, ammonia, nitrates and organic compounds. Health problems are associated with each of them, such as leukaemia, saturnism..., as well as modifications of the eco-system, e.g. entrophication of lakes and rivers. The nature of water treatments obviously depends on the kind of contaminants, and zeolite-based processes are of great concern in this field.

The main pollutants in gaseous emissions concern: volatile organic compounds (VOCs), greenhouse gases, NOx and SOx. We will only discuss the emissions which can potentially be treated by zeolites.

VOCs correspond to stable compounds of vapor pressure above 13 Pa under normal conditions, and belong to the different groups of organic chemistry (hydrocarbons, alcohol, ketones, chlorinated compounds ...). VOCs are charged for taking part in the greenhouse effect, the increase of ground level ozone, and photochemical smogs. VOCs can be toxic, carcinogenic ... VOCs emissions originate from transportation (≈ 30%), solvent uses (≈ 20%), industrial processes (≈ 3%) and various sources (≈ 47%). Solvent uses and industrial processes are the emissions sources which are easiest to control.

For 8,000 years, the temperature of earth's atmosphere has remained constant, but a sudden rise has occurred since the last century (+1°C) with the concurrent increase of $CO_2$ concentration from 280 (in 1860) to 350 ppm at present. This is due to the global warming by extra-emissions of greenhouse gases from anthropogenic activities: $CO_2$, $CH_4$, $N_2O$, $O_3$, CFC ... The contributions to the global warming effect, which integrates the emission flows and the global warming potential, are 81% for $CO_2$, 7% for $CH_4$ and 9% for $N_2O$. Policymakers have acknowledged the potential dangers of these emissions and implemented the Kyoto Protocol in 1997 to reduce emissions by 7% (US), 8% (EU) and 6% (Japan) below their 1990 level by the year 2008. The gases of main concern were $CO_2$, $N_2O$ and $CH_4$, but only emissions of the former two can be controlled by technologies using zeolites. Obviously, $CO_2$ comes first from fossil fuel combustion and can be primary controlled by more energy-saving processes or greener energy sources. Beside its contribution to global warming, $N_2O$ takes part in the depletion of stratospheric ozone layer. $N_2O$ emissions amount to 8 $10^6$ ton N year$^{-1}$ from agriculture (≈ 43%), biomass burning (≈ 18%), vehicles (≈ 13%), and power plants and industrial processes (≈ 23%) (1). These latter two are the easiest to control.

NOx ($NO+NO_2$) emissions are responsible for acid rain (deforestation), photochemical smog (health desease), and intensification of ground-level ozone. Total emissions of NOx amount to ca 45 $10^6$ ton N year$^{-1}$, with 75% from anthropogenic activities. On a global scale, the major sources are the combustion of fossil fuels (≈ 48%), biomass burning (≈ 16%), decomposition in soils (≈ 13%), and lightnings (≈ 10%). NOx emitted from man-made sources is distributed among mobile (≈ 70%) and stationary sources (≈ 30%) with 20% from industrial processes and 80% from large combustion plants. Obviously, emissions from stationary sources are the easiest to control and technologies with zeolites are currently in use.

Some issues regarding technological achievements, with or without zeolites, by catalysis for pollution abatment have been reviewed recently (2, 3), as well as an review of the best available technologies for reducing NOx and $N_2O$ emissions from industrial activities (4).

## 16.2 Waste water treatment

Water treatment by zeolites is essentially based on their remarkable properties as adsorbents and ion exchangers. Ion exchange properties are mainly characterised by sieving effects, cation exchange capacity (CEC), and ion exchange selectivities (5). Sieving effects depend on the hydrated and/or anhydrous diameter of the ions and the pore size of the zeolite. CEC is a function of Si/Al ratio and the cation affinity. Ion exchange selectivity is affected by the anionic field strength at exchange site, and the relative free energies of cation in zeolite or fully hydrated. These properties have been used for the ammonium and heavy metals removal from municipal and industrial sewages, and the abatment of radionuclides from effluents of nuclear plants, or catastrophe management. As regards the cost, a very large use of natural zeolites has been made for this purpose. This was well reviewed by Colella recently (6). Depending on the kind of cations to be extracted, various zeolitic-rich tuffs containing clinoptilolite, chabazite, phillipsite and mordenite can be used with the following selectivity patterns:

| | |
|---|---|
| Clinoptilolite (HEU): | Cs > Pb > $NH_4$ > Na > Sr > Cd,Cu,Zn |
| Chabazite (CHA): | Cs > $NH_4$ > Pb > Na > Cd > Sr > Cu > Zn |
| Mordenite (MOR): | Pb > Cs > $NH_4$ > Na > Cd |
| Phillipsite (PHI): | Cs > Pb > $NH_4$ > Na > Sr > Cd > Zn |

A particular aspect of 'water treatment' is the rehabilitation of accidentally contaminated soils by radionuclides. This is well illustrated by the works carried out after the Cernobyl catastrophe. The incorporation of clinoptilolite into contaminated soils reduced the transport of heavy metals and radionuclides from soils into ground water and biomass (7). Union Carbide's IONSIV IE-95 (CHA) and A-51 zeolites (LTA) with excellent $Cs^+/Na^+$ and $Sr^{2+}/Na^+$ selectivities, respectively, have also been employed for decontamination of high activity level water in the reactor containment building from $Cs^+$ and $Sr^{2+}$ after the accident at Three Miles Island (5). The radio-isotope loaded zeolites were then transformed into glasses for ultimate disposal.

Zeolites are very efficient for the removal of heavy metals from industrial waste water and drinking water (6, 8). There are particularly good prospects for the use of PHI- and CHA-rich tuffs for Pb and Cd removal. Excellent Cr removal was achieved with tuffs containing FAU and PHI (9). After saturation, the zeolites are generally hardened as cement pastes (6).

Regarding the removal of ammonia from waste and drinking water, this is the process of longer standing among those using zeolites as cation exchangers. HEU- and CHA-based materials are the most efficient and widely used for that purpose. Several units are operating worldwide with treatment capacities up to 100,000 $m^3$ $day^{-1}$ (6). In case of strong competition between ammonia and divalent cations, e.g. $Ca^{2+}$, the selectivity $NH_4^+/Ca^{2+}$ of HEU can be multiplied by ten upon treatment at 800°C (10). The process is usually carried out in fixed bed reactors with discontinuous regeneration steps. Ammonia stored in the zeolite is exchanged back with alkali and valorized as $(NH_4)_2SO_4$ fertilizer, or discharged. A process with in situ biological regeneration within the fixed bed reactor has been described recently (11). The process uses a nitrifying bio-agent for transforming ammonia. The driving force for ammonia desorption being the onward of the bio-chemical reaction.

Fe-MFI has been reported active as a Fenton-type catalyst in the wet oxidation of phenol by $H_2O_2$ (12).

Zeolite-based materials are also promising for the removal of organic compounds from industrial waste water (13). This is particularly true for chlorinated pollutants and the preferred process is based on adsorption/separation using hydrophobic molecular sieves (HMS). Compared to carbon adsorbents, HMS presents a good compromise between sorption capacities, selectivity to organics compared to water, and regenerability (vide infra, section 16.3.1.).

## 16.3 Treatment of gaseous emissions

The treatment of gaseous emissions from potentially noxious compounds involves abatment of VOCs, SOx, NOx and greenhouse gases mainly.

### *16.3.1 Treatment of volatile organic compounds (VOCs)*

The treatment of streams (gas or liquids) containing VOCs by zeolite-based materials belongs to adsorption/separation and/or catalytic oxidation. Regarding adsorption/separation alone, it concerns mainly industrial streams with relatively high VOCs concentrations. The VOCs taken up by zeolite are recovered and recycled. At low VOCs concentrations in complicated matrix, adsorption is often coupled with incineration or catalytic oxidation.

16.3.1.1 Adsorption/separation processes

Because VOCs are often present in streams with excess water, the preferred adsorbents for their selective removal are hydrophobic materials (13). Carbons were and still remain largely used, but HMS is attracting more and more attention. Due to incombustibility, they can be used in a wider temperature range than carbons for regeneration. Additionally, they offer opportunities for selective separation by pore size effects. In contrast, carbons have the efficiency advantage only at higher VOCs concentration and to be of lower cost.

At present, among HMS two main zeolitic types, available at a commercial scale, are receiving attention. The first one belongs to the MFI type and can be obtained directly by synthesis. The second is prepared by dealumination of FAU type zeolite in order to increase the Si/Al ratio above ca. 50.

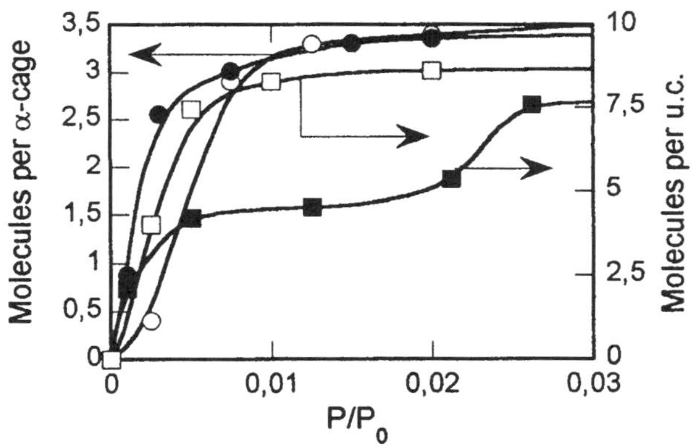


**Fig. 16.1** Adsorption at 298 K of trichloroethylene (○,●) and tetrachloroethylene (□,■) on FAU (open symbols) and MFI (full symbols) (from ref. 14)

Adsorption/separation processes are based on adsorption isotherms (thermodynamics) and intracrystalline diffusivity (kinetics). Figure 16.1 illustrates various shapes of adsorption isotherms depending on the VOC nature, trichloroethylene (TCE) and tetrachloroethylene (PCE), and of the zeolite, MFI with Si/Al > 500 and FAU (Si/Al > 100) (14). The isotherms of VOCs adsorbed on FAU present a more or less S-shape which corresponds to type V of the IUPAC classification. In contrast, the isotherms of VOCs on MFI are more of type I, with the additional particularity of a step at 4 molecules per u.c. for PCE adsorption. The adsorption useful domain of these isotherms is the abrupt step which corresponds to the filling of micropores. S-shape isotherms of FAU are not favourable to adsorption, but favourable for desorption, and the reverse is true for type I isotherms of MFI. The FAU will then be more easily regenerated than MFI.

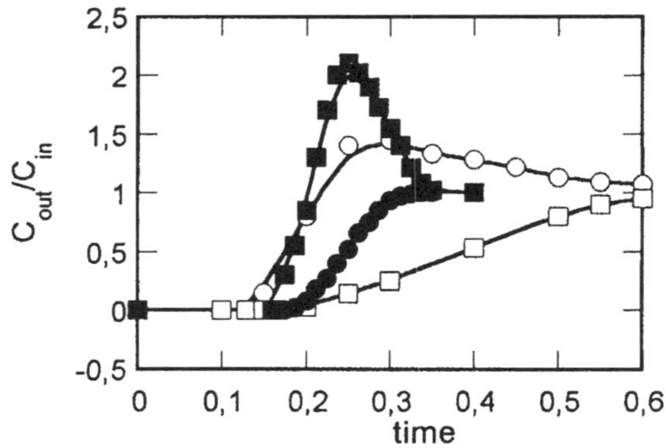


**Fig. 16.2** Breakthrough curves at 298 K of mixtures of trichloroethylene (○,●) and tetrachloroethylene (□,■) on FAU ( open symbols) and MFI (full symbols) (from ref. 14)

A second important point is the shape of the breakthrough curves in a multicomponent stream. Figure 16.2 shows such curves for a TCE+PCE mixture on FAU and MFI (14). The breakthrough curves can be predicted from the individual isotherms of pure compounds: on FAU, PCE displaces TCE in agreement with its higher boiling point, in contrast, PCE is displaced by TCE on MFI due to structural characteristics of the zeolite and relative size of channels and adsorbed molecules. In the case of binary mixture containing PCE and TCE, two columns connected in series and packed with FAU and MFI are enabled to completely clean polluted air via separation and eventually recovery of the pollutants. In a binary VOCs mixture with very close boiling points, the reversal of the breakthrough order can occur twice depending on the relative concentration of both components, as shown for TCE and isopropanol on FAU (15).

Regarding process, the VOCs adsorption is most generally performed using fixed beds of adsorbents or rotor concentrators, using steam or hot air for the desorption step (13). The fixed bed technology is based on a two adsorber scheme for processing gas or wastewater streams. One bed operates on the adsorption mode and the flux is shifted to the second one once the breakthrough occurs. The first bed is then regenerated.

The rotor concentrator technology is used in cases where short contact times must be achieved, for small VOCs concentrations and large volumes of gas streams. The gas to be processed flows through a corrugated wheel whose honeycomb structure has been coated with powdered adsorbent. The continuously rotating wheel is divided into three zones, one for adsorption, one for desorption and the third one for cooling the regenerated zone in order to maintain the operating temperature (Fig. 16.3). The concentration of VOCs is increased by ten in the outflowing air stream. This system can be coupled with incinerator facility. The size and cost of an incinerator are related to air flow and VOCs concentration, and the air flow from the concentrator is reduced by ca 90%.

The adsorption/separation process alone has long been used industrially for solvent recovery for cost saving by recycling. Retrofit of investment was gained in the very short term. However, as the cost of HMS is very high, zeolites are only used in applications for which carbons are not well suited.

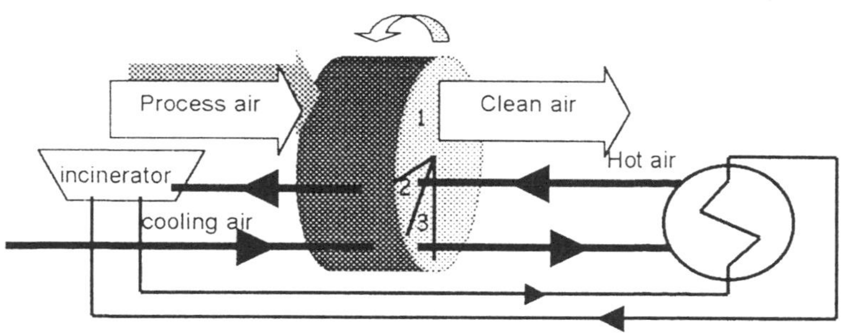


**Fig. 16.3** Flow sheet of the rotor concentrator technology

Another promising application was the temporary storage of hydrocarbons (HCs) emitted by vehicles under cold-start conditions (16). However, the volume of adsorber still necessary to take up HCs was found to be too large to be practical. The first reason for insufficient HCs storage is the competitive adsorption by the excess water vapor present in automotive exhaust.

16.3.1.2 Catalytic combustion of VOCs.

The catalytic combustion is an efficient way for VOCs abatment when present in dilute emissions and complex matrix. Both noble metals and transition metals (TM) based catalysts can be used (17, 18), and zeolites as support are receiving more and more attention.

Noble metals are preferred when non chlorinated VOCs have to be destroyed. For reasons of accessibility, FAU type zeolites are the most frequently employed catalysts and a parameter of utmost importance is the number of accessible metal sites and the reductibility. This was recently put in evidence for the *o*-xylene oxidation on Pd-FAU catalysts (19). $Pd^0$ was proposed as an active site, and the specific activity increased with the ratio $(n_{Pd})/n_{acid}$. A lower acidity (high Si/Al ratio) favors the $Pd^{2+}$ and PdO reductibility, and decreases the extent of coke formation.

TM-based materials are preferred when Cl-containing VOCs have to be processed (17). The activity would be related to the nature of TM and acidity. It was shown that TM-exchanged zeolites exhibit good performance for $CH_2Cl_2$ oxidation, in particular Cr-FAU was the most active (20). A handicap is the high selectivity to CO, which could be reduced by adding Pt. In $CH_2Cl_2$ oxidation high acitivity and selectivity to deep oxidation of $CH_2Cl_2$ to $CO_2$ was thus achieved on Pt-FAU with a great number of Pt sites and low Si/Al ratio (21).

*16.3.2 Control of greenhouse gas emissions*

Several potential control technologies have been proposed for controlling $CO_2$, $CH_4$, $N_2O$, CFCs, $O_3$ ... emissions, but only some of those concerning $CO_2$ and $N_2O$ are based on zeolites materials.

16.3.2.1 Recovery of $CO_2$ from flue gases

It is claimed that the recovery of $CO_2$ from flue gases emitted by power plants, steel mills, cement kilns and fermentation could become increasingly viable. $CO_2$ is not very diluted in these flue gases, with concentration ranging from 10 to 30%. For further utilisation, $CO_2$ needs first to be separated and captured, or at least concentrated. Membrane technology based on polymer materials could be promising, but needs much more selective materials.

A pressure swing adsorption process (PSA) has been described with high efficiency for separation and capture of $CO_2$ in $N_2$ at content from 16 to 25% (22). High purity $CO_2$ (> 99%) was recovered with efficiency ranging from 53% to 70% depending on $CO_2$ concentration. The selectivity and sorption capacity of zeolite 13X (FAU type) was much better than those of activated carbon. However, the influence of $H_2O$ on process efficiency was not reported. It is clear that $H_2O$, always present in flue gases from combustion, should first be separated to prevent inhibition of the zeolite.

From these scarce examples, the better perspectives to control $CO_2$ emissions will be the development of energy saving processes and alternative fuels.

### 16.3.2.2 Control of $N_2O$ emissions

Emission levels for $N_2O$ are expected to become regulated in the near future, and already imposed by taxes in France. There is therefore a strong incentive towards emission control. In principle, two methods are available, i.e. reducing $N_2O$ formation (primary measures) or after-treatment (en-of-pipe solutions). Regarding the end-of-pipe solutions, there are two very different situations depending on the $N_2O$ concentration in the tail gas. The treatment of high concentrated $N_2O$ stream (20-40%) coming from adipic acid plants, glyoxal plants ... is nearing completion, but does not involve zeolite based catalysts due to the high temperature of the process, 770-1070 K. In contrast, the treatment of low concentrated streams (< 1%) from nitric acid plants, power plants ... remains a challenge.

Catalysis is of vital importance in $N_2O$ removal technologies, which are based on the catalytic decomposition of $N_2O$ to $N_2$ and $O_2$ without or with the help of a reductant, and the valorisation of $N_2O$ as a strong and selective chemical oxidant at low temperature. The catalytic decomposition of $N_2O$: $2N_2O \rightarrow 2N_2 + O_2$ can be described in its simplest form by Eqs.16.1-3:

$$N_2O + * \rightarrow N_2 + O* \quad (1)$$

$$N_2O + O* \rightarrow N_2 + O_2 + * \quad (2)$$

$$2O* \leftrightarrow O_2 + 2* \quad (3)$$

In the presence of a reductant, e.g. CO, $C_3H_6$ or $NH_3$, the surface oxygen O* can also be removed according to:

$$CO + O* \rightarrow CO_2 + * \quad (4)$$

$$C_3H_6 + 9O* \rightarrow 3CO_2 + 3H_2O + 9* \quad (5)$$

$$2NH_3 + 3O* \rightarrow N_2 + 3H_2O + 3* \quad (6)$$

Among a huge number of catalytic formulations which have been evaluated for the reaction, TM-exchanged zeolites have shown high activities (see refs. 23, 24 for a review). Moreover, they are not, or are weakly, inhibited by the presence of excess $O_2$ (Eq.16.3 backward), as compared to TM oxides catalysts. This is the remarkable case of MOR, MFI, FER, BEA … exchanged with Co or Fe, which do not suffer any inhibition by $O_2$, and exhibit the highest activity. These materials have received particular attention, and it was shown that the preparation protocol employed for Fe-MFI has a very great influence on the catalytic performances with respect to activity and stability (25-28). Depending on whether the catalyst has been prepared by aqueous ion exchange, solid state exchange, chemical vapor deposition, incorporation of Fe in the framework, or dryness impregnation, various Fe species have been identified. There is not a general agreement about the exact nature of the active species, except on one point: the large iron oxide aggregates exhibit very poor activity for $N_2O$ conversion.

From spectroscopic studies (ESR, Mössbauer …), thermal analysis (TPR, TPO …) and quantum chemical modelling, several models have been designed for the architecture of the highly reactive oxo-cation formed by interaction of $N_2O$ with Fe-MFI. Panov et al. (29), prompted by biological model, designed a binuclear Fe complex with various peroxide bridges. Because of the sharp increase of rate with Fe loading in Fe-MFI, prepared from CVD or solid state exchange, El-Malki et al. (27) also concluded that binuclear sites and nanoclusters are more active for $N_2O$ decomposition than mononuclear sites. From ab initio quantum chemical calculations, Arbuznikov and Zhidomirov (30) confirmed the probability of binuclear oxo-cations with peroxide bridges and hydroxyl groups. Lazar et al. [31] also proposed an oxygen-bridge between two Fe atoms, one of the two belonging to the zeolite framework. In contrast, the reduction of $N_2O$+$O_2$ by $NH_3$ Fe-BEA exhibited a strong increase of TOF, mole of $N_2O$ transformed per mole of Fe, at low Fe content (32). It was proposed that the treatment by $N_2O$ of $Fe^{2+}$-BEA, obtained from reduction of exchanged and calcined samples, led to the formation of both mononuclear and binuclear Fe oxo-cations. The former are more easily reduced by $H_2$ and CO and their formations are preferred at low Fe content (33). From DFT computations, Yoshizawa et al. (34) proposed that active sites in Fe-MFI should have relevance to mononuclear iron-oxo species of the type $(FeO)^+$. It was also proposed that the reactivity of Fe-TON in $N_2O$ decomposition should be mainly attributed to Fe species of the framework origin (35). It is clear that the debate about the exact nature of these sites still remains open in Fe-zeolite catalysts. Moreover, compared to the rigidity of the MFI framework, the BEA is very flexible which allows i) reversible changes from tetrahedral to octahedral coordination of Al, and ii) the presence of defects, dangling bonds and partially coordinated aluminic fragments.

From these reported works, only one clear feature emerges about the role of the zeolite for the high activity of TM: to keep TM in a highly dispersed state as cation or oxo-cations of low nuclearity. Moreover, most of the attempts to identify the nature of Fe species as a function of zeolite structure or protocol of preparation have been carried out ex situ before the reaction. It will be of the utmost importance for future directions of works to look at the identification of Fe species in running conditions.

The catalytic decomposition of $N_2O$ is regulated by the cycle $Fe^{II}/Fe^{III}$. $N_2O$ interacts with reduced "$Fe^{II}$" site (Eq.16.1) to yield an extra-lattice oxygen '$Fe^{III}$-O' species, the so-called α-oxygen by Panov et al. (29). Most of studies have concluded that the removal of α-oxygen (Eq. 16.2) exhibits the lowest rate constant. The remarkable behavior of Fe exchanged in some zeolites could be ascribed to the occurrence of "easily reduced and completely reversible" Fe oxo-cation sites (28). TPR experiments by $H_2$ and CO of $N_2O$-treated Fe-MFI (27) and Fe-BEA (33) have shown that such '$Fe^{III}$-O' sites are much more reductible than those formed by $O_2$ treatment. The influence of the zeolite on the reductibility of Fe species could be understand from quantum chemical calculations (DFT method) on model clusters of FAU and BEA containing $Cu^{II}$, $Co^{II}$ and $Fe^{II}$ cations (36). The calculations indicate that a charge transfer from zeolite to TM ion occurs of ca. one electron, and that the TM-zeolite system behaves like a supermolecule. Obviously, the addition of a reductant such as hydrocarbons (25, 26, 37, 38), CO (39, 40) or $NH_3$ (41) boosts the $N_2O$ decomposition by a faster reduction of '$Fe^{III}$-O' species.

The structure of the zeolite has a clear influence on $N_2O$ decomposition (42). On a series of Fe-exchanged zeolite the order of reactivity identified by the light-off temperature (50% $N_2O$ conversion) was: FER > BEA > MFI,MAZ >> FAU. However, the order did not remain the same when $NH_3$ was added, since all the light-off temperatures were lowered by 70-120 K, except for Fe-FER which remained the same. The order of reactivity then became: BEA > OFF,MFI,FER > MAZ >> FAU.

Since the decomposition is regulated by the redox cycle $Fe^{II}/Fe^{III}$ represented by Eqs.1,2, the kinetics can be treated by the very classical Mars and van Krevelen model (23, 40). Moreover, it was experimentally demonstrated that inhibition by excess $O_2$ is of low extend for Co- and Fe-zeolite, the rate law can thus be expressed as:

$$r = k_1k_2P_{N2O}/(k_1+k_2) \qquad (7)$$

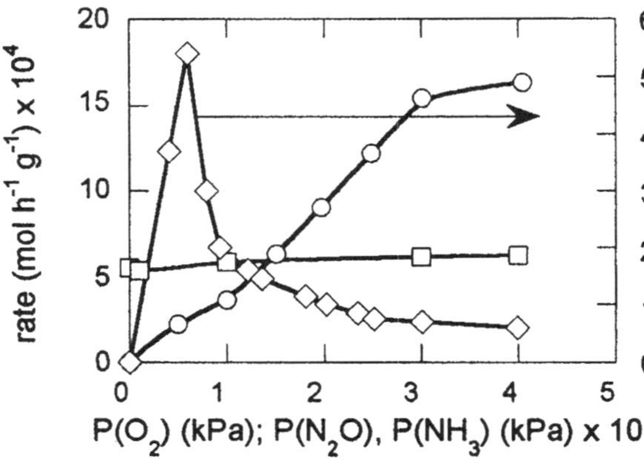


**Fig. 16.4** Rate of $N_2O$ reduction by $NH_3$ on Fe-BEA as a function of $N_2O$ (○), $NH_3$ (◇)and $O_2$ (□) pressures; $T_R$ = 618 K, VVH = 25,000-35,000 $h^{-1}$

A detailed kinetic study of $N_2O$ decomposition over Co-, Cu-, and Fe-zeolite has been reported by Kapteijn et al. (40) with analysis of the boosting effect of CO and NO by several rate laws. Regarding the kinetics of $N_2O + NH_3 + O_2$ reaction to $N_2$ over Fe-BEA (41), the dependency of the rate with respect to $N_2O$, $NH_3$ and $O_2$ pressures are presented in Figure 16.4. A rate law was proposed to account for the volcano-shaped dependency of the rate vs. $NH_3$ pressure, and based on the strong adsorption of $NH_3$ on '$Fe^{II}$' sites.

An interesting point is the promoting effect of NO on $N_2O$ conversion (40), and pointed out earlier by Li and Armor on Co-MFI (43). This is important because $N_2O$ and NO are both present in comparable amounts in the tail gases from acid nitric plants. That allows us to contemplate deNOx and de$N_2O$ in one step in a single SCR reactor. The simultaneous catalytic reduction of NO and $N_2O$ over Fe-BEA has proven to be more efficient than when processing NO and $N_2O$ alone (Figure 16.5) (44). The synergy between NO and $N_2O$ for their simultaneous removal is readily expressed by Eq. 16.1 and the following reaction:

$$NO + O^* \rightarrow NO_2 + * \qquad (8)$$

The SCR of $NO_2$+NO then proceeds easily (see section 16.3.4.2) according to:

$$NO_2 + NO + 2NH_3 \rightarrow 2N_2 + 3H_2O \quad (9)$$

Fe-BEA and Fe-FER catalysts have been tested in pilot plant conditions in derivation of a nitric acid tail gas. At ca. 673 K and real feed conditions (NOx = 1400 ppm, $N_2O$ = 1700 ppm, $NH_3$ = 2200 ppm (slip < 5 ppm), $O_2$ = 2.5%, $H_2O$ = 4%, P = 350 kPa, VVH = 18,000 $h^{-1}$), full NOx and 30-40% $N_2O$ conversions were achieved (45).

As stated above, when present in high concentration, e.g. exhaust gas from adipic acid plants, $N_2O$ can be valorized as a strong oxidant. The direct oxidation of benzene to phenol was thus demonstrated at 673 K with 97-98% phenol yield from benzene and 85% from $N_2O$ on a Fe-MFI catalyst (29). Due to potential interest in selective oxidation processes using $N_2O$, several incentives entail recovering $N_2O$ from $HNO_3$ tail gas ($N_2O$ < 1%). Several variously exchanged zeolites have been tested for the adsorption/recovery of $N_2O$ (24). Ba-MFI is the most efficient for $N_2O$ adsorption at ca. 323 K and desorption at 423-473 K with 90% $N_2O$ removal from the flue gas.

### *16.3.3 Removal of NOx in lean exhaust gases from mobile sources*

Whereas the use of a three-way catalyst is an established technology for the catalytic reduction of NOx produced by gasoline engine operating at stoichiometry, there is still no appropriate catalytic technology for NOx emission abatment in vehicles with diesel and lean burn engines. The following approaches based on catalysis and zeolites have been investigated for the NOx control in lean exhaust gases: (i) catalytic decomposition of NO; (ii) selective catalytic reduction of NOx (SCR) with urea or ammonia; (iii) SCR by hydrocarbon (HC), added or already present in the exhaust gas, and (iv) NOx removal through NOx storage, e.g. NOx storage and reduction system (NSR) and NOx storage and recirculation technique (SNR). These different NOx control technologies are presented hereafter.

#### 16.3.3.1 The catalytic decomposition of NO

Although initially promising, the catalytic decomposition of NO has proven to be difficult to realize. A large number of metal-zeolites formulations has been tested

(46-49) but since the discovery by Iwamoto (1) of the efficiency of Cu-exchanged zeolite, in particular Cu-MFI, no better achievement has taken place to date. The main conclusions regarding NO decomposition on Cu-MFI have been summarized by Fritz and Pitchon (47) as follows: (i) the decomposition is stoichiometric into nitrogen and oxygen; (ii) the reaction is inhibited by oxygen; (iii) the presence of over-exchanged copper species increases the activity; (iv) the zeolite can desorb oxygen, during which sequence some copper $Cu^{2+}$ species are reduced to $Cu^{+}$; (v) the activity is correlated with the concentration of $Cu^{+}$; (vi) the concentration of $NO^{\delta+}$-$Cu^{2+}$ increases with a concurrent decrease of $NO^{\delta-}$-$Cu^{+}$ and $(NO)_2{}^{\delta-}$ -$Cu^{+}$ concentrations during the formation of molecular nitrogen at room temperature; and (vii) the nitrogen formation stops when the concentration of $NO^{\delta-}$-$Cu^{+}$ and $(NO)_2{}^{\delta-}$ -$Cu^{+}$ species become constant.

Despite intensive works in this field, the most promising system, Cu-MFI, is subject to inhibition by water, very sensitive to poisoning by $SO_2$, efficient only at low space velocities, and the catalyst activity and selectivity are not satisfactory (49).

#### 16.3.3.2 The selective catalytic reduction of NOx with urea or ammonia

The SCR of NOx using ammonia or urea has been used for many years in power plants, industrial processes, as well as in stationary diesel engine applications (see section 16.3.4). Different SCR catalytic materials based on platinum, vanadium oxide or zeolites exhibit different operating temperature windows and must be carefully selected for a particular SCR application. However, The SCR of NOx with ammonia or urea has not yet been commercially used for diesel trucks, buses or cars due to their complexity, large size, safety concerns, isocyanic acid emissions, and ammonia/urea injection control issues. Nevertheless, SCR by ammonia is still the only control technology capable of reducing diesel NOx emissions to levels required by future emission standards. Numerous development programs attempt to adapt the SCR technology for mobile diesel engines (50).

#### 16.3.3.3 The selective catalytic reduction of NOx with hydrocarbons

The SCR of NOx by hydrocarbons has also been widely investigated since the initial reports of Held et al. (51), Iwamoto et al. (52) and Ritscher and Sandner (53). The potential of zeolite-based catalysts for practical applications, the mechanism and kinetics aspects of the SCR reaction over TM-zeolite, and the efficiency of different TM-zeolite formulations have been reviewed recently by Traa et al. (54), Parvulescu et al. (48) and Fritz and Pitchon (45). In particular, the influence of the zeolite structures, acidity, nature of the reductant and nature of the metal have been discussed. The following trends could be put forward:

1) The importance of protons in the efficiency of the catalysts

2) Co-, Pd-, Ga- or In- containing catalysts are efficient and selective catalysts in reduction of NOx by methane (55); methane (> 90% of natural gas) is widely used as a fuel source for many combustion processes and electric utilities and also as alternative fuels

3) Pt/zeolite operates at a significantly lower temperature than does Cu/zeolite, and exhibits relatively high resistance to deactivation, but the major drawback is the high selectivity to $N_2O$

4) Highly loaded Fe/zeolite catalysts show good activity and a reasonable stability under realistic conditions, but the light-off temperature is too high and a significant amount of CO is formed

5) Cu-MFI catalysts have been the most extensively studied. They are very active and selective towards nitrogen under certain experimental conditions, but subjected to hydrothermal deactivation

In the selective catalytic reduction, the hydrocarbon selectively reacts with NOx rather than with $O_2$ to form nitrogen, $CO_2$ and water. Traa et al. (54) have classified all the mechanisms proposed in the SCR of NOx by HC over TM-zeolites within three groups:

1) Catalytic decomposition of NO and subsequent regeneration of the active site from adsorbed oxygen by HC:

$$2NO \rightarrow N_2 + 2\,O_{ads} \quad (10)$$
$$HC + O_{ads} \rightarrow CO_x + H_2O \quad (11)$$

2) Oxidation of NO to $NO_2$ which acts as a strongly oxidizing agent of HC

$$NO + 1/2O_2 \rightarrow NO_2 \quad (12)$$
$$NO_2 + HC \rightarrow N_2 + CO_x + H_2O \quad (13)$$

3) Partial oxidation of HC to aldehyde, alcohol... which in turn reacts with NOx:

$$HC + O_2 \text{ and/or } NOx \rightarrow HC^* \quad (14)$$

$HC^* + NOx \rightarrow N_2 + CO_x + H_2O$ (15)

In agreement with the authors, the borderline between the different types of mechanisms are not rigid and, in practice, combinations of all three categories might operate.

The main causes of the deactivation of diesel catalysts are poisoning by lubrication oil additives (phosphorus), and by $SO_x$, and the hydrothermal instability. The SCR by HC is less sensitive to $SO_x$ than the NO decomposition. The Cu-based catalysts are slightly inhibited by water vapor and $SO_x$, and suffer deactivation at elevated temperature. Noble metal catalysts such as Pt-MFI undergo low deactivation under practical conditions, are active at temperatures below 573 K but the major and undesired reduction product is $N_2O$ (56).

The two major requirements in these MFI-based materials are an increase of hydrothermal stability, and a wider temperature window with high conversion under actual diesel exhaust gas. This still constitutes the most important challenges for a commercial application of these catalysts to reduce NOx from mobile sources emissions.

#### 16.3.3.4 NOx removal systems through NOx storage techniques (e.g. NSR and SNR)

Recent concepts to achieve NOx purification from lean exhaust gas are emerging. In the NSR system, the material favors the reactive adsorption of NO to $NO_2$ which is stored as a nitrate complex in $O_2$ rich atmosphere. The exhaust gas is then switched to a stoichiometric or HC rich environment in which the nitrate is thermodynamically unstable. The stored NOx is then released and catalytically reacts with excess HC species in the exhaust gas to form $N_2$. In the SNR system, the NOx are temporarily stored on an adsorbent and periodically recirculated to the combustion chamber to be decomposed in the combustion process. In these processes the key point is the adsorption/desorption of $NO_2$ which has been reviewed by Arai and Machida (57). The amount of reversible and irreversible adsorption of NO per Cu ions over Cu-zeolite decreased with the aluminium content of zeolite and linearly increased when increasing the exchange level (58). Over various TM-MFI (Si/Al = 23.3), reversible and irreversible adsorption of NO were the largest on Cu- and Co-MFI. More recently FAU type zeolites exchanged with alkali and alkaline earth metals have been studied for their NOx adsorption capacity which decreases in the following order: Ba-FAU > Cs-FAU > Na-FAU > Li-FAU (59). Preadsorption of $O_2$, $CO_2$, or $SO_2$ onto a Cu-MFI (Si/Al =23.3) has little effect on the reversible adsorption of NO, but poisoning by CO or $H_2O$ is serious (60). The problem of site inhibition due to coexisting gas is the major hindrance for the use of zeolite materials in NOx removal systems through NOx storage.

16.3.3.5 New concepts for NOx purification from lean exhaust gas

Based on the mechanism 2 reported above, Iwamoto et al. (61) developed the so-called IAR method (intermediate addition of reductant) which can be described as follows:

Conventional SCR by HC

$$NO + O_2 + HC \xrightarrow{\text{catalyst}} N_2 + CO_2 + H_2O \quad (16)$$

IAR process

$$NO + O_2 + \xrightarrow{\text{ox. catalyst}} NO_2 + O_2 \xrightarrow{\text{red. catalyst}} N_2 + CO_2 + H_2O \quad (17)$$

$$HC \longrightarrow \uparrow$$

where the hydrocarbon is added between an oxidation catalyst and a reduction catalyst. Pt-MFI was employed as an oxidation catalyst and various TM-exchanged MFI as reduction catalyst. The combination of Pt-MFI and Zn-MFI was found to be the most efficient and a gain of conversion of ca 50% was observed compared to a conventional SCR by HC on Zn-MFI and Pt-MFI alone. A similar concept has been applied for the plasma-assisted catalytic reduction of NOx (62, 63). In this process, a plasma first oxidises NO to $NO_2$ in the presence of a hydrocarbon:

$$\text{plasma} + NO + HC + O_2 \rightarrow NO_2 + \text{HC-products} \quad (18)$$

and then the catalyst reduces $NO_2$ to $N_2$:

$$NO_2 + HC \xrightarrow{\text{Catalyst}} N_2 + CO_2 + H_2O \quad (19)$$

Oxidation by plasma is selective for NO and does not oxidize $SO_2$ to $SO_3$.

The reduction of $NO_2$ by HC has been less frequently approached. Few works (64-66) essentially concern the role of $NO_2$ in the mechanism of SCR by HC of NO.

For deNOxing lean exhaust gas within a broad temperature range, methods using multi-stage catalysts active in different temperature domains have been claimed, e.g. by Toyota and Tosoh Corporation (67) for the combination of Co-Ba/H-MFI and Cu/H-MFI on a cordierite honeycomb and by Mazda Motor

Corporation (68) with Pt/Cu-MFI and Zr/Cu-MFI. This is close to the concept of dispersing several metals on a zeolite support: for example catalytic layers of zeolite with Cu ions and precious metals have been designed. The SCR by HC on zeolites containing more than one metal has been reviewed recently by Traa et al. (54); numerous combinations have been reported and essentially concern the patent literature.

### *16.3.4 Removal of NOx emitted from stationary sources*

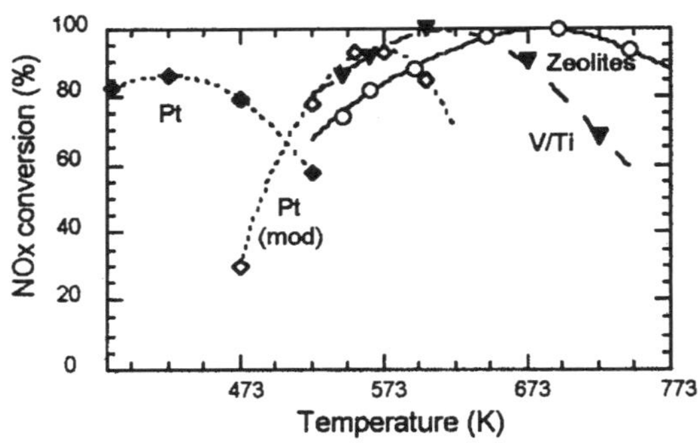


**Fig. 16.5** Windows of activity in the SCR of NOx by $NH_3$ for different catalyst types (from ref. 71)

There are three possible technologies (69) for reducing NOx in net oxidizing mixtures: (i) the non-selective catalytic reduction (NSCR), (ii) the selective non-catalytic reduction (SNCR), and (iii) the selective catalytic reduction (SCR). Only the SCR uses zeolite-based catalysts. The selective catalytic reduction of NOx with the help of a reductant (HC, $NH_3$ or urea) is the best control technology to achieve near full NOx removal from off-gas of power plants, nitric acid plants, stationary Diesel engines, marine vessels.... Major processes use $NH_3$ (or urea) as reductant. The catalyst is a combination of $V_2O_5$ and $TiO_2$ (sometimes promoted with $WO_3$ or $MoO_3$) as extruded monoliths or deposited on a plate structure (69). An increasing number of installations use a zeolite-technology at present (70, 71). Figure 16.5 presents the windows of activity of various formulations (72).

The basic reaction in the SCR process is:

$$4NH_3 + 4NO + O_2 \rightarrow 4N_2 + 6H_2O \quad (20)$$

Since Pence and Thomas (73) who first documented the activity of H-MOR as a catalyst for reducing NOx to $N_2$ by $NH_3$, there was significant interest in developing zeolite-based catalysts (70, 72, 74, 75): (i) wide SCR operating window, (ii) high NOx conversion, (iii) high selectivity towards $N_2$, (iv) very low $NH_3$ slip, (v) sulfur tolerant above 700 K, and (vi) less disposal problem for the spent catalyst.

In the following parts, the performance, the reaction mechanism and the stability towards water and $SO_x$ in the SCR of NO by $NH_3$ will be discussed according to the nature of the zeolite-based catalysts.

16.3.4.1 H-form of zeolites

*Effect of H and Al content.* MOR and MFI zeolites have been primarily studied. Sodium forms exhibit very low activity in comparison with H-forms (76). The catalytic activity is a strong function of the degree of proton exchange, i.e. with the acidity of the catalyst (76). One should therefore expect that SCR activity be dependent on the Al content. Andersson et al. (77) found the activity proportional to Al content on a series of acid-leached H-MOR samples with Si/Al ratio from 16.5.3 to 9.9. Eng et al. (78) do not observe any correlation in their study on H-MFI and H-MOR. But they found that a minimum exists for Al content below which the SCR activity is negligible (Si/Al = 10 and 35 for MOR and MFI respectively). More recently, Stevenson et al. (75) reported that H-MFI catalysts of varying Si/Al ratios from 12 to 350 exhibited the highest activity at the lowest Si/Al ratio.

*Kinetic and mechanism.* In the range 573-623 K, the reaction is generally characterized by first order in $O_2$ (75,78), 0.7-1 in NO (75,78,79) and negative order -0.55-0.6 in $NH_3$ (75,78).

$NH_3$ appears to inhibit the SCR reaction by limiting access of NO to the active sites, whereas the presence of $O_2$ favors the NO reduction. The main role of $O_2$ is to react with NO to form a reactive intermediate $NO^+$ or $NO_2$ (75, 80, 81). This step is generally proposed as rate-determining. In the mechanism reported by Brandin et al. (81) on H-MOR the first step is the formation of the nitrosonium ion:

$$4NO + O_2 \rightarrow 4NO^+ + 2\,O^{2-} \qquad (21)$$

$NO^+$ and $O^{2-}$ rearrange into structural nitrite and nitrate (limiting step) which further react with $NH_3$ to give $N_2$ and $H_2O$. Lewis sites formed from dehydroxylation of a Brönsted site are proposed as the active centres. According to Brandin et al. (81), in the SCR over H-MOR the NOx conversion goes through a maximum value at ca. $NO_2/NO$ = 1. From $^{15}$NMR studies of the $NO+NH_3+O_2$ reaction over H-MFI,

Filimonova and Mastikhin (82) proposed that adsorbed $N_2O_3$, formed from NO and $NO_2$ can react with ammonia according to:

$$(N_2O_3)_{ads} + 2\ NH_3 \rightarrow NH_2NO + NH_4NO_2 \quad (22)$$

the decomposition of nitrosamide and ammonium nitrite then leads to nitrogen and water. From kinetics in transient states, Eng and Bartholomew (80) conclude that a pair of $NH_3$ molecules adsorbed onto neighboring acid sites are necessary for reduction of NO to $N_2$. Thus, the initial step of SCR mechanism is the adsorption of $NH_3$ onto neighboring Brönsted sites and rearrangement of $NH_4^+$ complexes to the 3H structure. The next step involves reaction of adsorbed $NH_3$ with $NO_2$ to form an active complex, which reacts rapidly with NO to form $N_2$ and $H_2O$. On the other hand Stevenson et al. (75) suggest that Brönsted sites on extra-framework aluminium are more reactive than on framework aluminium

*Effect of $H_2O$.* In presence of $H_2O$ (10 and 20%) the $N_2$ formation rate has been found nearly the same as in absence of $H_2O$ over H-MFI (78), while the inhibiting effect of $H_2O$ was reported by Ramachadran et al. (83).

### 16.3.4.2 Cu-containing zeolite

*Effect of metal content and nature of the host structure.* The activity of Cu-zeolites has been studied in the temperature range 298-873 K. Whatever the zeolite structure, the SCR activity at temperatures below 573 K increases quite linearly with the copper content (76, 77, 84-89). Thus, Kieger et al. (84) proposed an active site composed of two neighboring Cu ions due to the increase of TOF with Cu content. Copper oxide aggregates are less active and selective to $N_2$ (84, 87, 90) than Cu cations or oxo-cations, due to their high activity for the side reaction of $NH_3$ oxidation (84, 90). The TOF at 573 K for Cu-FAU was low compared to Cu-MFI and Cu-MOR which show similar values (86). Kieger et al. (84) proposed that the lower activity of Cu-FAU (84, 86) is due to the high proportion of Cu located in non-accessible sites (sodalite cages and hexagonal prisms). When considering the Cu species only located in the supercage, Cu-FAU exhibits similar specific activity as Cu-MOR and Cu-MFI (84). The 'blocking' of sodalite cages of FAU by alkaline earth (Ca, Ba,...) or rare earth (La, Pr, ...) strongly enhances the proportion of accessible copper ions and leads to Cu-M-FAU catalyst highly efficient in NOx reduction by $NH_3$ to $N_2$ (84,92).

**Table 16.1** Reaction order with respect to NO, $NH_3$ and $O_2$ in the SCR by $NH_3$ on various catalysts

| Catalysts | Temp. (K) | Reaction order in | | | Ref. |
|---|---|---|---|---|---|
| | | NO | $NH_3$ | $O_2$ | |
| Cu-MFI | 573 | 1.0 | 0 | 0.6 | 86 |
| Cu-NaFAU[a] | 383 | 1.0 | 0.5 | - | 93 |
| Cu-NaFAU[a] | | 1.0 | 0 | - | 87 |
| Cu-NaFAU | 503 | 0.6 | 0.4 ($NH_3$/NO < 1)<br>0 ($NH_3$/NO = 1) | 0.4 | 94 |
| Cu-MOR | 588 | 1.0 | | | 79 |

[a] without $O_2$

*Kinetic and mechanism.* Table 16.1 presents the reaction order with respect to NO, $NH_3$ and $O_2$ for the SCR on various Cu-zeolites. On Cu-exchanged zeolites it is generally accepted that the rate of the SCR of NO by $NH_3$ obeys to a Langmuir-Hinshelwood type mechanism involving reaction on copper-containing sites of an adsorbed species derived from NO with an adsorbed species derived from $NH_3$ (76, 84, 87, 88, 95), with stronger adsorption of $NH_3$ than NO (96). Moreover, the mechanism is regulated by a redox cycle of copper ions (84, 85, 87, 97). From temperature programmed oxidation (TPO), SCR of NO by $^{15}NH_3$ and previous elements suggested by Mizumoto (85, 87), Kieger et al. (84) proposed the mechanism reported in Figure 16.6.

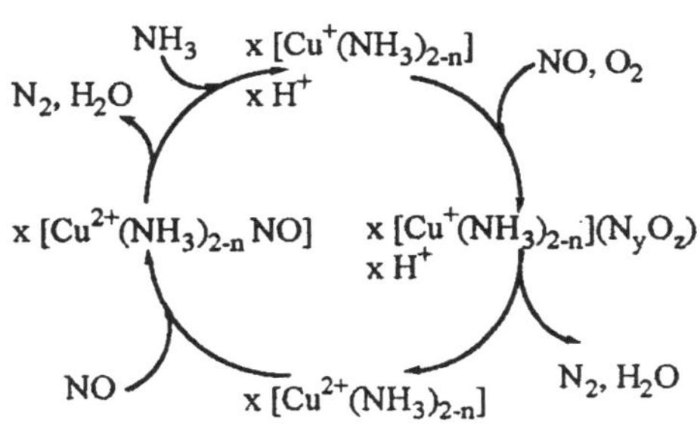


**Fig. 16.6** Catalytic cycle of the SCR of NO by $NH_3$ on Cu-FAU (from ref. (84))

In this mechanism $Cu^{2+}$ is reduced by $NO+NH_3$ to $Cu^+$, which in turn is oxidised to $Cu^{2+}$ by $NO+O_2$. The original feature of the process demonstrated by TPO and SCR with $^{15}NH_3$ concerns the oxidation of $Cu^+$ complexes which is faster by $NO+O_2$ than by $O_2$ alone. In view of the correlation between copper content and TOF, the active sites below 550 K were proposed as Cu ions in close vicinity, may be $[CuOCu]^{2+}$, stabilized by $NH_3$ and located in the supercages. Above 600 K, all Cu ions become active. Similar arguments were developed by Komatsu et al. (86) who proposed a nitrate-like species (Fig. 16.7) as intermediate:

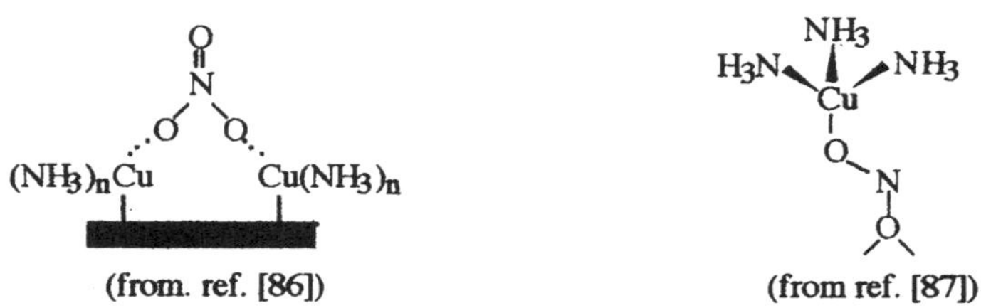


**Fig. 16.7** Copper active species proposed in the SCR of NO by $NH_3$

Based on IR and ESR evidences Willamson and Lunsford (87) have suggested mononuclear Cu species (Fig. 16.7) active at low temperature (283 K) and in absence of oxygen.

*Effect of $H_2O$ and/or $SO_x$.* According to Salker and Weisweiler (98), water vapor (5%) enhances NO conversion on Cu-MFI catalyst. Using gas mixture containing 3 or 9% $H_2O$ Sullivan et al. (95) observed an increase of the NO conversion to $N_2$ over Cu-MFI in the temperature range 373-823 K. Cu-MOR catalysts also showed a pronounced positive effect of having moisture in the gas mixture (83). From TPD studies, Salker and Weisweiler (98) have shown that water makes the strong chemisorption of ammonia easier and may explain its beneficial role on the reaction.

Ham et al. (99) have studied the effect of $SO_2$ in the temperature range 523-723K. Below 573 K a decrease of activity by 60% at 523 K and 20% at 573 K occurs, while above 573 K no inhibition was observed. During the tests at low temperature with $SO_2$, accumulation of ammonium sulfate and/or ammonium bisulfate occurs and may cause severe blocking. At higher temperatures these salts decompose and this may explain the absence of inhibition. Complete regeneration of deactivated catalysts can be achieved by heat treatment at 773 K. Over Cu-H-MOR bound with silica and deposited on a cordierite honeycomb, only 12% decrease of NO conversion was observed at 673 K after 350 h passivation in presence of 200 ppm $SO_x$ and 10% $H_2O$ (100). All these results emphasize the stability of copper zeolite catalysts under reaction gas mixture containing water and/or $SO_x$.

16.3.4.3 Fe-containing zeolite

MFI and MOR host structures for Fe ions were found preferable to FAU for the SCR of NO by $NH_3$ in the presence of oxygen (101). On Fe-FAU catalysts, Amiridis et al. (102) have shown that only a fraction of Fe cations are active due to species located in inaccessible sites. A change of Si/Al ratio from 2.4 to 4.4 has only a modest effect on the SCR catalytic activity of Fe-FAU (102). Upon the increase of the Si/Al ratio from 10 to 100 of Fe-MFI with the same Fe exchange level, the NO conversion decreased rapidly (101); the TOF thus decreased with the increase of Fe content (101). From this result Yang et al. (101) underline the importance of a good balance between Brönsted and iron sites. The methods of preparation of Fe-zeolite appears as a crucial point since overexchanged Fe-MFI prepared by sublimation exhibited higher activity than overexchanged Cu-MFI prepared by conventional technique and this specially at temperatures above 600 K (103).

In the temperature range 490-545 K, the kinetic dependence of the SCR on Fe-FAU was first order in NO and zero order in $NH_3$ (102). The positive role of $O_2$ on the reaction rate was also shown.

Long and Yang (104) proposed a mechanism similar to that described above for H-MFI. It involves the reaction between an $NO_2$ adsorbed species (NO oxidised by $O_2$ on $Fe^{3+}$) and two $NH_4^+$ ions to form an active intermediate which subsequently reacts with another gaseous or weakly adsorbed NO to produce $N_2$ and $H_2O$. According to Schmidt et al. (105) a redox cycle $Fe^{2+}/Fe^{3+}$ is involved in the mechanism and binuclear Fe oxo-cations were suggested as active species.

In presence of 5% $H_2O$ and 0,05% $SO_2$ in the gas mixture, 15-30% loss of activity is observed at 573 K over Fe-MFI (101). The addition of cerium ions increases the stability towards the presence of $H_2O+SO_2$ (106). Above 573 K Yang et al. (104) reported that $H_2O+SO_2$ have little effect on the activity. According to these authors (104), water and sulfates increase the Brönsted acidity of the catalysts and consequently the adsorption of ammonia (104). On overexchanged Fe-MFI prepared by sublimation, the presence of water (2.5%) in the gas mixture also promotes the NO conversion. $SO_2$ (0.02%) or $H_2O+SO_2$ inhibits the reaction only at temperatures below 650 K (103). On the other hand, upon addition of 8% $H_2O$ at 573 K, Ramachandran et al. (83) observe a drop of 40% in the NO conversion over a pelletized Fe-MOR containing an alumina binder.

16.3.4.4 Ce-, V-containing zeolite, and other zeolitic type catalysts

Ce-exchanged zeolite catalysts have been studied in the SCR of NO by $NH_3$ by van den Bleek and co-workers (107-113). This group has reported the high activity and selectivity of Ce-exchanged zeolites (NOx conversion > 80%) at 573-773 K and a VVH of 30,000 $h^{-1}$ compared to Cu-, Co- and Fe-MFI (107). In several reports the following aspects were discussed: effect of cerium content (108), reaction kinetics and mechanism (109), influence of $H_2O$ and $SO_2$ (110, 111) and catalyst packing for large scale applications (112, 113). Briefly, the SCR activity strongly increases with the cerium content up to ca 0.75% (108). The reaction is regulated by the redox cycle $Ce^{III}/Ce^{IV}$ (109). In the low temperature region (573 K), the reaction mainly proceeds through a $NO_2$ intermediate, while at high temperature (783 K) a nitrosation reaction mechanism with involvement of $N_2O_3$-like species is assumed to contribute substantially (109). In simulated exhaust gas with 12% $H_2O$, an 11% decrease in NO conversion was found at 723 K after 116 h reaction on Ce-MFI/$Al_2O_3$ extrudates ($Al_2O_3$ used as binder) (110); the deactivation was attributed to dealumination. In real diesel exhaust gas (stationary diesel engine) the decrease of activity was due in large part to sulfur poisoning and in a lesser part to the deposition of elements (Ca, P and Zn) contained in the lubricating oil (110).

**Table 16.2** Some zeolite-based catalysts reported active in the SCR of NOx by $NH_3$.

| Catalyst | References | Catalyst | References |
|---|---|---|---|
| H-MOR | 76,77,79,80,81 | Ce-zeolite | 107-113 |
| H-MFI | 75,78,80,82 | V-MFI | 115-116 |
| H-FAU | 80 | V-FAU | 114 |
| Cu-FAU | 74,84-88,90,91,117 | Exchanged natural zeolites | 118 |
| Cu-MOR | 79,81,86,97,99,100 | Cu-MFI/Raney composite | 119 |
| Cu-MFI | 83,86 | Cr-MFI | 98 |
| Cr-MFI | 98 | $NH_4^+$-zeolite | 120,121 |
| Fe-FAU | 102,105 | Cu-MFI-ferrisilicate | 89 |
| Fe-MFI | 83,98,101,103,106 | | |

The NO reduction by $NH_3$ over oxo-vanadium species encapsulated in FAU was investigated by Adams et al. (114) for mechanistic aspects. In presence of oxygen, Grunert and co-workers (115, 116) have observed that vanadyl-exchanged MFI exhibits comparable conversion as a $V_2O_5/TiO_2$ reference catalyst but at a higher temperature (115). Isolated $VO^{2+}$ ions were proposed as the active sites for the SCR reaction. At temperatures below 623 K, the deactivation is slow in dry conditions but intensifies in the presence of water (4.5%) and $SO_2$ (200 ppm), while above 623 K a better stability is observed (116). Similar instability in the presence of $H_2O$ and $SO_x$ was reported by Ramachandran et al. (83). Other zeolite-based catalysts have been evaluated for the SCR of NOx by $NH_3$ (Table 16.2).

16.3.4.5 An example of industrial application

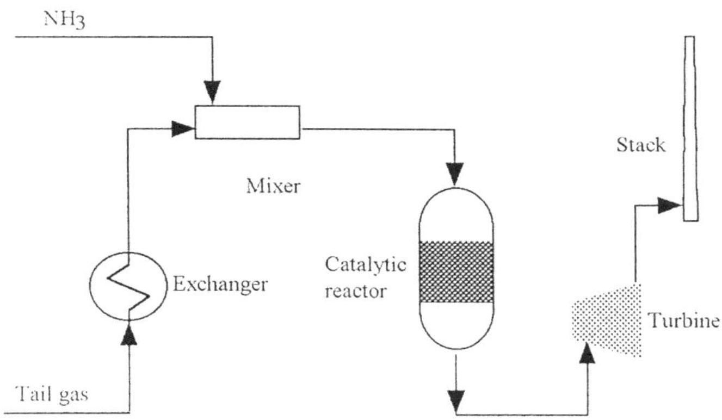


**Fig. 16.8** A SCR process scheme (from ref. 70)

For a decade, the fertilisers producer Grande-Paroisse Ato Fina (GP) has developed a zeolite based catalyst in the SCR of NOx by $NH_3$ for the NOx treatment of exhaust gases emitted by nitric acid plants (70). To date 27 deNOx installations assembled by GP are operating world-wide with a zeolite-base catalyst. In the SCR process (Fig. 8), the tail gas (NOx: 0.01-0.2% with $NO/NO_2 \sim 1$, $N_2O$: 0.05-0.2%, $O_2$: 1-4%, $H_2O$: 0.3-0.7% balance with $N_2$) and ammonia are mixed before reacting on a pelletized Cu-FAU catalyst (binder: bentonite). The operating conditions are the following: temperature = 443-773 K (preferably 473 K), pressure = 3-5 kPa and VVH = 10,000-50,000 $h^{-1}$ (standard: 25,000 $h^{-1}$). In the standard conditions, less than 0.005 % of NOx are unreacted with low ammonia slip (< 5 ppm). The selectivity to $N_2$ is better than 90%, whatever the running conditions with the improved formulations Cu-M-FAU (M: Ca, Ba, La,...) (92). This example shows the great interest in using zeolite based catalysts.

## 16.4 Concluding remarks

There are several issues for cleaning noxious emissions and wastes using zeolites.

Ion-exchange of ammonia is a mature technology but the selective removal of heavy metal and radionuclides in complex media, which can be achieved with highly selective organic ion-exchange resins, still remains challenging with zeolites. However, the use of inorganic ion-exchangers will be much more attractive for glazing as ultimate disposal.

In VOCs, the major issues concern $CH_4$ and short-chain HCs which are difficult to adsorb selectiviely. Moreover, $CH_4$ is the most abundant HC emission and the most refractory to low cost oxidation. Another problem deals with the removal of hydrophilic organic compounds (alcohol, ketone ...) in waste water.

The SCR of NOx by $NH_3$ is the best control technology but a new breakthrough would be achieved in power plants by the SCR of NOx using methane as reductant. Regarding deNOx from mobile sources, new concepts are appearing, and NOx trap and plasma-assisted catalytic reduction seem promising.

The removal of $N_2O$ in low concentration in flue gases will be fully efficient with the development of catalysts active below 570 K. Moreover, the simultaneous removal of NOx and $N_2O$ in the same SCR reactor will also constitute a significant step forward.

## References

1. Lee D.S. et al., *Atmospheric Environment* **31** (1997) 1735-1749.
2. Ertl. G. et al., eds., "Environmental Catalysis", in *Handbook of Heterogeneous Catalysis*, (Wiley, VCH, Weinheim, 1997), 1559-1696.
3. Janssen F.J.J.G. and van Santen R.A., eds., "Environmental Catalysis", (Imperial College Press, London, 1999).
4. "NOxCONF 2001, La Pollution Atmosphérique d'Origine Industrielle", Paris, 21-22 March 2001.
5. Shermann J.D., in *Zeolites: Science and Technology*, Ed. Ramoa Ribeiro F. et al., (NATO ASI Series, Martinus Nijhoff Publishers, The Hague, 1984), 583-623.
6. Collela C., *Stud. Surf. Sci. Catal.* **125** (1999) 641-655.
7. Sobolev W. et al., *Stud. Surf. Sci. Catal.* **135** (2001) 31-P-05.
8. Bailey et al., *Wat. Res.* **33** (1999) 2469-2479..

9. Collella C. et al., *Stud. Surf. Sci. Catal.* **135** (2001) 01-O-05.
10. Taborsky P., *US Pat.* (1994) 5,304,365.
11. Lahav O. and Green M., *Wat Res.* **32** (1998) 2019-2028.
12. Fajerweg K., and Debellefontaine H., *Appl. Catal. B: Envirol* **10** (1996) L229-L235.
13. Fajula F. and Plee D., *Stud. Surf. Sci. Catal.* **85** (1994) 633-651.
14. Garrot B. et al., *Stud. Surf. Sci. Catal.* **125** (1999) 683-690.
15. Chihara K. et al., *Stud. Surf. Sci. Catal.* **135** (2001)18-P-15.
16. Otto K. et al., *Ind. Eng. Chem. Res.* **30** (1991) 2333-2340.
17. Spivey J.J., *Ind. Eng. Chem. Res.* **26** (1987) 2165-2180.
18. Armor J.N., *Appl. Catal. B: Environmental* **1** (1992) 221-256.
19. Dégé Ph. et al., *Appl. Catal. B: Environmental* **27** (2000) 17-26.
20. Chatterjee S., and Greene H.L., *J. Catal.* **130** (1991) 76-85.
21. Pinard L. et al., *Stud. Surf. Sci. Catal.* **135** (2001) 30-O-05.
22. Chue K.T. et al., *Ind. Eng. Chem. Res.* **34** (1995) 591-598.
23. Kapteijn F. et al., *Appl. Catal. B: Environ.* **9** (1996) 25-64.
24. Centi G. et al., *Chemtech.* **December** (1999) 48-55.
25. Centi G. and Vazzana F., *Catal. Today* **53** (1999) 683-693.
26. Kögel M. et al., *J. Catal.* **182** (1999) 470-478.
27. El-Malki El-M. et al., *J. Catal.* **196** (2000) 212-223.
28. Pérez-Ramírez J. et al., *Stud. Surf. Sci. Catal.* **135** (2001) 30-O-02.
29. Panov G. I. et al., *Catal. Today* **41** (1998) 365-385.
30. Arbuznikov A.V. and Zhidomirov G.M., *Catal. Lett.* **40** (1996) 17-23.
31. Lázár et al., *J. Phys. Chem. B* **102** (1998) 4865-4870.
32. Delahay G. et al., *J. Catal.*, in press.
33. Mauvezin, M. et al. *J. Phys Chem. B* **105** (2001) 928-935.
34. Yoshizawa K. et al., *J. Phys. Chem. B* **104** (2000) 734-740.
35. Kasture M. et al., *Stud. Surf. Sci. Catal.* **125** (1999) 579-586.
36. Berthomieu D. et al., *Stud. Surf. Sci. Catal.* **135** (2001) 15-P-23.
37. Pophal C. et al., *Catal. Lett.* **44** (1997) 271-274.
38. Kameoka S. et al., *Catal. Lett.* **69** (2000) 169-173.
39. Leglise J. et al., *J. Catal.* **86** (1984) 392-399
40. Kapteijn F. et al., *J. Catal.* **167** (1997) 256-265.
41. Coq B. et al., *J. Catal.* **195** (2000) 298-303.
42. Mauvezin M. et al., *Catal. Lett.* **62** (1999) 41-44.
43. Li Y. and Armor J. N., *Appl. Catal. B: Environ.* **3** (1993) 55-60.
44. Coq B. et al., *Appl. Catal. B: Environ.* **27** (2000) 193-198.

45. Kieger S. et al., *Proceedings* NOxCONF 2001, La Pollution Atmosphérique d'Origine Industrielle, Paris, 21-22 March 2001.
46. Iwamoto M., in *Stud. Surf. Sci. Catal.* **54** (1990) 121-143.
47. Fritz A. and Pitchon V., *Appl. Catal. B: Environ.* **13** (1997) 1-25.
48. Parvulescu V.I. et al., *Catal. Today* **46** (1998) 233-316.
49. Iwamoto M.et al., *J. Phys. Chem.* **95** (1991) 3727-3730.
50. Koebel M. et al., *Catal. Today* **59** (2000) 335-345.
51. Held W. et al., *SAE Trans., Section 4* N° 900496 (1990) 209-216.
52. Iwamoto M. et al., *Shokubai* **32** (1990) 430-438.
53. Ritscher J.S. and Sandner M.R., *US Pat.* (1981) 4,297,328.
54. Traa Y. et al., *Microporous Mesoporous Mat.* **30** (1999) 3-41.
55. Armor J. N., *Catal. Today* **26** (1995) 147-158.
56. Gilot P. *et al*, *Fuel* **76** (1997) 507-515.
57. Arai H. and Machida M., *Catal. Today* **22** (1994) 97-109.
58. Zhang W.X. et al., *Chem. Lett.* (1993) 851-854.
59. Monticelli O. et al., *Appl. Catal. B: Environ.* **21** (1999) 215-220.
60. Zhang W.X. et al., *Langmuir* **9** (1993) 2337-2343.
61. Iwamoto M. et al., *Appl. Catal. B: Environ.* **17** (1998) 259-266.
62. Vogtlin G.E. et al., *US Pat* (1996) 5,711,147
63. Penetrante B.M. et al., *SAE Technical Paper Series* N° 982508 (1998) 1-10.
64. Chajar Z. et al., *Catal. Lett.* **28** (1994) 33-40.
65. Bethke K.A. et al., *Catal. Lett.* **31** (1995) 287-299.
66. Shimokawabe M. et al., *React. Kinet. Catal. Lett.* **52** (1994) 393-397.
67. Mizuno T. et al., *US Pat.* (1995) 5,443,803.
68. Takemoto T. et al., *US Pat.* (1995) 5,409,671.
69. Bosch H. and Janssen F., *Catal. Today* **2** (1988) 369-532.
70. Gry P., in "NOxCONF 2001, La Pollution Atmosphérique d'Origine Industrielle", Paris, 21-22 March 2001
71. Kiovsky J.R. et al., *Ind. Eng. Chem. Prod. Res. Dev.* **19** (1980) 218-225.
72. Byrne J.W. et al., *Catal. Today* **13** (1992) 33-42.
73. Pence D.T. and Thomas T.R. in *Proceedings of the AEC Pollution Control Conferenc,* Oak Ridge (USA) 1972, 115-121.
74. Heck R.M., *Catal. Today* **53** (1999) 519-523.
75. Stevenson S.A. et al., *J. Catal.* **190** (2000) 228-239.
76. Choi E.Y. et al., *J. Catal.* **161** (1996) 597-604.
77. Andersson L.A. H. et al., *Catal. Today* **4** (1989) 173-185.
78. Eng J. and Bartholomew C.H., *J. Catal.* **171** (1997) 14-26.
79. Medros F.G., et al., *Ind. Eng. Chem. Res.* **28** (1989) 1171-1177.

80 Eng J. and Bartholomew C.H., *J. Catal.* **171** (1997) 27-44.
81. Brandin J.G.M. et al, *Catal. Today* **4** (1989) 187-203.
82. Filimonova S.V. and Mastikhin V.M., *React. Kinet. Catal. Lett.* **54** (1995) 419-426.
83. Ramachandran B. et al., *Catal. Today* **55** (2000) 281-290.
84. Kieger S. et al., *J. Catal.* **183** (1999) 267-280.
85. Mizumoto M. et al., *J. Catal.* **55** (1978) 119-128.
86. Komatsu T. et al., *J. Catal.* **148** (1994) 427-437.
87. Williamson W.B. and Lunsford J.H., *J. Phys. Chem.* **80** (1976) 2664-2671.
88. Seiyama T. et al., *J. Catal.* **48** (1977) 1-7.
89. Komatsu T. et al., *J. Chem. Soc. Faraday Trans.* **94** (1998) 949-953.
90. Delahay G. et al., *Catal. Today* **54** (1999) 431-438.
91. Kieger S. et al., *Appl. Catal. B: Environ.* **25** (2000) 1-9.
92. Coq B. et al., *EU Pat* (1999) 0914866 A.
93. Arakawa T. et al., *Bull. Chem. Soc. Jpn.* **50** (1977) 1431-1436.
94. Kieger S., *PhD dissertation* (1997)
95. Sullivan J. A. et al., *Appl. Catal. B: Environ.* **7** (1995) 137-151.
96. Choi E.Y. et al., *J. Mol. Catal.* **69** (1991) 247-258.
97. Ham S.W. et al., *Catal. Lett.* **42** (1996) 35-40.
98. Salker A.V. and Weisweiler W., *Appl. Catal. A: Gen.* **203** (2000) 221-229.
99. Ham S.W. et al., *Catal. Today* **11** (1992) 611-621.
100. Nam I.S. et al., *Catal. Today* **38** (1997) 181-186.
101. Long R Q. and Yang R. T., *J. Catal.* **188** (1999) 332-339.
102. Amiridis M.D. et al., *J. Catal.* **142** (1993) 572-584.
103. Ma A.Z. and Grunert W., *Chem. Comm.* (1999) 71-72.
104. Long R. Q. and Yang R. T., *J. Catal.* **194** (2000) 80-90.
105. Schmidt R. et al., *J. Phys. Chem.* **96** (1992) 8142-8149.
106. Long R.Q. and Yang R.T., *J. Am. Chem. Soc.* **121** (1999) 5595-5596.
107. Ito E. et al., *Stud. Surf. Sci. Catal.* **96** (1994) 661-669.
108. van Kooten W.E.J. et al., *Appl. Catal. B: Environ.* **21** (1999) 203-213.
109. Ito E. et al, *Appl. Catal. B: Environ.* **4** (1994) 95-104.
110. van Kooten W.E.J. et al., *Appl. Catal. B: Environ.* **25** (2000) 125-135.
111. van Kooten W.E.J. et al., *Catal. Lett.* **63** (1999) 227-231.
112. Rebrov E.V. et al., *Appl. Catal. A: Gen.* **206** (2001) 125-143.
113. Ito E. et al., *Catal. Today* **27** (1996) 123-128.
114. Adams R.C. et al., *Catal. Today* **33** (1997) 263-278.
115. Wark M. et al., *J. Catal.* **175** (1998) 48-61.
116. Piehl G. *et al*, *Catal. Today* **54** (1999) 401-406.

117. Mizumoto M. et al., *J. Catal.* **59** (1979) 319-324.
118. Mishima H. et al., *Appl. Catal. B: Environ.* **19** (1998) 119-126.
119. Ma A.Z. et al., *Appl. Catal. B: Environ.* **27** (2000) 37-47.
120. Richter M. et al., *Catal. Today* **54** (1999) 531-545.
121. Richter M. et al., *Appl. Catal. B: Environ.* **15** (1998) 129-146.

# SUBJECT INDEX